W9-BEE-141

Living with the Coast of Alaska

Living with the Shore

Series editors, Orrin H. Pilkey and William J. Neal

The Beaches Are Moving: The Drowning of America's Shoreline
New edition
Wallace Kaufman and
Orrin H. Pilkey Jr.

Living by the Rules of the Sea
David M. Bush, Orrin H. Pilkey Jr.,
and William J. Neal

Living with the
Alabama-Mississippi Shore
Wayne F. Canis et al.

Living with the California Coast
Gary Griggs and Lauret Savoy et al.

Living with the Chesapeake Bay and Virginia's Ocean Shores
Larry G. Ward et al.

A Moveable Shore: The Fate of the Connecticut Coast
Peter C. Patton and James M. Kent

Living with the East Florida Shore
Orrin H. Pilkey Jr. et al.

Living with the West Florida Shore
Larry J. Doyle et al.

Living with the Georgia Shore
Tonya D. Clayton et al.

Living with the Lake Erie Shore
Charles H. Carter et al.

Living with Long Island's South Shore
Larry McCormick et al.

Living with the Louisiana Shore
Joseph T. Kelley et al.

Living with the Coast of Maine
Joseph T. Kelley et al.

Living with the New Jersey Shore
Karl F. Nordstrom et al.

From Currituck to Calabash: Living with North Carolina's Barrier Islands
Second edition
Orrin H. Pilkey Jr. et al.

The Pacific Northwest Coast:
Living with the Shores of Oregon and Washington
Paul D. Komar

Living with the Puerto Rico Shore
David M. Bush et al.

Living with the Shore of Puget Sound and the Georgia Strait
Thomas A. Terich

Living with the South Carolina Coast
Gered Lennon et al.

Living with the Texas Shore
Robert A. Morton et al.

Living with the Coast of Alaska

Owen Mason, William J. Neal, and Orrin H. Pilkey
with chapters by Jane Bullock, Ted Fathauer, Deborah Pilkey, and Douglas Swanston

Duke University Press Durham and London 1997

The Living with the Shore series is funded by the Federal Emergency Management Agency

Living with the Shore Series

Publication of 20 volumes in the Living with the Shore series has been funded by the Federal Emergency Management Agency.

Publication of the Alaska volume was assisted by funding from the National Science Foundation Earthquake Hazard Mitigation Program.

Publication has been greatly assisted by the following individuals and organizations: the American Conservation Association, an anonymous Texas foundation, the Charleston Natural History Society, the Office of Coastal Zone Management (NOAA), the Geraldine R. Dodge Foundation, the William H. Donner Foundation, Inc., the George Gund Foundation, the Mobil Oil Corporation, Elizabeth O'Connor, the Sapelo Island Research Foundation, the Sea Grant programs in New Jersey, North Carolina, Florida, Mississippi/Alabama, and New York, The Fund for New Jersey, M. Harvey Weil, Patrick H. Welder, Jr., and the Water Resources Institute of Grand Valley State University. The Living with the Shore series is a product of the Duke University Program for the Study of Developed Shorelines, which was initially funded by the Donner Foundation.

Printed in the United States of America on acid-free paper ∞
Library of Congress Cataloging-in-Publication Data
appear on the last printed page of this book.

Contents

Figures, Tables, and Risk Maps

Figures

Tables

Risk Maps

Preface

Thirty years have passed since the Good Friday earthquake of 1964; a whole generation has lived without personal knowledge of this sobering experience of injury and loss. As a whole, the members of this same generation—and indeed, most people today—are not particularly aware of the frequent damage and loss caused by natural hazards in Alaska. And yet, Nome was struck by the "storm of the century" in 1992. Airline traffic has been inconvenienced by activity from Augustine, Redoubt, and Spurr volcanoes over the last decade. Coastal communities such as Nome are experiencing serious coastal erosion, in part because of their own mismanagement. Even small villages experience problems, significant in proportion to their size. Shishmaref, for example, built a seawall to stop erosion, but the wall failed within weeks and the town's buildings are threatened still.

Today's residents and future settlers in Alaska's coastal zone create a conflict with nature. Seismic events and coastal processes are normal and natural geologic events, neither unusual nor threatening until humans occupy this dynamic zone. Then, natural processes become natural hazards. The multifarious factors at work mean that geologic hazards in Alaska often assume a multiple character. The threat is not just from a single hazard, but from a series of interacting processes. For example, high rainfall can produce river flooding, glacial outburst floods, or landslides and debris flows. Earthquakes cause buildings to collapse due to ground shaking, but seismic effects may also induce landslides, debris flows, ground liquefaction, submarine landslides, and tsunamis.

A great opportunity now exists for Alaska and its people to design for and live with natural processes rather than confront them head-on. The goal of this book is to provide a starting point for recognizing geologic multihazards and then coping with them and reducing their impacts. Our risk evaluation adopts a multihazards approach that is also multidisciplinary. The authorship of this book reflects a wide range of applied disciplines in defining and mitigating natural hazards. Dr. Owen Mason is a geologist and anthropologist with more than 20 years of experience in coastal geology and archaeology, primarily in Alaska. His work has taken him to almost every region of Alaska's coast, and his multidisciplinary training brings an important perspective to human occupation of the coastal zone, land use, and the related hazards. Dr. Douglas Swanston is also an experienced Alaskan scientist with cross-discipline expertise. His work throughout Alaska as a forest hydrologist and his knowledge of landslides and related processes yield special insight into the hazards associated with unstable slopes. Ted Fathauer, of the National Weather Service Forecast Office, wrote the chapter on climatic hazards. Ted's firsthand observations and voluminous memory of climatic events in Alaska over the past 20 years provide a sound basis for the explanation of the coastal hazards associated with wind, sea, and ice. Geologists Dr. Orrin Pilkey and Dr. William Neal have collaborated for more than 20 years in coastal studies and are the coeditors of the 20-volume Living with the Shore series. Deborah Pilkey authored the chapter summarizing construction methods for mitigating wind, snow, and permafrost hazards; and Jane Bullock, Senior Policy Advisor to the Director, Federal Emergency Management Agency, provided the chapter on earthquake-resistant construction. Chapter 3 reproduces, with permission, much of the material published in *The Next Big Earthquake,* a pamphlet written by Peter Haeussler, U.S. Geological Survey, that was distributed to all Alaskan newspaper subscribers in March 1994.

Four components form the core of this multidisciplinary approach. Part 1 discusses the physical framework and human history of Alaska. Part 2 reviews the various hazards to which Alaska is subject. Part 3 presents hazard risk assessments for each coastal community in Alaska with more than 50 residents. Part 4 outlines frameworks for adapting to hazards and mitigating their potential impact.

Part 1 begins by describing the geography, climate, and geologic history responsible for natural hazards in Alaska (chapter 1). The history of human occupation and development in Alaska is covered in chapter 2. Part 2 discusses each major hazard from the perspective of the process and effects on the coastal zone. The outline progresses from seismic events such as earthquakes, volcanoes, and tsunamis (chapter 3); to physically unstable slopes, including avalanche, debris flows, and snow load (chapter 4); to climate-related processes, including storm surge and ice override (chapter 5); coastal erosion (chapter 6); river and glacier outburst flooding (chapter 7); and human-caused hazards such as oil spills and fires (chapter 8). Chapters 9 through 12, in part 3, provide hazard summaries for individual coastal cities and towns, including past experiences, present mitigations, and suggestions for improvement. Chapters 13 through 15, in part 4, summarize mitigation techniques through construction and coastal zone management in Alaska as determined by federal and state programs. The appendixes provide response checklists to individual hazards, a list of contact agencies, and useful references.

We are pleased to acknowledge the numerous institutions and individuals who provided the assistance that led to the completion of this project. Funding was provided by the Federal Emergency Management Agency (FEMA) and the National Science Foundation. The authors volunteered their time and provided illustrations and photographs from their personal collections. Stefanie Ludwig provided photographs. During the authors' travels across Alaska, numerous individuals, from taxi drivers to coastal property owners to museum staff members, shared their experiences, observations, and "war stories." Knowledgeable and helpful governmental and institutional representatives included Nancy Bird, Prince William Sound Science Center, Cordova; Marlene Campbell, City and Borough of Sitka; Jan Caulfield, City and Borough of Juneau; Paul Day, City Manager, Nome; Linda Freed, Kodiak Island Borough; Melanie Fullman, Coastal Coordinator, Ketchikan Gateway Borough; Richard Glenn, Ukpeagvik Inupiaq Corporation–Naval Arctic Research Laboratory, Barrow; Christy Miller, State Disaster Coordinator, Anchorage; and David Salmon, Prince William Sound Science Center. Important data on ashfalls were drawn from conversations with James Begét, Department of Geology, University of Alaska. Mary Anne Perez and Debbie Gooch provided administrative assistance and secretarial support. Amber Taylor deftly drafted the figures. We sincerely thank all of these individuals and apologize to the many we cannot name who also contributed to this effort.

William J. Neal
Orrin H. Pilkey

Living with the Coast of Alaska

PART I

The Lay of the Land and Its Occupation by Humans

1 Nature's Dangerous Brew: Geology, Climate, and the Coast

Americans from the lower 48 consider Alaska a frontier, perhaps the nation's last; even the state's license plates say so. As with the frontiers of the past, Americans view Alaska's land and seas as a challenge, a bronco to be tamed and broken—or, in this case, developed and exploited. The seeds of the earliest development were planted in the coastal zone first by Native Alaskans, then by Russians. Commerce in natural resources (furs, gold, fish, timber, and oil) spurred the growth of coastal villages, towns, and cities. History set the stage, and the coastal zone will bear the brunt of Alaska's growth in the twenty-first century. Two-thirds of the state's population now lives near the coast. Anchorage, Juneau, Ketchikan, Kodiak, Sitka, Kenai City, Valdez, Nome, Barrow, and scores of other towns and villages are all located on the coast. These communities face an arsenal of nature's hazards as daunting as those seen anywhere in North America.

Residents of the states that border the Atlantic Ocean and the Gulf of Mexico are becoming increasingly sensitized to such coastal hazards as storm surge, shoreline erosion, flooding, and hurricane winds. Alaska's coast faces all these plus additional processes driven by the climatic extremes of Arctic conditions. In straight distance, Alaska's seaward coast is 6,640 miles long, but when all of the corrugations and convolutions of bays, channels, fjords, and islands are counted, there are more than 33,000 miles of shoreline (fig. 1.1). This vast coastline faces two oceans and three seas, and accounts for more than half of the entire shoreline of the United States!

Alaska extends over 50 degrees of longitude, east to west, and over 20 degrees of latitude, south to north, and encompasses several climatic zones. The Alaska coast ranges from the windswept volcanic cliffs of the Aleutian chain to the temperate mossy rain forests of southeast Alaska—all carved in a variety of substrates, including sandy barrier islands, eroding bluffs of sand or silt, sheer bedrock cliffs, and silty river deltas.

The Alaskan shore presents a unique challenge to its residents. Virtually

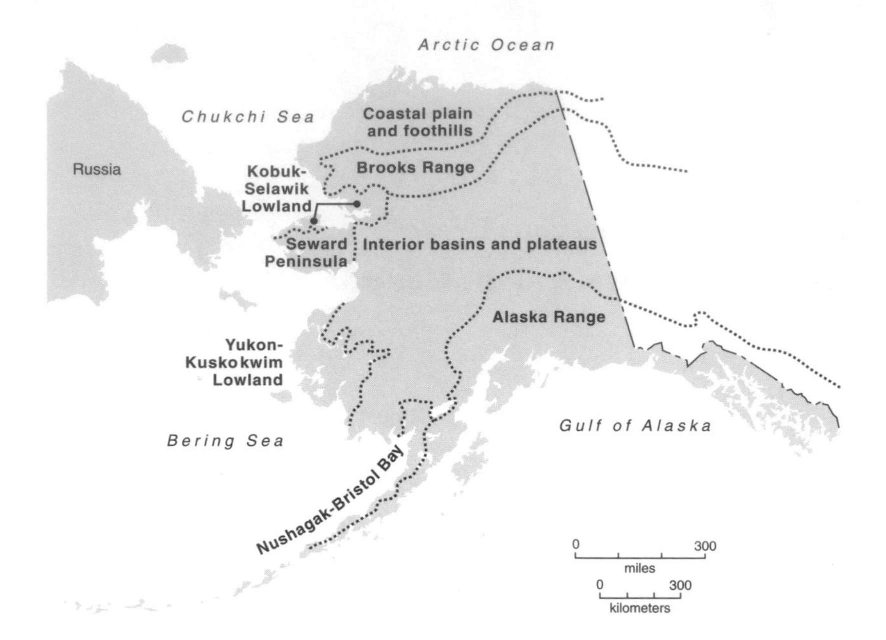

Figure 1.1 Map of Alaska showing the geographic subdivisions used in this book. The Alaska shoreline, some 33,000 miles in total length (including bays and fjords), encompasses a huge variety of coastal types, each with its own set of hazards.

the entire coast is subject to multiple hazards. Nearly all of Alaska may feel the effects of earthquake activity. In southern Alaska, the shore may drop or rise 3 feet or more in a single day as a result of seismic forces. The same earthquake, or one far distant in the Pacific, can send a towering tsunami wave hundreds of feet up coastal slopes. On the western, low-lying coast, communities face coastal erosion, storm surge, flooding, ice override, and possibly tsunamis as well. Areas with steep slopes have a host of other hazards to consider: debris flows, avalanches, wind channeling, and flash floods. And coastal processes constantly gnaw at Alaska's shores, reshaping the coast through erosion and deposition.

Living with the hazards of nature is a way of life, even a source of pride, for Alaskans. By necessity, portions of many communities must be located in hazardous zones. But the majority of private residential, commercial, and public property development can either avoid these hazards or greatly reduce their impact. Development is increasing, not always wisely, and communities are growing, putting more property and more lives at risk. The goal of this book is to provide the individual property buyer or owner with the fundamentals of hazard recognition and mitigation. The nature of Alaska's coastal hazards is the product of a high-energy brew of geology, climate, and the sea.

The Geologic Amphitheater

The geology of Alaska is perhaps the most complex in North America; the state is a jigsaw puzzle of varied parts that future generations of geologists will continue to unravel. This complexity is in part the result of a long history of upheavals along the margins of earlier crustal plates (table 1.1). Present-day Alaska owes its configuration and the frequency of its seismic and volcanic events to the competing movements of two major crustal

Table 1.1 Geologic and Historical Time Line of Events in the Development of Alaska (see figs. 1.2 and 1.3)

250 Ma	Alaska not yet assembled as land mass
240 Ma	Wrangellia terrane forms near Equator off modern-day Peru
>200 Ma	Alexander terrane forms in South Pacific
150 Ma	Arctic Coastal Plain rotating apart from northern Canada
130 Ma	Alexander superterrane and Wrangellia terrane join near Baja California and continue norhward
60 Ma	Assembly of Alaska nearly complete; Alaska Peninsula volcanic arc begins to emerge from sea
<5 Ma	Formation of Alaska Range
3–2 Ma	Early glaciation of Saint Elias Mountains
85–11 Ka	Bering Land Bridge intermittently exposed
15 Ka	Rapid melting of continental ice, increase in sea level
12 Ka	End of last glacial expansion
12–11 Ka	Earliest evidence of settlement in interior Alaska
11 Ka	Flooding of Bering Strait
5 Ka	Sea level stabilizing (close to modern level)
5–4 Ka	Earliest known semisubterranean pit houses in Alaska
3 Ka	Earliest known pottery in Alaska
3–2 Ka	Alpine glacier expansion and storms on coast
2 Ka	Complex societies develop in Bering Strait region, Kodiak and southeast Alaska; whaling at Point Hope
A.D.	
1500–1800	Little Ice Age
1741	Bering Strait discovered by Europeans
1799	Russian American Company established
1799	Russians establish fortress near Sitka
1816	Otto von Kotzebue explores Kotzebue Sound
1867	Alaska sold to U.S. by Russia
1897–98	Klondike gold rush
1899	Nome gold rush
1914	Anchorage founded
1942	Japanese invade Aleutians
1959	Alaska achieves statehood
1964	Good Friday Earhquake
1968	Oil discovered at Prudhoe Bay
1971	Alaska Native Claims Settlement Act passed
1977	Alyeska Pipeline completed

Note: Ma=millions of years ago; Ka=thousands of years ago.

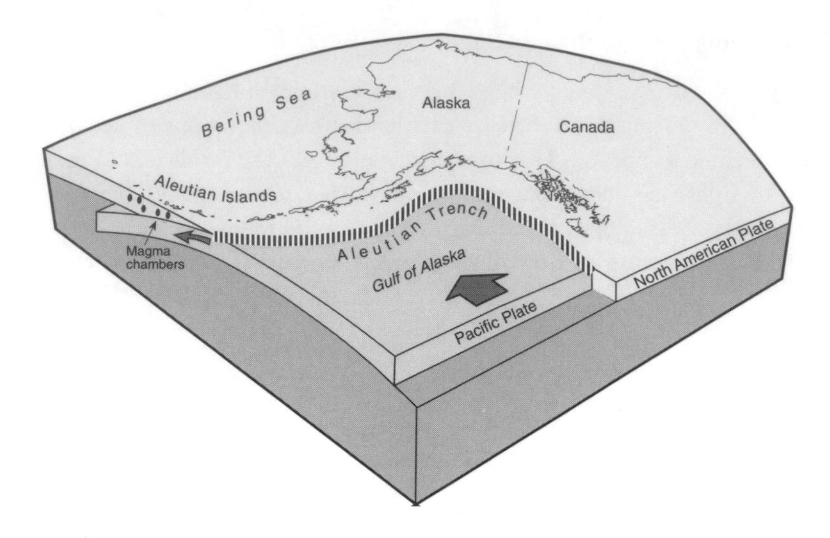

Figure 1.2 The crustal plates that make up Alaska. Subduction of the Pacific Plate beneath the North American Plate produces much of the earthquake and volcanic activity of Alaska.

plates (fig. 1.2). The Pacific Plate is moving northwestward against the North American Plate, and in the process the denser oceanic rocks are being pushed or pulled beneath the edge of Alaska. The tremendous energy expended along plate boundaries is expressed as earthquakes and volcanic activity. Faults marking earlier boundaries between former oceanic and continental blocks make up the quiltwork of Alaska's structural geology (fig. 1.3). Many of these faults are still active, placing much of the state in the highest risk category for earthquakes. The same events account for the rugged nature of the state's topography. An average uplift rate of a fraction of

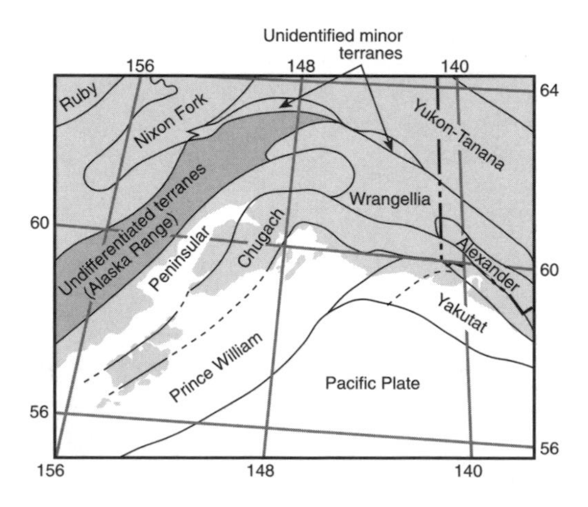

Figure 1.3 Geologic terranes of southern Alaska. The lines between the geologic terranes are fault zones, some of which are active.

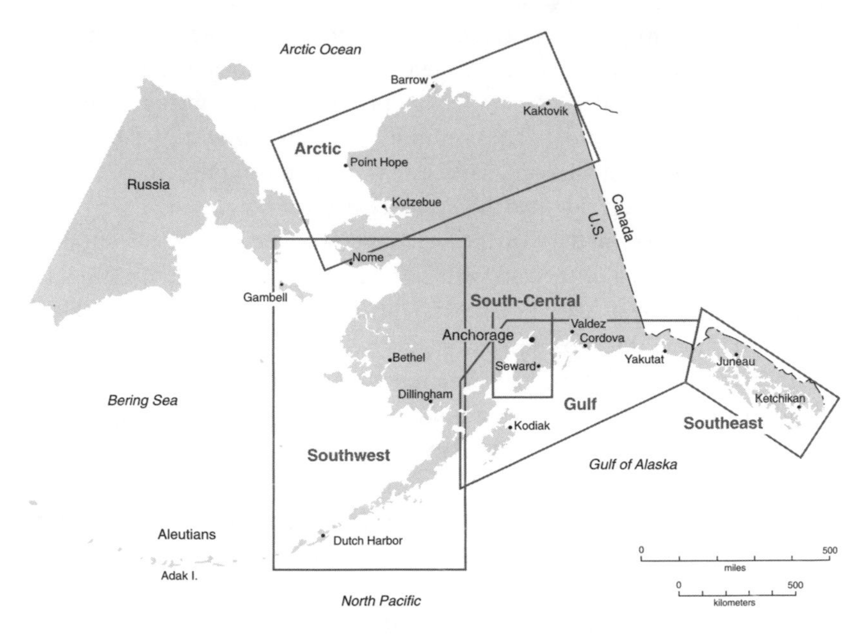

Figure 1.4 The major physiographic provinces of Alaska. Modified from *The Geology of Alaska,* ed. G. Plafker and H. Berg.

an inch per year doesn't sound like much, but it produces mountains over the span of millions of years.

The crustal plates are still moving. South-central Alaska is moving generally west along the Denali and Tintina faults, while southeast Alaska is moving in a northerly direction different from the rest of the state. The geologic evolution of Alaska is an ongoing process, as residents are frequently reminded when the peninsular volcanoes erupt or seismic waves vibrate the coast.

Two vastly different landforms provide the geologic backdrop for Alaska's coastal communities: the mountains and uplands of the southern arc, and the lowland plains of the west and north (fig. 1.4). Although Alaska's coast is famous for its rocky headlands, rugged slopes, and spectacular glacial fjords, more than half the coast is bounded by flat coastal plains with low bluffs of unconsolidated sand or silt. Most of Alaska's population centers lie on the rocky shores of the southern coasts; only small towns such as Nome, Barrow, and Kotzebue are located on the flat northern shores.

The contrast between the north and south coasts is largely one of hard bedrock in the south and soft, unconsolidated sediment in the north. The bedrock, the brittle component of the mountain massifs, is a geologic patchwork of different rock types and geologic ages stitched together along myriad fault lines as the continental blocks shift on the tectonic conveyor belt (fig. 1.3).

The mountains of Alaska impress even the casual observer with their youthfulness. Continued uplift has left staircases of former river terraces and, along with glacial erosion, maintains the fabled abrupt slopes of Denali (Mount McKinley). The continuing upward growth of the mountain chains can be felt all along the coast, as it was during the tremendously powerful 1964 Good Friday earthquake (see chapter 3).

In an earlier period, the Pleistocene, Alaska was part of the subcontinent of Beringia and was connected with eastern Siberia across the continental shelf that skirts west Alaska offshore. During the Great Ice Age, ice sheets thousands of feet thick formed and spread over half of North America from northern Canada to Iowa, and over much of northern Eurasia from England to Russia. A tremendous amount of seawater was locked up in this glacial ice, and sea levels fell dramatically as a result. As the sea level fell, Beringia emerged.

In Alaska, the glaciers of the Ice Age were limited largely to the mountains rather than being one continuous icecap. Even in the south the ice cover was not total, and the glacial shore of southern Alaska probably resembled the modern coast of Greenland (fig. 1.5). Most likely, there was enough land along that shore to allow the first colonization of America by human beings (see chapter 2).

Figure 1.5 Extent of glaciation in Alaska during the last major ice age. Note that large areas, including the North Slope, were not covered by glaciers. The general location of Beringia, the land area connecting Alaska and Siberia when the sea level was lower, is also shown.

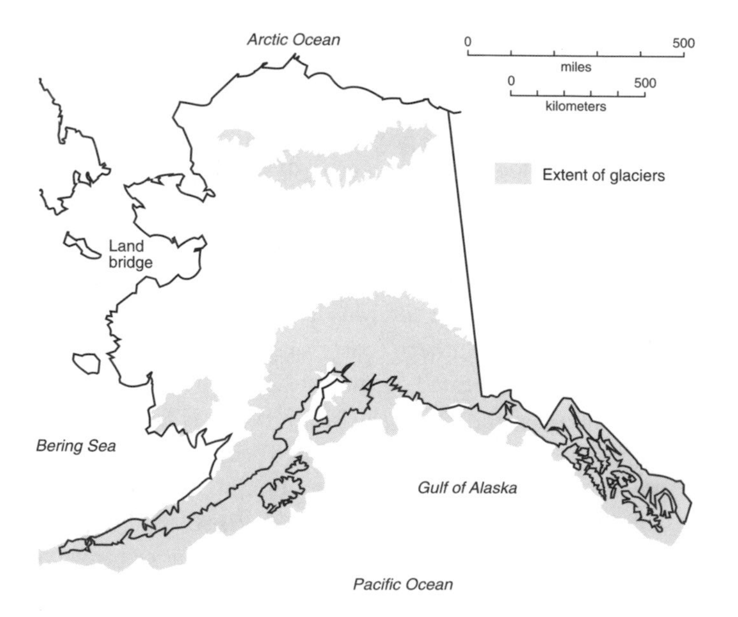

Just exactly how many times glaciers waxed and waned, carved and planed the margins of Alaska is still unknown. The seafloor's sedimentary record indicates that 15 to 20 glaciations scoured the land. About 2–3 million years ago the first Alaska Range glaciers dropped scattered glacial tillstones to the bottom of the Gulf of Alaska. Glacial periods followed in cycles of 23,000–110,000 years in response to the tilt of earth's axis, the shape of its orbit, the timing of its closest approach to the sun, and other factors. Worldwide, the last continental glacial expansion ended about 12,000 years ago, but glaciers in Alaska have continued to expand and contract since that time. Glaciers do not all respond to climate changes at the same time, and southeast Alaska glaciers are notoriously fickle, depending more on their own geometry and ice balance than on the big picture of worldwide climate.

Several major ice advances out of the mountains did occur throughout Alaska: one occurred about 2,000–3,000 years ago, and a major glacial expansion took place quite recently during the Little Ice Age of the sixteenth through nineteenth centuries. Mountain uplift due to seismic energies continued concurrently with the glacial erosion of higher valleys and the filling of lower valley bottoms by rivers charged with glacier-scoured sediments.

In terms of human settlements and land use, the land-altering effects of the last glaciation are paramount. The sediments exhumed and emplaced by those glaciers are today's engineering challenges and often complicate other geologic hazards such as earthquakes, mass wasting (slow downslope movement of rock debris), coastal erosion, and storm surge. Glaciers vacated portions of south-central and southeast Alaska as recently as 200–400 years ago. In fact, actively calving glaciers are only 3 miles away from housing subdivisions in Juneau (fig. 1.6).

On a global scale, climatic warming and the wastage of glaciers translates into higher sea levels. As sea level rises, glacial valleys are flooded, forming fjords; stream gradients are lowered; and the loci of sediment deposition shifts inland. Dry coastal plains become wetlands and marshes, and the energy of the sea transforms bluffs and cliffs into beaches and spits. Coastal and shoreline evolution contributes in large measure to Alaska's ongoing changes and the kaleidoscope of resulting hazards.

Sea Level and Coastal Changes

Nearly all of us are familiar with short-term changes of sea level due to the daily tides, storms, and fluctuations in atmospheric pressure (see chapter 5). In 1964, some Alaskans witnessed the power of seismic movements to lift or depress the sea level up to several tens of feet within minutes.

Figure 1.6 The Mendenhall Glacier viewed from the air. This glacier's retreat during the last few decades may be responsible for the slight sea level drop experienced by Juneau. Removal of the weight of the ice has caused the land to rebound, raising its elevation relative to sea level. One of this Juneau suburb's hazards is flooding caused by irregular rates of water discharge from the glacier.

In the long term of millennia or tens of thousands of years, the greatest single factor controlling sea level is the worldwide change in ice volume. The sea level drops as water is frozen into glaciers. Conversely, sea level rises as water is added to the sea during glacial melting or as surface seawater expands due to atmospheric (climatic) warming. Human activities may be responsible for increasing the temperature of water and increasing the rate of glacial melting.

Sea level change is usually a function of both the land and the sea. Worldwide, the sea level is slowly rising in most shoreline localities. Along many of Alaska's shores, however, the relative sea level is dropping as the land continues to rebound after the retreat of the glaciers. The earth responds to pressure much as a sponge does. First, the pressure of the ice mass compresses and deforms the earth's surface; then, as the weight of the ice is removed, the earth recovers its integrity. This process, termed isostatic rebound, occurs mostly within a few thousand years of the ice layer's removal, but slower rebound may continue for many thousands of years afterward. Rates of isostatic rebound in Alaska today depend on former ice thickness and the recency of the last glacial advance. In Skagway and Haines, more than an inch of uplift every year may possibly be attributed to isostatic rebound.

Vicissitudes of Sea Level in Alaska

During the late Pleistocene, the sea level was considerably lower (by nearly 300 feet) than it is today and the shoreline was a considerable distance out to sea from its present limits. The continental shelves surrounding Alaska were fully exposed, creating the large land bridge of Beringia. As continental glacial ice melted rapidly about 15,000 years ago, the sea level rose. Most of this rise was rapid, occurring in the first several thousand years after the glaciers retreated. Rates of rise may have averaged 1 foot per 20 years, many times the present rate. Rising seas covered the Alaska continental shelves from both the Pacific and Arctic Oceans. The sea encroached up river valleys, too, forming long estuaries.

About 11,000 years ago, the rising sea levels sundered the connection between Siberia and Alaska, flooding a low pass and initiating the Bering Strait at a depth of around 160 feet below current sea level. Barely 1,000 years later, the sea level had risen another 100 feet, flooding the low plains that are now Norton and Kotzebue Sounds. River deltas across the state shifted in both position and elevation, keeping pace with the position of the sea level.

The sea level rose in southern Alaska, too, but the land rose even more rapidly in isostatic rebound. For that reason, former shoreline features such as deltas or storm ridges are now several hundred feet *above* the present sea level. The elevation of former beach and shoreline features depends largely on the former thickness of glacial ice, which thinned toward its outlet at the former sea margin—in many cases now the outer coast of southeast Alaska. In general, the highest uplifted shoreline features lie within the inner fjords, and the lowest are on the outer coast. Portions of the outer coast of southeast Alaska were not glaciated and consequently underwent little or no isostatic rebound.

The sea level continued to rise rapidly until about 5,000 years ago, then slowed significantly. As a result, shoreline features (e.g., spits and beach ridges) that were present at that time are still preserved along many Alaskan shores. The preservation of coastal features requires a surplus of sediment input relative to the rate of sea level rise, a comparatively low tidal range, and the absence of subsidence. According to the age of now-submerged peat layers, the sea level in the Chukchi and Beaufort Seas was about 4.5 feet below present levels in 3000 B.C.

Coastal Evolution

Shorelines provide an excellent lesson on the interplay between geologic, climate, and oceanic processes. The dynamism of the shoreline is readily apparent to even casual observers, who can easily distinguish a gravelly,

Figure 1.7 Eroding permafrost bluff in Barrow, September 1991. The gradual retreat of this bluff clearly threatens some of the community's buildings. The famous Mound 44 archaeological site, where the "Frozen Family" was found, is in the background.

storm-raked winter beach from a sandy summer beach. The bruised and barren scarp associated with an eroding bluff is another familiar type of coast (fig. 1.7). Unconsolidated, unvegetated bluffs are unstable and readily collapse. Bluff failure can be initiated from above or within if the sediments are suddenly saturated with rainwater (or even water from lawn watering) and the increase in weight exceeds the holding capacity of the sediments. The resulting slumps can be extensive and catastrophic in areas with steep slopes (see chapter 5).

Bluff erosion is more commonly incremental, however, and is often triggered by the onslaught of storm waves at the base of the bluff. Over the long term, beaches may form at the base of bluffs as sediments erode, depending on the persistence and strength of the waves available to remove the eroded material. The presence of a beach provides natural insulation from erosion; the beach must be removed before the bluff is completely open to wave attack. Often, we tend to forget that eroded bluff sediments are the source of the beaches so valued for recreation or access, as along the Kenai Peninsula, where beaches provide work space for seiners and parking for recreational vehicles. Seawalls erected to prevent bluff erosion starve beaches (see chapter 6).

Bluff erosion can occur under a variety of conditions—some catastrophic, some gradual. A major factor in bluff erosion is the type of sediment in the bluff. Hard-rock bluffs, or cliffs, may change little during human life spans, although an earthquake or tectonic uplift can drastically affect them, as happened in Prince William Sound in the space of minutes after 5:36 P.M. on March 27, 1964. Sand bluffs are usually more erodible than other types. In Alaska, where most bluffs are frozen for all or part of the year, bluff collapse rates can also be significant in fine sediments such as silt. On the North Slope of Alaska, storms produce elevated sea levels and high-energy waves that can carve substantial notches into ice-rich bluffs. Such notches can penetrate several yards under the bluffs, and eventually the cavity causes the collapse of a whole stretch of bluff line.

Earthquakes sometimes trigger large collapses in bluffs composed of fine sediments, such as those west of Homer. Earthquake-induced subsidence can also initiate wave attack on bluffs by effectively removing the protective beach or dunes, as occurred on the Kenai Peninsula.

Along many parts of the Alaska coast, especially in areas surrounding river deltas, the land is growing seaward. This process, termed accretion, is sporadic and is often associated with spits and barrier islands. Most commonly, linear accumulations of sand or gravel are added to beaches or along tidal inlets in fair-weather periods or following low-energy storms that carry sediment landward (fig. 1.8). During the last 5,000 years of sea level stabilization, more than 100 beach ridges, 3 miles in width, were added to the Cape Krusenstern foreland on the north shore of Kotzebue Sound. Similar additions of land occurred across the sound at Cape Espenberg, near Wales, and at nearly a dozen other locations along the Bering and Chukchi Seas from the Alaska Peninsula to Point Barrow. Locations where beach ridges are actively accreting to the coast provide records useful for reconstructing past climates and are helpful in locating archaeological sites.

Dates obtained from Alaska beach ridges show that storms were particularly intense about 3,300–1,700 years ago and 1,200–200 years ago. These periods correlate with major glacial expansions worldwide. The latter period includes the Little Ice Age, associated with widespread advances of glaciers in the mountains across the Northern Hemisphere.

Rocky coasts may develop characteristic erosional features if the sea level is maintained at the same position for a lengthy period (fig. 1.9). Erosion on rocky shorelines depends on the strength and resistance of the underlying rocks, the distribution of planes of weakness within the rock, and, in Alaska especially, the number of freeze-thaw cycles, which can "soften" up the rocky faces and unleash a multitude of rock clasts as gravel. When storms and tidal forces act on less resistant rocks, a shore platform, or wave-cut terrace, is produced as cliffs are cut back. Shore platforms slope gently seaward and may be exposed as rocky flats during low tide. Wave energy concentrated at the position of the low- to mean-tide range will produce a distinct

Figure 1.8 Beach ridges such as these on the Espenberg Spit are linear accumulations of sand and gravel that accrete to the shoreline. Usually they indicate that the shoreline is growing seaward.

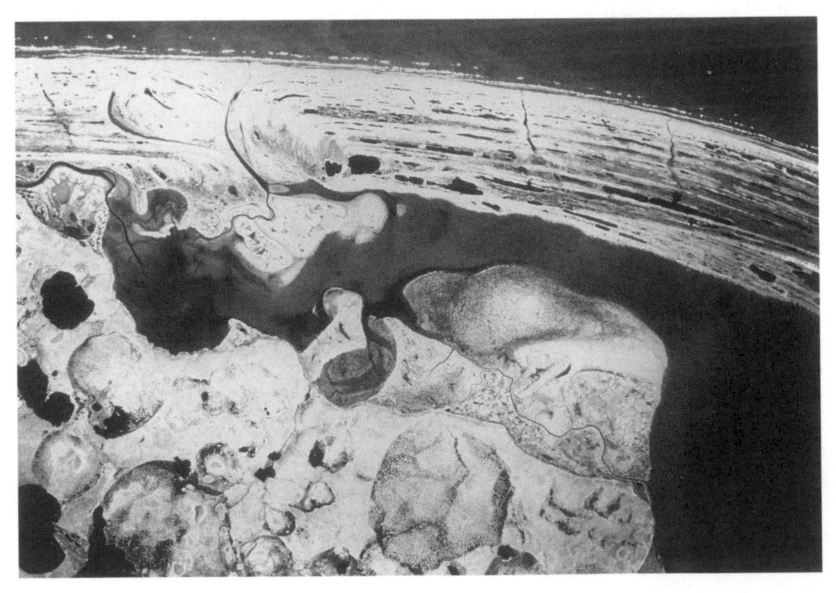

Figure 1.9 The rocky coastline of Amchitka Island in the Aleutian chain. The flat surface above the sea cliff is a wave-cut platform that was once the seafloor before being uplifted. The isolated rocks offshore are sea stacks.

tidal bench. Waves and other erosive forces produce a variety of other landforms in rocky coasts, including caves, arches, sea stacks, and other residual masses. If subjected to rapid uplift due to tectonics or longer-term isostatic rebound, a series of older shore platforms, wave-cut cliffs, caves, or other features may be preserved at elevations above the present sea level.

Ice and the Alaska Coast

In the summer, the Alaska coast behaves like most coasts: sand and gravel move onshore, offshore, and alongshore; offshore bars form and move parallel to the coast (see chapter 6). In general, the summer beach, even in northwest Alaska, is a lethargic beast fattening itself with sediment. But as soon as autumn arrives, the shores of northern Alaska begin to resemble the "Icebox" decried by Secretary of State William Seward's detractors. Once the ice foot has crusted over the swash zone, the coast is insulated against erosion, and beaches and bluffs are largely stable for the duration (fig. 1.10).

At Point Barrow, on the Arctic Ocean coast, ice formation typically starts in early October, and ice melt or breakup is not complete until mid to late July. Along the Bering Sea coast, near Nome, freeze-up is usually in early November, and breakup occurs during late May. In the southern Bering Sea from the Yukon Delta to Bristol Bay, ice usually forms later and breaks up earlier. Of course, extreme weather conditions can add a few weeks either way, as do variations in latitude, wind direction, and duration. In addition, the warming associated with El Niño sometimes nudges the ice margin up the southeastern shore of the Bering Sea, north of its usual position.

The greatest energy available for shoreline transformation comes from the storms produced during the open-water months, as is also the case elsewhere in the world. In Alaska, ice plays only a comparatively minor role in shoreline evolution, principally by slowing down or preventing other processes. Because no waves hit the shore when the surf zone is frozen, shorefast ice counteracts the effect of strong winter winds. Pack ice and pan ice reduce the amount of fetch available to generate or transport waves within the sea. However, the presence of ice can also mean that less time is available for recovery after storms, and beaches and bluffs may thus be more susceptible to the lasting impact of a single erosive storm. This is the

Figure 1.10 The maximum and minimum areal extents of pack ice or sea ice around Alaska.

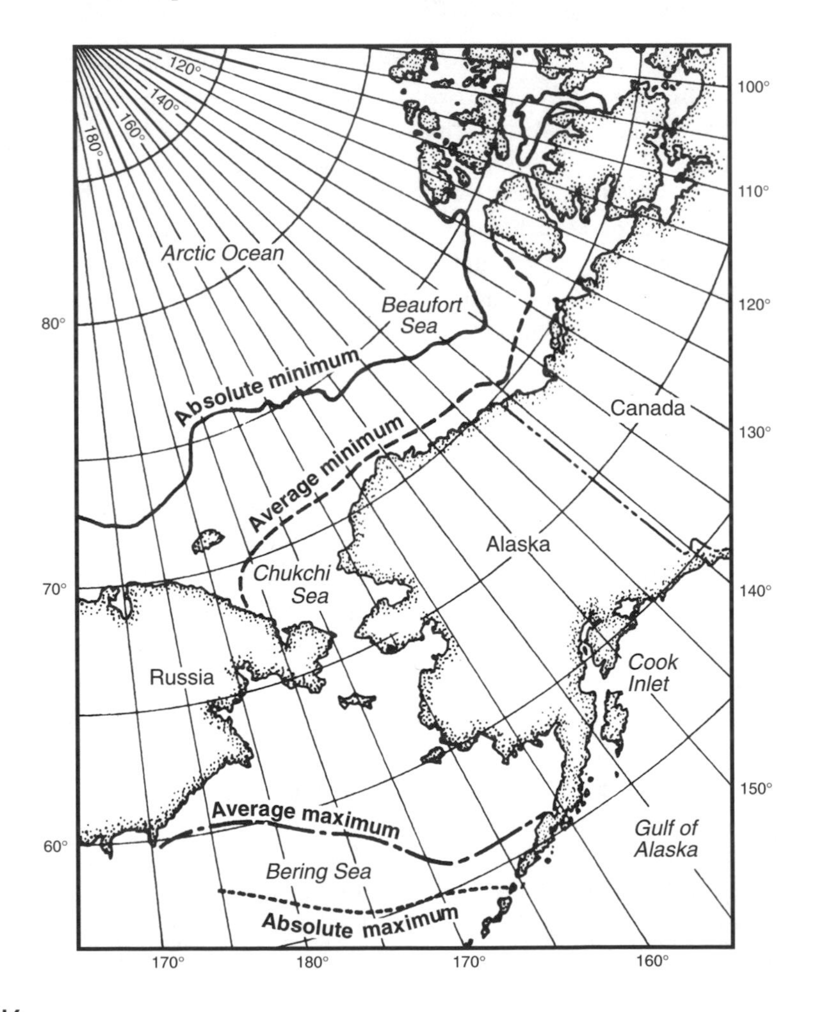

case on the north shore of Alaska, where the erosion rate is five times that of temperate zones—nearly 200 feet of bluff lost in 30 years. Alternatively, ice can add material to beaches by bulldozing the lower shoreface and pushing cones of sediment several feet high onto the upper beach. Sediment incorporated into the ice foot during freeze-up may be released on the upper beach during breakup as waves hit the swash zone. On the North Slope, large blocks of ice on beaches form wallows, erosional depressions 6–9 feet deep, when subjected to wave action.

Three types of ice affect Alaskan shores: shorefast ice forms as a deposit plastered to the shore; pack ice, or sea ice, forms as floating, shifting pans of ice; and berg ice (fig. 1.11) detaches, or calves, from glaciers. All three types of ice can offer both benefits and hazards to communities. During the fall, drifting pack ice can function as a storm barrier by decreasing the ability of the wind to generate waves. Of course, large chunks of ice can easily become lethal weapons if they are rafted onshore during a storm, as happened in the Nome surge of November 1974. In fact, many urban Alaskans are unaware that the shore is not necessarily "locked in" from November to June. On occasion, strong winter storms wrench away the shorefast ice and allow seawater to surge onshore.

The hazards of ice to shipping and transport are familiar, but its benefits are not. In winter, the shorefast ice in stable ice areas such as Kotzebue Sound and the Yukon Delta becomes a superhighway. Government authorities maintain veritable roads over the ice, and snow machine travel between islands and the mainland is frequent. Knowledge of shorefast and pack ice is part of the cultural heritage of the Inupiat (north Alaska Eskimo) people, replete with a detailed terminology and vocabulary. The Iñupiat hunt seals at their breathing holes in shorefast ice, and they use opening and closing breaks, or leads, in pack ice to hunt bowhead whales.

Although shorefast ice is a yearly occurrence on the northern half of the Alaska coast, ice affects some of the southern shores along the Gulf of Alaska as well. Floating ice is common farther south in Prince William Sound, Cook Inlet, and southeast Alaska. In south-central and southeast Alaska, icebergs detach from fjord-head glaciers during summer.

Along the south shore of the Alaska Peninsula, ice may form in sheltered embayments, but it does not form a continuous surface. Ice in Cook Inlet, with its high tides, forms mostly as pan ice and rarely covers the entire embayment. Floating ice and beached ice pans are frequent winter sights along the tidal flats of Anchorage. Close ice pans (covering more than 70–90 percent of the water surface) dominate the upper inlet from early January to early March. Generally, Cook Inlet is ice free by April, although in record years thin ice has persisted until late April. Kachemak Bay and lowermost Cook Inlet are generally open, although Homer Spit has had pack ice offshore in record ice years. Shorefast ice is restricted to the west coast of

Figure 1.11 Berg ice, a common scene near the glaciers of southeast Alaska.

Cook Inlet, a rocky margin with numerous sheltered bays. Typically, shorefast ice has formed along the entire west coast of Cook Inlet from the Susitna River mouth to Kamishak Bay by early February.

In Prince William Sound, shorefast ice regularly forms at the heads of fjords in cold weather. The detachment of large icebergs off the Columbia glacier can be a serious navigational hazard near the Valdez Arm, at the marine terminus of the Trans-Alaska Pipeline. Bergs are driven into the sound under the influence of northerly winds during winter. Berg-rich waters are found in Aialik Bay, down from the Harding ice field, into Knight Island Pass, down from the Sargent ice field. Little or no ice reaches the open areas of Prince William Sound.

In southeast Alaska, icebergs are limited to areas in the vicinity of calving glaciers. Waves up to 10 feet high can be generated when a chunk of ice breaks away from a glacier. The greatest amount of this float ice is found in spring; it decreases with the onset of summer. If the glaciers continue their present retreat, the amount of ice may increase significantly in the next few years. At present, icebergs are confined to the passage east of Petersburg, down from the LeConte Glacier, and three fjords south of Juneau (Taku, Tracy, and Endicott Arms). Glacier Bay and appropriately named Icy Strait contain many icebergs throughout the year that often prove hazardous to navigation up to 20 miles out to sea. At times, ice piles up along the shores of Icy Strait. Otherwise the coasts of the Alexander Archipelago are ice free throughout the winter.

Nothing Stays the Same

Alaska's geologic setting is a youthful complex of dynamic features that include zones of seismic and volcanic activity, widespread unstable slopes, and an incredibly long coastline subject to an impressive array of natural forces. The surface of the land is overlain by climatic and oceanographic regimes that are just as dynamic, multiplying the hazards for Alaskans. The dynamism of its geologic and biotic environments, the minerals in its rocks, and the heavy rains that nourish its immense trees have produced bountiful resources that make Alaska irresistibly attractive for economic development. Abundant minerals, timber, furs, and fish have enticed humans into harm's way.

2 Settling Alaska: 11,000 Years of Human History

The descendants of the first colonizers of Alaska, its native peoples, constitute nearly 15 percent of the present population, although many coastal villages are 80–90 percent Native American. Most villagers continue to participate in the perceived natural order of cycling between animals and humans through hunting and fishing for subsistence.

Peopling the New World

As soon as the 50-mile-wide Bering Strait was discovered by Europeans in 1741, it became clear that Native Americans had entered the New World either by walking over the ice-covered strait or by boat. Geologists studying the history of the strait found that it closed and opened as continental ice sheets formed and melted. The first entry of Bering Strait water into the Arctic Ocean occurred about 4 million years ago, as dated from Bering Strait shells found in Arctic Ocean sediments. Old shorelines near Nome, now uplifted by tectonism, provide details of the subsequent history. The Bering Strait was dry ground from 85,000 years ago until about 11,000–20,000 years ago. The rise of sea level from its Ice Age minimum to its present level took nearly 6,000 years to complete, and a sea level very near modern levels was attained about 5,000 years ago.

Most archaeologists believe the first Americans made a land crossing. Until recently, however, most of the evidence from geologic cores of the continental platform indicated that the Bering Strait last flooded about 14,000–15,000 years ago. This created a contradiction, because the earliest archaeological remains in Alaska are only 12,000 years old. To cross open water, hunters would have needed watercraft, and they probably would have focused on hunting maritime animals rather than on large terrestrial mammals. New radiocarbon dates from an extensive peat bed 150 feet below the surface of the Chukchi Sea show that the land bridge remained exposed until 11,000 years ago, however, which allows plenty of time for a land crossing. On the other hand, the minority of archaeologists who favor the maritime pathway were given a boost by recent geological evidence that the coast of Alaska and British Columbia had a less forbidding climate and more resources than was previously thought.

The language and ethnic background of the very first settlers cannot be ascertained on the basis of the few stone tools that were left at interior Alaska campsites 11,000–12,000 years ago. Rising sea levels have overrun the old coastal village sites, making it difficult to find evidence of early maritime peoples along Alaska's shores. The earliest known coastal sites, 8,000–9,000 years old, are in the Aleutians and southeastern Alaska. Clearly, some of the earliest Alaskans had watercraft, either kayaks or dugouts, from which they hunted marine mammals or speared fish; they probably depended on seals for sustenance.

Human populations remained relatively low and dispersed over the landscape for several thousand years, from about 8,000 to 5,000 years ago. Archaeologists have found belowground pit houses, sealing harpoons, throwing boards, and arrowheads 4,000–5,000 years old that resemble those used by modern-day Iñupiat and Yup'ik Eskimos and Athapaskan Indians. By 3,000 years ago (1000 B.C.), sizable winter villages comprising multiple families had grown up at salmon streams. These people made pottery, traded over long distances, and accumulated substantial stores of food. In southeast Alaska, the people built elaborate wooden fishing weirs in the intertidal zone to trap salmon.

About 2,000 years ago, the time of the Roman Empire, complex societies developed in the Bering Straits region, on Kodiak Island, and in southeast Alaska. Settlements on coastal promontories provided dependable access to sea mammals (walrus and whales), and success in hunting led to class differences, especially in communities that hunted whales.

The oldest firm evidence of whaling is from Saint Lawrence Island and dates to about A.D. 500. Whaling communities, led by wealthy whaling captains, had an organized political life. These communities grew, along with the political power of their leaders, leading to the development of elaborate art styles and battles for the control of strategic resources. At times, religious cults flourished throughout the region. Bering Strait communities engaged in trade for iron, clay, skins, and other perishable goods. Climatic changes, usually colder temperatures, caused declines in animal availability that led to famine, adversity, and increased warfare. By A.D. 1700, the population of Alaska's coastal areas had grown to levels only recently exceeded. Villages like neighboring Nuwuk and Utqiagvik in the Barrow vicinity had more than 300 people each. The Kodiak Islands were home to several thousand Koniag Eskimos.

Motivations for choosing a location most often involved the acquisition of subsistence foods or resources such as whales, seals, and salmon. By necessity, settlements were located in high-risk areas, but moving or rebuilding was considerably less complicated back then.

Patterns of Modern Settlement

Around 1750, European seamen focused their exploratory and mercantile zeal on the northwestern Pacific, which constituted the last remaining geographic unknown. The names of Alaska shoreline features are a cartographic record of this progress, from the first voyages of Vitus Bering in 1741 to that of the searchers for the lost Franklin party in the 1840s, when the basic outline of Alaska was filled in on European maps. The Russians, who had been on Kamchatka since the 1640s, had proximity on their side, although the Spanish were the early birds in the Pacific Ocean. After several brief forays north of the Bering Strait into Kotzebue Sound, the Russians focused their attention southward toward the Aleutians and the Gulf of Alaska. Spaniards led by João Cabrilho (called Juan Cabrillo by the Spanish) had first visited California in the 1540s. The Spanish founded San Francisco in 1776 and sent Juan de Fuca as far north as Washington in the 1780s. A group of Spaniards led by Captain Malaspina reached the glaciated Alaska coast near Yakutat in 1792; a piedmont glacier the size of Switzerland is named in his honor.

The general outline of Alaska was mapped in the late 1770s by the Englishman James Cook, who skirted the coasts of south-central and west-to-northwest Alaska, naming many embayments along the way, including Cook Inlet, Norton Sound, Bristol Bay, Cape Lisburne, and Icy Cape in the Chukchi Sea. In 1816, the Russian-sponsored explorer Otto von Kotzebue outlined the shallow shores of Kotzebue Sound, lending the names of his crew members Shishmaref, Espenberg, Krusenstern, and Choris to the Alaskan shore. Numerous other Russians, including Golovnin and Alexander Kashevarov, recorded the contours of the southern Bering Sea coast from 1810 to 1825. The Englishman Frederick William Beechey established the first contact with Point Barrow people in 1826.

The European explorers were fueled by economic self-interest. The primary target was the valuable fur-bearing sea otters found in the Aleutians, the Kodiak Archipelago, and southeast Alaska. The scale of Russian depredations on sea otters is legendary, and the millions of fur seals of the Pribilof Islands faced extinction by 1805, scarcely 20 years after the Russians found them. The Russians were the first to build permanent settlements in Alaska; the British and Americans were content to trade with native inhabitants from ships. The first Russian traders, Grigory Shelekhov and Ivan Golikov, settled near the native village of Unalaska in the 1770s and at Three Saints

Bay on Kodiak in 1784. By 1790 Shelekhov had relocated at Pavlovsk Harbor, the present site of Kodiak, which boasts a good harbor and a series of defensible bedrock promontories.

The major period of Russian expansion followed the incorporation of the Russian American Company in 1799. The Russians founded a fortress near Sitka in 1799 but faced stiff resistance from the Tlingit people, who destroyed the fort in 1802. The Russians returned in 1804, destroyed a Tlingit village, and built the town of Novo-Arkangelsk, or Sitka, which was declared the capital of Russian America in 1808. Russian settlements remained fairly small and thinly spread over the landscape, although forts were established along the southwest Alaska coast at Saint Michael on Norton Sound, at the mouth of the Nushagak, on the Kuskokwim, and several hundred miles up the Yukon River. Perhaps the most profound effects of the Russian colonial occupation were the introduction of Christianity by Russian Orthodox priests and, inadvertently, epidemic diseases such as smallpox. Native villages suffered substantial declines in population during the nineteenth century. Other profound changes followed from the coercion by Russian traders that led to the abandonment of many villages, forcing many Native Americans into trade and otter hunting.

In 1867, with fur revenues declining, the Russians sold their interest in Alaska to the United States, this despite the fact that nearly all the land and its people were sovereign and independent of Russian authority. In fact, the American purchase and occupation of Alaska were not ratified by any treaty with Native Alaskans until the passage in 1971 of the Alaska Native Claims Settlement Act (ANCSA).

The U.S. occupation of Alaska involved several phases of economic and military endeavors: (1) military occupation, 1867–1912; (2) territorial status, 1912–59; and (3) statehood, since 1959. Few outsiders came to stay permanently during the first phase; most visitors were looking for short-term profits. As elsewhere in the West, the military explored, mapped the interior, and staffed the forts. As furbearer numbers declined, mining, timber, and fishing gained in economic importance. The first cannery opened in 1878 near Sitka, and more than 55 canneries were in operation by 1900. Petersburg, for example, originated as a cannery town in 1897. The discovery of gold in 1880 led to the founding of Juneau, named for Joe Juneau, one of the lucky miners. Other towns thrived as supply depots for miners; the Klondike gold rush of 1897–98 led to the establishment of Valdez and Skagway as supply centers and entry points. After 1906, Cordova served as the terminus for a 196-mile railway up the Copper River to the Kennecott Mine near McCarthy. The discovery of gold in 1899 on the beaches of the Seward Peninsula led to the founding of the city of Nome in 1902. By that time 25,000 prospectors had swarmed onto the beaches in search of riches.

Presbyterian missionaries founded several towns in the late 1890s, including Haines and Metlakatla. The efforts of missionaries and educators fostered the centralization of the Native American population in settlements after 1920.

The population of Alaska remained predominantly Native American until 1930, despite the sudden influx of fortune hunters during the gold rush decade of 1900–10. The establishment of roads and railroads into the interior led to the founding in 1914 of the state's largest city. Anchorage started out as a construction camp on the northernmost shore reachable by sea traffic. The town remained relatively small (about 4,000 people) until World War II, when the erection of Fort Richardson tripled the population in several months. The military presence was in response to the Japanese invasion of the western Aleutians in 1942—the first violation of U.S. soil by a foreign power since the War of 1812.

Americans' perception of Alaska shifted during and after World War II, when cold war foreign policy makers touted Alaska as a strategic outpost. The federal government plowed more than $1.25 billion (1950s value) into military base construction throughout Alaska during the postwar period. By 1960, the military comprised nearly half of the workforce. It was not the military that catalyzed the dramatic development that has taken place in Alaska since the 1960s, however; it was oil—discovered first in 1957 near Kenai and in 1968 at Prudhoe Bay on the North Slope.

Although it had been a state since 1959, Alaska's value to the Union was firmly established with the completion of the Alaska Pipeline in 1977. The tax income generated from petroleum spawned intense construction in nearly every Alaskan village and town. The ANCSA and other initiatives led to the decentralization of education and a boom in local government. The land and cash settlement set up 10 regional native corporations, each charged with the requirement of profitability. This drive for profit within the native-owned corporations has led to intense development pressures, both on the oil-rich North Slope and in the timbered southern areas where logging on Native American land is extensive.

Booms in other economic activities have been more sporadic. Fishing and logging peaked in the 1970s, while retailing continued to expand into the Alaskan market in the 1980s. The fortunes of Alaska shifted again when the price of oil declined in the early 1990s, following the Gulf War. Alaska now faces a dramatic decrease in state income and investment, which will be exacerbated by the declining viability of the Prudhoe oil fields over the next 20 years. Increasingly, communities will be faced with hard choices and severe constraints on development. This makes it even more important that coastal development be undertaken with prudence.

Long-Term Records of Catastrophes in Alaska

Oral records of some natural catastrophes have been passed down from generation to generation in the native tradition, but firsthand scientific observations are limited to the historic record and the period of Russian and American occupation. Native American elders report that lava flows occurred several hundred to 1,000 years ago on the Seward Peninsula, and native observations confirm the long-term trends of erosion and subsidence along the coast from the Yukon Delta to Barrow. Yup'ik elders in the Yukon-Kuskokwim Delta warn that the land is thinning with age and that one is likely to "fall through," a good description of the hazards of subsidence. Complete records generally cover a period of less than 100 years.

Research and dating of geological deposits are necessary to gain a long-term perspective of the land's history. Geologists can probe, drill, and excavate buried surfaces subsequently flooded or covered by sediment (e.g., a sand layer) and link these to earlier tsunamis. Buried organic deposits can be assigned ages using radiocarbon (^{14}C) dating, a method based on the predictable relationships of decay of unstable radioactive carbon atoms.

Past earthquakes can be inferred by digging trenches or drilling cores into subsidence areas. When the land subsides, the coastal marsh vegetation is flooded and eventually is covered by marine sediments transported during successive tidal cycles. The alternating succession of land and marine peats allows geologists to reconstruct the history of earthquakes in Alaska. Using this approach, researchers from the Alaska Department of Geophysical and Geological Survey (DGGS) argue that at least during the past 5,000 years (given the uncertainties of radiocarbon dating), earthquakes as large as the 1964 Good Friday event (fig. 2.1) have occurred about every 600–800 years. Other researchers argue that the interval between massive earthquakes during the last 5,000 years is only 400–500 years, within a range of 200–800 years. The last very large earthquake occurred more than 800 years ago; another could occur in the near future.

Although the Russians occupied the area for nearly 200 years, seismic records for southeast Alaska are limited. The Lituya Bay quake of September 1899 on the Yakataga seismic zone of the Queen Charlotte fault (see fig. 3.1) was a large earthquake, estimated at 8.4 on the Richter scale. Much of the land surface from Yakutat to Skagway was uplifted by this quake, up to 18 feet in places. An earthquake along the Yakataga seismic zone in 1979 produced a reading of 7.7 on the Richter scale. Using radiocarbon dating and maps of raised shoreline features, geologists infer that large earthquakes have uplifted the land along the northern margins of the Queen Charlotte fault system at intervals of 1,000–1,500 years. Several large earthquakes were reported off southern southeast Alaska during the 1940s and 1950s. Seismic

Figure 2.1 Damage in downtown Anchorage after the 1964 earthquake. The land subsidence here occurred when the Bootlegger Cove formation (fig. 3.2), which underlies much of Anchorage, slumped. Slippage on this formation was particularly severe along the bluffs facing Knik Arm. From the Alaska Steamship Company Collection (acc. 87-175-641) in the Alaska and Polar Region Department Archives, University of Alaska at Fairbanks.

activity is particularly high in the vicinity of Glacier Bay and the northern Lynn Canal. Literally dozens of small to medium-sized earthquakes were mapped between 1970 and 1987.

Volcanic Eruptions

Volcanic eruptions in Alaska are the subject of considerable interest, and many volcanoes have well-documented histories extending over several thousand years. One of the largest eruptions to affect Alaska, named Old Crow after a Yukon Territory locality, occurred more than 140,000 years ago when a submarine eruption spewed a thin layer of ash over nearly the entire state. Around 3,500 years ago, the Aniakchak caldera collapse generated a widespread ashfall that covered much of western Alaska as far north as Cape Espenberg on the Seward Peninsula. Few subsequent eruptions have covered such a wide area, although a considerable number of ashfalls over the last 4,000 years have issued from the Hayes vent of Mount Spurr and from Mount Katmai. The direction, and hence areal extent, of ashfalls is variable. Hayes vent ashfalls affect the Cook Inlet area inland to the upper Susitna River, while Katmai ash is found primarily on the Alaska Peninsula and Kodiak Island. The 1912 eruption of the Novarupta vent unleashed a rain of ash that blanketed much of Kodiak to a thickness of nearly 1 foot (fig. 2.2). The ashfall forced some villages to relocate.

In fact, some archaeologists argue that repeated volcanic eruptions caused resource failures in southern Alaska that led to widespread human migrations. In the 1980s, eruptions from Mount Redoubt and Mount Spurr produced similar fears in villages from Kenai to Anchorage. Almost two-thirds of the volcanic centers on the Aleutian Peninsula have erupted over the last 300 years. The Aleutian volcanoes are also prolific producers of ash; archaeological data indicate repeated eruptions in the eastern Aleutians from 8,000 to 4,000 years ago. In southeast Alaska, Mount Edgecombe near Sitka has had several large eruptions in the last 12,000 years; the most recent was more than 4,000 years ago.

A veritable wall of volcanoes faces the communities along Cook Inlet (fig. 2.3). The eight volcanoes on its western shore are known to erupt with considerable regularity. Three volcanoes are Alaska's most prolific sources of ash: Mount Redoubt, Mount Spurr, and Mount Saint Augustine. The traces of tephra layers deposited in Kenai Peninsula ponds and lakes indicate to geologists that volcanoes in the Cook Inlet area have erupted every 10–35 years during the twentieth century and at least once every 50–100 years dur-

Figure 2.2 Volcanic ash blankets a cemetery outside Kodiak immediately after the June 1912 eruption of Mount Katmai. A number of Alaskan communities face potential ashfall accumulation thick enough to disrupt community life and cause health problems. Rainwater runoff, a process that often clogs drains, gutters, and sewers, has already removed some of the ash here. Photo courtesy of the Kodiak Historical Society.

Figure 2.3 The location of volcanoes in Alaska that have been active over the last 2,000 years. Clearly, volcanic activity is closely related to the interactions between the Pacific and North American Plates (see fig. 3.1). Modified from *Volcanic Hazards from Future Eruptions of Augustine Volcano, Alaska,* by J. Kienle and S. Swanson.

ing the last 500 years. It is unclear whether the number of volcanic eruptions is on the rise, because the ash layers from low-level eruptions are not always preserved in lake sediments. The longest period without eruptions (23 years) followed the 1912 Katmai event and preceded the 1935 Mount Saint Augustine ashfall.

Ashfall is not the only volcanic hazard that residents of south-central Alaska must face. Eruptions from Mount Redoubt can be accompanied by debris avalanches, lahars (clay-rich mudflows), and floods. The timing of these events in the past can be estimated from the age of plant remains buried under lahar deposits. Around 3,600 years ago (1600 B.C.), a massive slope failure produced two large lahars down the Crescent River valley. The volcanic activity continued for 2,000 years and led to several more lahars, with a very large one occurring sometime between A.D. 1000 and 1600. In the last 200–300 years there have been at least five or six eruptions large enough to produce lahars and floods that could have affected the Drift River oil terminal, had it been in place.

Eruptions of Mount Saint Augustine, in lower Cook Inlet, are well documented by the debris fans that cascade down its slopes to sea level; wave erosion exposes cross sections of these deposits. From the study of these fans, geologists have established that over the last 2,000 years, the island volcano has erupted somewhat regularly every 100 years or so. Debris fans from Mount Saint Augustine are not without considerable danger for humans. As a fast-traveling debris flow hits Cook Inlet, its energy is transferred to the sea surface, creating a tsunami that can reach the opposite shore of the inlet in a matter of minutes.

Tsunamis

Sea waves generated by slides or flows into the water, earthquakes, and submarine volcanic eruptions pose an enormous threat to coastal zones. The geologic record attests to their frequency on the Alaskan scene. The record of debris fans produced by Mount Saint Augustine volcanism is also a record of tsunamis, like the one in 1883, which was observed to strike the opposite shore of the inlet. One of the largest tsunamis of record occurred in Lituya Bay in 1958 when a large landslide fell into the bay. Several ships in the bay were carried hundreds of feet aloft by the force of the wave. Tsunamis accompanied the 1964 earthquake and were responsible for most of the widespread damage, particularly at Kodiak, Chenega Bay, and Seward. Geologic evidence from Bristol Bay indicates that the Aniakchak caldera explosion 3,500 years ago may have produced a tsunami large enough to overtop the high bluffs on the northern margin of the bay.

Debris Slides and Avalanches

Alaska's steep slopes, high precipitation, and seismic shaking aid gravity in causing downslope failure. Historic records ranging from newspaper stories to tree rings make it possible to reconstruct the frequency of debris flows and slides. Slides in the Juneau area were relatively common around 100–200 years ago. The history of recent landslides and avalanches is detailed in chapter 5.

Glacial Advances

Few towns in Alaska are immediately threatened by the expansion of glaciers, although detailed maps of the Juneau area indicate that nearly half of the Mendenhall Valley was covered by ice as recently as the late eighteenth century. The retreat of this ice produced localized flooding and left substantial deposits of gravel and sand. As glaciers expand, they sometimes impound glacial lakes, which may drain catastrophically when the glacial "dam" fails. Geologic evidence suggests that the Taku glacier near Juneau drained in this manner at least two times, once more than 1,200 years ago and once in the last several hundred years. South of Anchorage, the Portage glacier and the Kenai Peninsula glacier were also considerably larger during the eighteenth century, a period termed the Little Ice Age.

An earlier glacier expansion during the ninth to twelfth centuries is recorded in the growth rings of trees killed by the glaciers. Similarly, tidewater glaciers in Prince William Sound expanded during the Little Ice Age but retreated during the warmer twentieth century. Floating bergs present a continual spectacle to tourists and a hazard to navigation. On a smaller scale, the breakup of shorefast ice can create problems for communities built too close to the beach (fig. 2.4). Some archaeologists propose that Prince William Sound and Kachemak Bay were abandoned by people when the cooling associated with glacial advances severely curtailed critical resources.

History of Storm Surges

Records of storms that have struck northwest Alaska are preserved in ancient geologic deposits as well as in newspaper archives, although the latter extend back only 100 years. Storms in Alaska intensify when worldwide climates cool and the difference between tropical heat and Arctic cold increases. During the Little Ice Age and earlier glacial advances, the coasts of the Bering and Chukchi Seas witnessed massive storms in repeated succession. During the warmer intervals, storms arrived less frequently. The timing and sequence of these events can be reconstructed by examining beach ridge deposits (see chapter 1).

The sequence of twentieth-century storms in the Bering Sea can be reconstructed by reading the newspapers from Nome. As the boomtown of the gold beaches, Nome depended on high-seas transport for supplies, and anything that disrupted sea transport was newsworthy. Even today the arrival of the supply barge is a great event. Newspaper accounts of storms in Nome are thus particularly detailed and reveal that about 60 storms have hit the Nome area in the last 95 years (fig. 2.5). Storms occur with some regularity; about every 10–11 years a big storm has hit the Nome area, with smaller events occurring every 3–7 years. Comparisons with temperature data show that storms at Nome occur during colder periods, just as the prehistoric records from geologic sections indicate. Several decades were nearly storm free, especially the years from 1917 to 1928 and from 1950 to 1960. By contrast, two periods witnessed heavy storms: 1900–15 and 1936–46. The intensity and frequency of storms in the latter period persuaded community leaders in Nome that shoreline protection was important, and in 1950 the U.S. Congress appropriated $2 million to build the Nome seawall.

Reconstruction of the history of storms in southern Alaska is limited to historic records, although a considerable number of prehistoric storm ridges can be found on the rocky shores of Prince William Sound, Kodiak Island, and even along Lynn Canal. Beach ridges in Lynn Canal show in-

Figure 2.4 Ice floes damaging the waterfront in Nome before seawalls were constructed (ca. 1920s). From the Lulu Fairbanks Collection (acc. 68-69-2950), Alaska and Polar Region Department Archives, University of Alaska at Fairbanks.

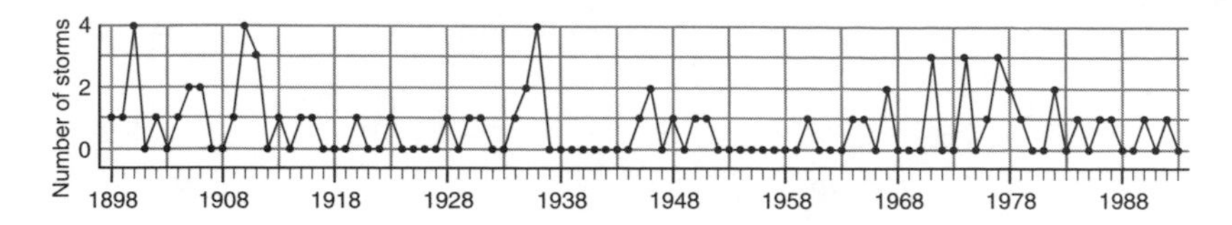

Figure 2.5 Storm history of Nome, Alaska. The extent of damage of the associated storm surges (or rise in sea level) depends on the level of the tide and the size of the waves on top of the storm surge. The higher the level of the lunar tide, the more damaging the storm surge. Assembled from the *Nome Nugget.*

creased storm activity during the Little Ice Age as intense northerly winds raked across southeast Alaska from the interior.

Conclusions

The bounty of resources along Alaska's shores places its residents in a double bind: to use the sea and its shores, it is necessary to live in a geological hot seat. The next calamity may be tomorrow or a generation away, but it will come. The silent volcanoes are sentinels of that inevitability, just as the placid sea is an illusion, as all mariners know. For all of its recorded history and for millennia before, Alaska has witnessed countless earthquakes, volcanic eruptions, tsunamis, storm surges, floods, and slope movements. Some archaeological evidence suggests that areas were abandoned in response to such natural events. Casualties and property loss may have been just as important for early cultures as for people today. In prehistoric times, however, recovery from disaster meant reassembling tools and wardrobe and replacing stored food, and perhaps moving to a new, safer site. Early settlements could be moved with little difficulty, but modern industrial technology enforces a more fixed commitment to locations and renders people more inflexible in their responses to disaster. When locating buildings and other structures, our choice must be the best not only in terms of utility and economics, but also in terms of low risk with respect to natural hazards. In the long run, low-risk locations will provide economical solutions, minimize costs, and save lives.

Part II
Hazards in Alaska

3 Earthquakes, Volcanoes, and Tsunamis: A Deadly Trio

As the Pacific Plate collides with the North American Plate (fig. 1.3), the next great geologic jigsaw piece is being emplaced into the Alaskan puzzle. This collision involves the expenditure of enormous energy, especially along the plate boundaries. The energy released in such tectonic processes creates the greatest hazards on earth (fig. 3.1). Most of Alaska's great earthquakes and all of its volcanoes originate deep beneath the plate boundary, at the Aleutian Megathrust–Queen Charlotte fault.

Earthquakes are characterized by the following:

1. Great release of destructive energy
2. The probability that areas well beyond the location of the event will be affected
3. The likelihood that multiple secondary hazard events such as tsunamis and landslides will be triggered
4. Long-term certainty, but short-term unpredictability

Three of the world's 10 most destructive earthquakes between 1904 and 1992 occurred in Alaska, including the 1964 Good Friday earthquake; and of seven notable volcanic eruptions in the United States, three were in Alaska. Alaska's volcanoes have been restless through the 1980s and 1990s, and the potential exists for a Mount Saint Helens or Mount Pelée type of eruption somewhere in the state. Faults along the sutures between older terranes and associated with the mountain ranges also can generate earthquakes that affect the coastal zone (fig. 1.4). Further, great quakes on other parts of the Pacific rim may shake Alaska or send tsunamis to its shores.

Every resident and visitor to Alaska should be aware of the great hazards, particularly earthquakes and associated phenomena. The following discussion of earthquakes is reprinted with permission and minor modification from the March 1994 pamphlet *The Next Big Earthquake,* written and compiled by Peter Haeussler of the U.S. Geological Survey, which was distributed throughout Alaska.

Earthquakes

More than 30 years ago, on March 27, 1964, the second-largest earthquake ever recorded on earth shook the heart of southern Alaska. Scientists have long recognized that Alaska is one of the most seismically active areas of the world. "Great" earthquakes (larger than magnitude 8.0) have rocked the state on an average of every 13 years since 1900. Most of these large earthquakes have occurred far from heavily populated areas, but it is only a matter of time before another earthquake affects many Alaskans.

It does not necessarily take a major earthquake to inflict hardship. For example, the January 1994 earthquake in southern California was almost 1,000 times weaker than the 1964 earthquake in Alaska, but it caused billions of dollars in damage and claimed many lives because of its proximity to a populated area.

Alaska has changed significantly in the last 30 years. Its population has more than doubled. Many new buildings are designed to withstand intense shaking, some older buildings have been reinforced, and development has commonly been discouraged in particularly hazardous areas. Despite these precautions, future earthquakes may still cause damage to buildings, displace items within buildings, and disrupt basic utilities that we take for granted.

Fortunately, we can prepare for earthquakes. By identifying the greatest hazards, we can set priorities for using our resources most effectively to reduce damage. By becoming aware of the hazards posed by earthquakes and by taking actions such as those described in this chapter and in appendix A, we can drastically reduce loss of life and property and make Alaska a safer place to live.

Measuring an Earthquake

The energy suddenly released during an earthquake can produce a terrifying jolt, one often measured in terms of thousands of Hiroshima-size atomic bombs. The size of an earthquake is commonly stated in terms of its *magnitude*; its effects are measured in terms of *intensity*.

An earthquake's magnitude can be expressed in several ways. The most famous, the Richter scale, was devised in 1934 by Charles F. Richter. On the Richter scale and other magnitude scales, each whole-number step represents a tenfold increase in the size of seismic waves measured on a seismograph—a machine that measures how much the ground moves in an earthquake. However, a single step on the Richter scale (table 3.1) corresponds to

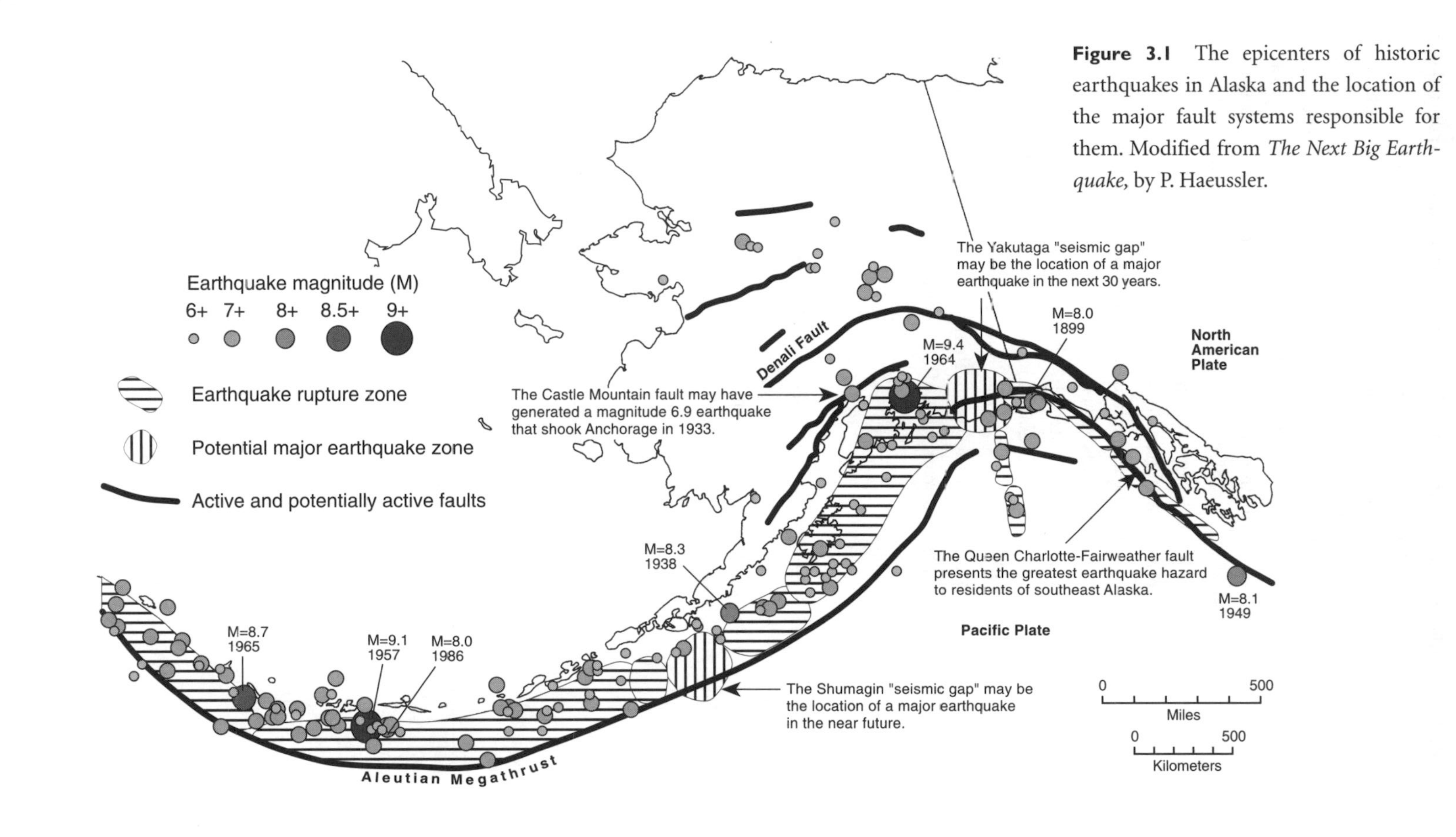

Figure 3.1 The epicenters of historic earthquakes in Alaska and the location of the major fault systems responsible for them. Modified from *The Next Big Earthquake*, by P. Haeussler.

a thirtyfold increase in the amount of energy released in an earthquake. The 1964 Alaska earthquake had a Richter scale magnitude of 9.2.

The intensity of an earthquake is not measured using seismography, as Richter scale magnitudes are, but is based on the earthquake's effects on man-made structures. The intensity of an earthquake can be very different in places only 100 feet apart because the amount of shaking, and therefore the damage, depends on the kind of soil or rock beneath a particular spot. A modified version of the Mercalli intensity scale, developed in 1902 by an Italian geologist, is often used to measure earthquake intensity (table 3.1). This scale ranges between an intensity of I, which is rarely felt, and XII, which results in damage to nearly all structures. Most residents in earthquake-prone parts of Alaska have experienced intensities up to IV. Mercalli intensities were X near the epicenter of the 1964 Alaska earthquake in Prince William Sound; VII in Kodiak, Homer, Seward, Valdez, Cordova, and Anchorage; and V–VI in Fairbanks, Fort Yukon, Yakutat, and Sitka. Regardless of how the magnitude or intensity of an earthquake is measured, any earthquake is important if it affects your community.

Earthquake Damage

Damage during an earthquake results from several factors:

1. *Duration of shaking.* The longer buildings shake, the greater the damage. The duration is determined by how the fault breaks during the earthquake. The strongest shaking during the 1964 earthquake lasted 3–4 minutes. During a magnitude 7 earthquake in Alaska, the shaking may last 30–40 seconds.

Table 3.1 Earthquake Magnitude and Intensity Scales Compared

Richter magnitude	Equivalent energy in weight of TNT (in tons)	Equivalent energy in Hiroshima-size atomic bombs	Mercalli intensity near epicenter	Witnessed observations
3–4	15	1/100	II–III	Feels like vibration of nearby truck
4–5	480	3/100	IV–V	Small objects upset, sleepers awaken
5–6	15,000	1	VI–VII	Difficult to stand, damage to masonry
6–7	475,000	37	VII–VIII	General panic, some walls fall
7–8	15,000,000	1,160	IX–XI	Wholesale destruction, large landslides
8–9	475,000,000	36,700	XI–XII	Total damage, waves seen on ground surface

2. *Strength of shaking.* Many damaging earthquakes occur within 15 miles of the earth's surface. These are the most common earthquakes in central and southeastern Alaska. In such events, shaking decreases rapidly with increasing distance from the fault that produced the earthquake. Deeper earthquakes are common beneath southern Alaska and the Aleutian Islands. Because of their greater depth, the shaking directly above such shocks is reduced, but the shaking decreases only gradually with increasing distance from the epicenter of the earthquake.

3. *Type of soil.* Shaking is increased in soft, thick, wet soils. In certain soils the ground surface may settle or slide. Shaking is reduced on bedrock, as is damage to structures built on a bedrock foundation.

4. *Type of building.* Type of construction is critical in building survival. The rules for earthquake-resistant construction are discussed later in this book.

Alaska Earthquake Statistics

—Eleven percent of the world's earthquakes occur in Alaska.

—Alaska has 52 percent of all the earthquakes in the United States.

—Three of the 6 largest earthquakes in the world in this century were in Alaska.

—Seven of the 10 largest earthquakes in the United States in this century were in Alaska.

—Since 1900, Alaska has had an average of:

1 magnitude 8 or larger earthquake every 13 years

1 magnitude 7–8 earthquake every year

4.5 magnitude 6–7 earthquakes per year

20 magnitude 5–6 earthquakes per year

90 magnitude 4–5 earthquakes per year

Why Earthquakes Are Inevitable in Alaska

The surface of the earth is made up of a dozen or so large fragments called plates. Most of these plates are more than a thousand miles across and more than 40 miles thick. They have been moving about for at least a few billion years, and will continue to shift in the future. The plates move steadily but slowly, passing one another at rates currently up to 4 inches per year. Most earthquakes occur at boundaries where plates are sliding past each other, but some are internal to plates. If the movement of the plates could be stopped, there would be no earthquakes, but that isn't possible, and future earthquakes are inevitable. Southern and southeastern Alaska lie at the boundary between two plates, the North American Plate on the north and east and the Pacific Plate on the south and west (fig. 3.1).

Plates can move relative to each other in three ways:

1. The plates can move toward each other. This is the situation in south-

ern Alaska and along the Aleutian Islands where the Pacific Plate dives beneath the North American Plate (fig. 1.3). The earth's largest earthquakes, such as the 1964 Alaska quake, generally occur where plates are moving toward each other. Volcanoes, such as those on the Alaska Peninsula and on the Aleutian Islands, also occur in areas of plate convergence.

2. The plates can slide by each other. This is the geologic setting off the coast of southeastern Alaska, where the North American Plate and the Pacific Plate slide past each other on the Fairweather–Queen Charlotte fault (fig. 3.1). The same type of movement on the San Andreas fault in California is what causes many damaging earthquakes there.

3. The plates can move away from each other. This situation occurs mostly in the deep oceans and does not typically generate large earthquakes.

These immense plates move at a steady rate, but at their edges the sliding motion is neither smooth nor constant. The motion of the plates strains or deforms the rocks at the boundaries until a point is reached at which the rocks can no longer withstand the strain. Then, a sudden slip releases energy that causes earthquake shaking at the earth's surface. When the plates are not slipping by each other, but are "locked" together, no earthquakes occur. Eventually, however, enough strain builds up, the locked section breaks, the two plates slide past each other, and we experience an earthquake. The whole process is much like pulling a concrete block with a bungee cord. At first, you pull on the bungee cord and it stretches out and the block does not move. Eventually, the pull on the cord is enough to get the block moving, and it slides forward with a jerk, then stops. If you keep pulling, the cycle repeats itself, just like the earthquake cycle.

Sudden slip during earthquakes occurs on different parts, or segments, of faults at different times. Figure 3.1 shows the times when different parts, or segments, of Alaska's faults broke and produced earthquakes. The highest earthquake hazards are at "seismic gaps"—spots on major faults where there has not been a recent large earthquake. The Shumagin and Yakutaga seismic gaps are considered very likely to have a major earthquake in the next few decades. Scientists are studying these areas in order to understand what happens before a large earthquake.

Liquefaction

When loosely packed, very wet sand is shaken during an earthquake, it may flow like a liquid; this is called liquefaction. Anyone who has walked on a beach may have seen a small-scale version of this process. Stamp your foot in the sand near the water's edge, and suddenly the area of your footprint vibrates like shaky gelatin.

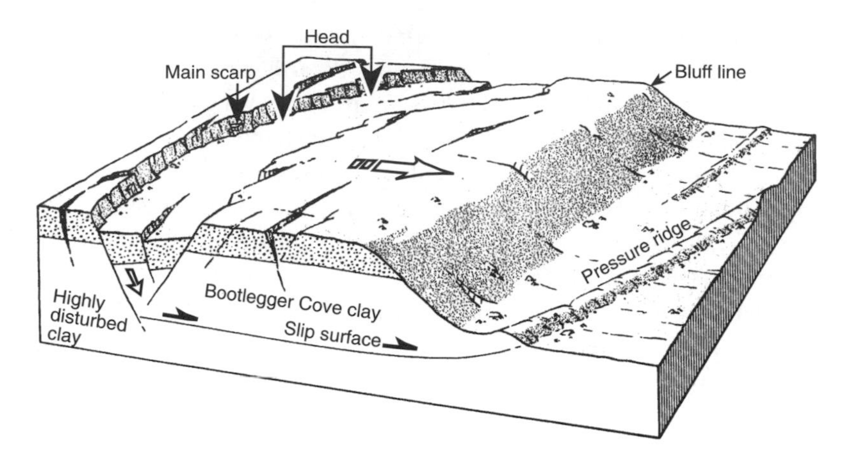

Figure 3.2 Diagram showing how the bluff around Anchorage failed during the 1964 earthquake, and how it will fail in future earthquakes. The Bootlegger Cove clay formation acts as a lubricant to the process of slumping, which is also aided by liquefaction. These combined processes caused houses to tumble about and formed the jumbled terrain preserved in Earthquake Park. Most of the failed bluff face has been smoothed over by bulldozers, but the underlying geologic potential for slumping remains unchanged. From USGS Professional Paper 542-A.

Earthquake-induced liquefaction is often accompanied by cracks in the ground surface and small eruptions of sand and water called sandblows. Liquefied soil is unable to support the weight of any ground or structures above it. Bridges and buildings may settle and tilt even though they were designed to withstand strong ground shaking. If the liquefied area is on a slope, massive landslides may result.

The Bootlegger Cove formation is the soil type that underlies much of Anchorage (see chapter 11). Liquefaction of a part of this formation caused much of the destruction in the Anchorage area during the 1964 earthquake (fig. 3.2). Formations or soils that liquefy are present in many low-lying parts of Alaska where there is a high water table.

Determine If You Live or Work in Particularly Hazardous Areas

Earthquake damage is typically concentrated in areas that can be identified in advance. The amount of shaking experienced in an earthquake can be very different in locations less than 100 yards apart. You can determine for yourself if the places where you live and work are particularly dangerous (see chapters 9–12). Asking yourself the following questions, and seeking the answers, is an excellent start.

Do you live where the ground can settle, slide, or shake violently? Landslides are likely to be triggered by significant earthquakes, especially on steep slopes and in areas underlain by soft ground. During the 1964 earthquake, much of the Turnagain Heights area of Anchorage slid toward Knik Arm. The area is underlain by a kind of soft clay (the Bootlegger Cove formation) that is particularly hazardous because it is full of water and will slide easily if shaken violently. Clay and other types of soft ground can intensify the shaking of an earthquake.

Regional planning reports include generalized maps that illustrate earthquake risk and show how risk can vary within a relatively small area. Even reasonably detailed maps give only an overview of the potential for shaking, liquefaction, landsliding, faulting, and damage, however. To investigate a particular building site, you should consult an engineering geologist, geotechnical engineer, or foundation engineer.

Do you live on a fault? If you took two books, put them side-by-side, and then pushed one of the books while keeping the other still, you would be imitating the process that makes earthquakes. The surface that the two books slipped along is called a fault. In the earth's crust, huge blocks of rock can move past each other along faults. When these blocks get hung up and then suddenly slip, an earthquake is produced. Faults are common in the earth's crust, but only some, called "active faults," produce earthquakes.

Particularly severe damage is likely where structures are built directly on top of active faults. The 1989 Loma Prieta earthquake in California is an example of a moderate-sized earthquake on a fault near a populated area causing a considerable amount of damage. The location of active faults is usually known. Buildings that lie on top of, or very close to, active faults should be considered especially dangerous.

Many earthquake-producing faults in Alaska lie far beneath the earth's surface. In fact, most residents of southern coastal Alaska live above a huge fault called the Aleutian Megathrust (fig. 3.1). The 1964 earthquake, and most major earthquakes in southern Alaska, are related to movement along this fault. Ground shaking caused by earthquakes along deep faults is a more widespread hazard than actually straddling a fault. Nevertheless, a small earthquake nearby can be just as destructive as a big earthquake farther away.

Do you live where the ground may subside? About 55,000 square miles of Alaska subsided up to 8 feet during the 1964 earthquake, submerging low-lying parts of Portage, Whittier, Homer, Seldovia, and Kodiak. Although no one was killed by the subsidence itself, regional subsidence can intensify the effects of tsunamis as well as render property useless. The subsidence that affects Alaskans is limited to coastal areas of southern Alaska and is caused by regional warping of the earth's crust—a different process from local subsidence caused by landsliding.

Future large earthquakes will inevitably cause regional subsidence. Such subsidence may not occur in the same area as the 1964 earthquake, but it will probably affect areas of the Alaska Peninsula and the Aleutian Islands. In view of the fact that the largest subsidence in the 1964 earthquake was about 8 feet, it would be wise to place structures more than 10 feet above sea level in coastal areas. Although a very large area was also uplifted in the 1964 quake—Cordova in particular was affected—uplift generally had a less harmful impact on people than subsidence.

Other Hazards Associated with Earthquakes

The failure of submarine slopes can be as devastating as slope failures on land, and as costly in lives and property. Poorly consolidated sands and silts commonly accumulate on deltas where rivers meet the sea. Because deltas are level ground, home builders, businesses, and engineers are often tempted to build on them. The Alaska Railroad invested considerable money and energy in the 12 lines of track in its terminal facility at Seward. Within minutes of the long-duration Good Friday quake of 1964, however, the deltaic sands and silts fractured and slid seaward. Several factors added to the submarine failure, including high artesian water pressure in the delta and the absence of confining pressure offshore because it was low tide. The old Valdez waterfront experienced six seismically induced submarine landslides in conjunction with the 1964 quake, leading to severe destruction. Submarine landslides that lower the land surface also make areas more vulnerable to the tsunamis that may follow earthquakes, as occurred in Seward.

Earthquakes are not the only force that initiates submarine slope failures in fjord-head deltas. If delta sediments are unstable and are under high pressure, the mere removal of the adjacent water during the daily tidal cycle may lead to submarine landslides. Three nonseismic submarine landslides damaged dock facilities in Valdez in the 1920s and 1940s.

Developers and home buyers should examine geologic logs and study aerial photographs to determine the stability of delta property. The most prudent course of action is to avoid such property altogether, but this alternative is not always easy in areas with restricted amounts of flat land. For this reason, the former rail yard in Seward is still used, albeit for seasonal camping by recreational vehicles. The waterfront in Skagway is also located on a potentially hazardous delta. In October 1994, a slide from the delta front in Skagway produced a localized tsunami that destroyed the ferry dock and adjacent areas. Fortunately this event did not occur during the summer season, when thousands of passengers might have been on the dock. Under similar conditions, Valdez was moved, and new waterfront and port facilities constructed.

Getting Your Building Inspected

How do you locate a professional to advise you on the resistance of your building to earthquake shaking? Civil and structural engineers and architects are trained and licensed to provide such information about structures. Geologists, foundation engineers, and geotechnical engineers are trained and licensed to evaluate the soil conditions and recommend appropriate action.

When you hire a consultant, you are asking an experienced professional to review a potential problem and perhaps to provide plans and specifications for correcting the problem. The amount of work that will be required is not known initially, and thus it is important to select someone you trust and to develop a scope of work as you proceed.

Start by calling a professional organization. Ask for information about the different types of work that might be required; for information about how to select an engineer, geologist, or architect; and for a list of members in your area.

Contact several firms or individuals and determine if they do the sort of work you need to have done. Ask for information that explains the firm's areas of competence and ask for the names of others it has served. Check with other clients to see how satisfied they were.

Recognize that the quality of the advice given and the work performed—as well as the price you pay—may depend entirely on the care you take in making a selection.

Become informed. Even if you do not understand the technical details, ask enough questions to understand the concepts and relative importance of the issues involved. Do not be afraid to ask questions that you fear might appear stupid. Your money is going to be spent, your life and belongings are at risk, and you have a right to understand what needs to be done and why.

For projects more complex than inspecting a single-family home, you should meet with a representative of the selected firm and discuss the options. In almost every case there will be a number of approaches for solving any given problem. Get the consultant to explain the pros and cons of each, as well as the dollars and risks involved. Once this is done, you will have defined the work the consultant will do for you. Then a fee can be set and you can discuss how changing the work would change the fee.

State and federal agencies do not inspect individual buildings. Your city building department may be willing to inspect your building, but its employees are not authorized to recommend corrective actions.

Determine the Safety of Your Home and School

Most people in Alaska are safe at home if they live in a one- or two-story wood-frame building. These buildings are not likely to collapse during earthquakes. The most common damage to wood-frame buildings is light cracking of interior walls, cracking of masonry chimneys, and cracking and possible collapse of brick or masonry veneer on exterior walls. A cracked chimney should be inspected by a qualified professional before the woodstove or fireplace is used.

Nevertheless, some one- or two-story wood-frame buildings are hazardous. Buildings that are not adequately bolted to the foundation may fail at or near ground level. Chapter 14 has information on adding foundation bolts and bracing the cripple walls found in some older homes. Correcting these problems will drastically reduce the earthquake risk for many residents. Chimneys in older homes may need bracing to prevent toppling during earthquakes.

Modern public elementary and high school buildings have generally performed well during earthquakes, with the exception of the Government Hill school and West High School in Anchorage in the 1964 earthquake. However, knowledge about proper seismic design and where to locate buildings has increased dramatically since then. Older school buildings may need to be reassessed in light of modern building codes.

Mobile homes, portable classrooms, and modular buildings can slide or bounce off their foundations during earthquakes. Their supports need to be braced to resist vertical and horizontal forces. If portable classrooms are used at your local school, you should ask school officials whether the buildings are properly braced.

Determine the Safety of Other Buildings You Use

Buildings designed and constructed according to modern codes generally perform very well during earthquakes. Certain types of buildings, especially older ones, are potentially hazardous. Unreinforced brick buildings pose a particular hazard even in moderate earthquakes. Unbraced railings and walls that are inadequately anchored to the floors and roof can topple onto sidewalks or adjacent buildings.

Major damage often occurs in buildings with a "soft" first story—usually an open space with stand-alone columns rather than interior walls supporting the building above. Such spaces are usually used as garages, stores, or large offices. A soft first floor does not have enough strength to resist the horizontal shaking force of the upper parts of the building. Similarly, rooms added over garages of private homes or older split-level homes may not be adequately supported.

Damage to these types of buildings poses a threat to both life and property during earthquakes, but losses can be significantly reduced if the structures are strengthened before an earthquake. Strengthening offices and commercial buildings will reduce structural and nonstructural damage and may allow continuation of business after severe earthquakes.

If you believe a structure that you or your family uses is hazardous, check the references on earthquake-resistant design and construction in appendix C. Ask the building owner what consideration has been given to seismic design and strengthening. Many civil and structural engineers and architects are trained and licensed to investigate the strength of a structure and to recommend appropriate action to reduce earthquake risk.

If you live in a single-family home, ask a licensed engineer or architect to look at your home while you are present and to discuss the seismic issues with you. This typically involves only a few hours of work. A written report or plans and specifications for corrective action may involve more time. If you live in south-central or southeast Alaska, you should ask for such an inspection before buying a new home.

Estimate Your Risk

Earthquakes are a risk that we accept as part of living in Alaska. We face many risks in our lives, and we routinely take precautions to reduce our losses from them; for example, we wear seat belts to reduce the risk of injury during automobile accidents. Earthquake hazards can also be reduced significantly if individuals, businesses, and governments take appropriate action. The basic actions described in the earthquake checklist (appendix A) are reasonable precautions that should be taken by all residents of Alaska. Other actions—such as strengthening or replacing a dangerous building or even choosing to live in a safer building or in a safer part of your city—may involve significant expense and some disruption. Yet, damage to buildings and other structures is the primary cause of death, injury, and financial loss during a large earthquake.

To decide how much action is required to reduce earthquake hazards, you must estimate your risk. Earthquake risk varies from location to location, from structure to structure, and from person to person. Ask the following questions:

1. Is there a risk of serious injury or even death for occupants of a specific building?
2. What would be the cost of repairing or replacing a building after a large earthquake?
3. What would be the cost of not being able to use a building after a large earthquake?

4. What are the odds that time and money spent on action today will prove cost-effective within your lifetime and within the lifetimes of existing structures?
5. If a structure will be replaced by normal development within 10 years, is strengthening it to resist earthquake damage cost-effective?
6. Is such strengthening required by a governmental agency? Is it legally reasonable? Is it morally necessary?

We can live more safely with earthquakes by understanding the risks and taking reasonable precautions. In addition to evaluating your site and structure, you should develop a disaster plan and stock emergency supplies (see the earthquake checklist in appendix A).

Earthquake Insurance

Standard homeowners' insurance does not cover damage and destruction caused by earthquakes. Many homeowners are unaware that their existing fire insurance does not cover fires caused by earthquakes. The 1989 Loma Prieta earthquake in California caused more than $6 billion in damage, but insured property damage accounted for only 16 percent of the loss. In Alaska, about 18 percent of homeowners have purchased earthquake insurance, and about 12 percent of business owners have it.

The most common type of earthquake insurance is normally added as an endorsement on a standard homeowners' insurance policy. Typically, there is a deductible of 5–10 percent of the value of the home. This means that if your home is currently insured at $150,000, you would have to pay $7,500–15,000 on damages before the insurance company would pay anything. Separate deductibles may apply to the contents of the house and the structure. An important coverage is temporary living expenses; that pays for motel and meals if you have to move out of your home. There is usually no deductible on this coverage. The yearly cost of residential earthquake insurance is normally about $1.50–$2.00 per $1,000 of coverage on a conventional frame home. However, the rate may rise to $6.80–$12.80 per $1,000 of coverage on structures with brick or masonry veneer on the outside. Clearly, the insurance industry considers homes without brick or masonry to be better risks in an earthquake. To find out more about earthquake insurance, ask your insurance agent.

Volcanoes

As "cold" near-surface crustal rocks are dragged or pushed down along the plate boundary (fig. 1.2) into zones of higher thermal energy (a process known as subduction), melting may occur, forming magma. Masses of molten magma, less dense than the surrounding rock, tend to rise or intrude into the overlying crust. Faults and fractures provide avenues of opportunity that route the magma all the way to the earth's surface, generating volcanic activity. The composition of these magmas, or lavas, as they are termed when they reach the surface, determines their behavior. Lava rich in iron and magnesium is more fluid and emerges as a lava flow that crystallizes into a fine-grained dark rock known as basalt. Magmas rich in silica tend to be more sticky or viscous; they resist flow, creating explosive eruptive activity that produces cinder cones. The volcanic lavas that develop along the subduction zones at plate boundaries vary in composition between these two extremes. The result is alternating flow and pyroclastic (volcanic debris) production that builds into the great mountain peaks of composite or stratovolcanoes. The great volcanoes of the world, including Alaska's giants, are of this type, and history is peppered with their names (Vesuvius, Krakatoa, Pelée, Mount Saint Helens, Pinatubo). Some have even altered the short-term global climate. The importance of understanding Alaska's volcanoes is reflected in the fact that three major agencies cooperate in mapping, monitoring, and working to mitigate the risks from volcanic hazards (the U.S. Geological Service Alaska Volcano Observatory, the Geophysical Institute of the University of Alaska, and the Alaska Geological and Geophysical Surveys).

Volcanic Hazards

Figure 2.3 shows the distribution of more than 40 volcanoes in Alaska that have been active in the past 2,000 years (table 3.2). New eruptions from volcanoes that have been inactive for hundreds or even thousands of years, particularly composite cones, cannot be ruled out, because these systems are dormant rather than dead. The arcuate distribution of the volcanoes closely matches the earthquake distribution (fig. 3.1) and reflects the position of the subducting Pacific Plate. The Aleutian Islands are completely volcanic in origin (volcanic island arc), and all of the volcanic hazards discussed here should be expected on these islands. No volcano should be ignored, but the Aleutian volcanoes are very remote and less threatening to population centers of mainland Alaska.

In contrast, the composite cones of Cook Inlet and the Alaska Peninsula (table 3.2, fig. 3.3) are a significant threat both to their immediate surroundings and to communities well beyond them, particularly coastal communities (e.g., Anchorage, Hope, Tyonek, Kenai, Anchor Point, Homer, and Kodiak). The Katmai group, Mount Saint Augustine, Mount Redoubt, and Mount Spurr have all been active in the twentieth century; Novarupta erupted massively in 1912. Mount Saint Augustine has erupted at least five times since 1812, and Mount Redoubt has erupted five times in the twentieth century alone. Both of these volcanoes have disrupted Anchorage air traffic, produced tephra (ash and coarser material) falls, and triggered most of the primary and secondary hazards associated with volcanoes (table 3.3).

Lava flows are generally localized near the summit crater or flank vents of volcanoes. Flows might be a threat locally on some of the Aleutian Islands, and lava flows can dam stream valleys and cause flooding. The escape of heat during eruptions, whether flows or explosive, creates threatening secondary hazards that include glacial melting, glacial surges, flooding, and outburst flooding. Eruptions of Redoubt have generated such flooding on the Drift River (see chapter 7). Unconsolidated layers combined with steep slopes and water saturation create conditions that favor slope failure, which generates debris flows, mudflows, and avalanches. Like floods, de-

Figure 3.3 Major volcanoes in the Cook Inlet region and their associated hazards. These are the most dangerous volcanoes in Alaska because of their proximity to population centers, especially Anchorage. Mount Saint Augustine creates a significant tsunami hazard for Homer and Homer Spit. Modified from *Volcanic Hazards from Future Eruptions of Augustine Volcano, Alaska,* by J. Kienle and S. Swanson.

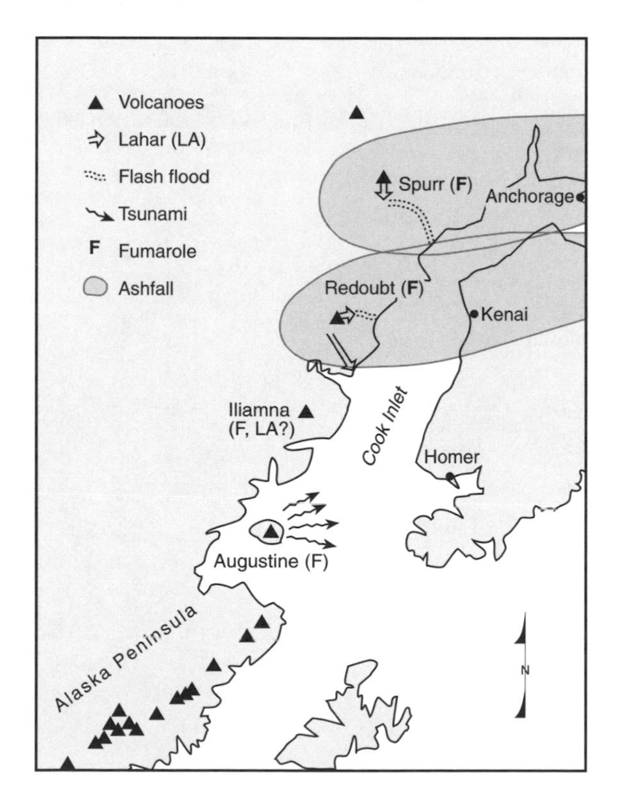

Table 3.2 Active and Potentially Active Volcanoes in Alaska (through 1990)

Volcano	Eruption type(s)	Number of eruptions in the past 200 years	Latest activity (in years before present or years A.D.)
	EASTERN ALASKA		
Wrangell[1]	Ash	1?	1902?
	COOK INLET		
Mount Hayes	Ash	0	450?
Mount Spurr	Ash	1	1953
Redoubt volcano[2]	Ash, dome	4	Ongoing
Iliamna Volcano	Ash	?	?
Augustine Volcano[3]	Ash, dome	5	1986
	ALASKA PENINSULA		
Katmai group	Ash, lava		
Mount Mageik		4	1946
Mount Martin		7	1990
Novarupta[4]		1	1912
Trident Volcano		1	1964
Mount Ugashik,[5] Mount Peulik (Ukinrek Maars)	Ash	1	1977
Yantarni Volcano, Mount Chiginagak	Ash	2	1971
Aniakchak Crater	Ash, dome	1 (or 2?)	1931
Mount Veniaminof	Lava, ash	7	1983–84
Mount Emmons,[6] Pavlof Volcano	Ash, lava	30	1987
Mount Dutton	Ash, lava	0	?
	ALEUTIAN ISLANDS		
Kiska Volcano[7]	Ash, lava	7	1990
Little Sitkin Volcano	Ash, lava?	1?	1900?
Mount Cerberus[8] (Semisopochnoi)	Ash, lava?	1	1987
Mount Gareloi	Ash, lava	6	1987
Tanaga Volcano	Lava, ash	1	1914
Kanaga Volcano	Lava	2	1933
Great Sitkin Volcano	Ash, dome	6	1974
Kasatochi Island	?	1	1828
Korovin Volcano	Ash	7	1951?
Pyre Peak (Seguam)[9]	Ash, lava	5	1977

Table 3.2 cont.

Volcano	Eruption type(s)	Number of eruptions in the past 200 years	Latest activity (in years before present or years A.D.)
Amukta Volcano	Ash, lava	3	1987
Yunaska Island[10]	Ash	2	1937
Carlisle Volcano[11]	?	1	1987
Mount Cleveland[12]	Ash, lava	10	1987
Kagamil Volcano	?	1	1929
Mount Vsevidof	Ash	1	1957
Okmok Caldera	Ash, lava	11	1988
Bogoslof Volcano	Ash, dome	6	1951
Makushin Volcano	Ash	7	1980
Akutan Peak	Ash, lava	21	1988
Westdahl Peak	Ash, lava	2	1978
Fisher Dome	Ash	0	1830
Shishaldin Volcano	Ash, lava	About 18	1987
Isanotski Peaks	Ash, lava	0	1845

Source: Appendix 1 and table 1 from *Living with Volcanoes*, by T.L. Wright and T.C. Pierson (USGS Circular 1073, 1992).

[1]Emission of gases and vapors from vents (fumarolic activity).
[2]Eruptions began December 1989. Debris flows caused temporary closing of the Drift River Oil Terminal. A 747 jet aircraft temporarily lost power in all 4 engines when it entered the volcanic ash plume and would have crashed had its engines not been started just 3,800 feet above the mountain peaks toward which it was heading.
[3]Ash plume disrupted air traffic and deposited ash in Anchorage. A dome built in the crater led to fear of dome collapse triggering a tsunami along the east shore of Cook Inlet, as happened in 1883.
[4]Largest eruption of this century; produced 29,000 cubic yards of volcanic material, which is equivalent to 230 years of eruption at Kilauea. Pyroclastic flow filled Valley of Ten Thousand Smokes, and as much as a foot of ash fell 100 miles away.
[5]Ukinrek Maars formed 0.9 miles south of Becharof Lake, 8 miles from Peulik.
[6]Pavlof is the most frequently active volcano in Alaska.
[7]Steam and ash emission.
[8]Possibly from Sugarloaf, satellitic vent on south flank.
[9]Eight lava fountains, as high as 95 yards.
[10]Minor ash emission
[11]Probably small steam and ash eruption, possibly from Cleveland.
[12]1945 eruption resulted in the only known fatality from Alaska volcanism.

Table 3.3 Hazards Associated with Alaska's Volcanoes and Volcanic Eruptions

Primary hazards	Secondary induced hazards
Lava flow	Glacial melting/surges
Lava dome	Flooding
Cone collapse	Damming/outburst flooding
Lateral blast	Debris flows
Pyroclastic flows	Mudflows/lahars
Ashflow	Avalanches
Basal avalanche	Fire
Glowing ash/gas cloud	Tephra/acid rain result in:
Tephra/ashfall	weight effect
	abrasion
Volcanic gases	corrosion
Tsunami	irritation
	suffocation
	contamination
	reduced visibility

bris flows and mudflows are common events on the flanks of Alaska's volcanoes, and they tend to follow preexisting stream valleys. One prehistoric debris flow off Mount Redoubt traveled more than 19 miles and deposited sediments more than 15 feet thick. Lava domes may plug the crater, creating conditions favorable for explosive events.

Mount Saint Helens in Washington State is a classic example of a major explosive event in which a lateral blast displaced a huge volume of volcanic material and destroyed everything in its path. The calderas (the bases of former volcanoes) present within Alaska's volcanic chain indicate that either cone collapses or volcanic blasts have occurred in the past. These events as well as routine explosive eruptions may give rise to one of the deadliest of volcanic processes, the pyroclastic flow, in which an air-cushioned avalanche of hot, dry pulverized rock debris and tephra, mixed with hot air and noxious gases, rushes down the mountain at very high speeds, engulfing everything in its path. The ash cloud portion of the flow may be so hot that it is incandescent, a glowing cloud (*nuée ardente*). Such a cloud killed 30,000 people on Martinique in the Caribbean when Mount Pelée erupted in 1902. Mount Saint Augustine volcano is very similar to Mount Pelée. Coarse debris, or a basal avalanche, makes up the basal portion of a pyroclastic flow. The density of this debris generally keeps pyroclastic flows following valleys down the sides of the volcano, but their high speed and vertical thickness allow some flows to move uphill or to overtop obstructions. Nevertheless, high ground away from valley floors should be the preferred site if you have a valid reason to be on a volcano.

Most Alaskans are more likely to experience volcanic hazards away from the actual volcano. Tephra and ash falls or associated acid rains may create an inconvenience or a real threat to health and property. Ash is abrasive and can damage engine and compressor parts, something particularly threatening to airplanes in the vicinity or downwind of an eruption. A dense ashfall can collapse roofs, clog drains, and reduce visibility. When ash is mixed with acrid gases or acid rain, corrosion and contamination of water and plants can occur. Breathing ash causes bronchial irritation and, in extreme cases, suffocation. As noted earlier, recent eruptions in Alaska (Augustine, Redoubt) have threatened air traffic and deposited ash in communities around Cook Inlet (figs. 2.4, 3.3). Mount Spurr ash fell in Anchorage in 1953 (fig. 3.4), and the 1912 Katmai eruption deposited ash in Kodiak, killed vegetation over a wide area, and resulted in the loss of livestock (fig. 2.2). The extent of the area affected and the degree of impact from ashfalls is determined by the length and intensity of the eruption, the height of the tephra plume, and the strength and direction of the prevailing wind. Most of the time, ash will be carried from the Cook Inlet volcanoes to the northeast, potentially affecting one-half of the state's population and disrupting the major air traffic corridors.

Like earthquakes, volcanoes can generate tsunamis, either as the result of undersea eruptions or when a large pyroclastic flow plunges into the sea and displaces a large volume of water. Such an event occurred off Mount Saint Augustine in 1883. It generated a 25–30-foot wall of water that struck Port Graham, flooding houses and carrying boats aloft. Natives immediately moved their cabins to higher ground. Property owners in low-lying areas of coastal communities along the eastern shore of Cook Inlet (e.g., Ninilchik, Happy Valley, Anchor Point, Homer, Seldovia, and English Bay) should take a lesson from these events more than a century ago.

Tsunamis (Tidal Waves)

Residents of Alaska's coastal communities should be particularly concerned about tsunamis and similar destructive sea waves. Again, *The Next Big Earthquake* presents a clear discussion of the tsunami hazard, short-term prediction, and how to heed the warnings.

Tsunamis are ocean waves produced most commonly by earthquakes, although not all earthquakes produce tsunamis. The word *tsunami* comes from Japanese and means "harbor wave," after the devastating effects such waves have had on low-lying Japanese coastal communities. Tsunamis are often incorrectly referred to as "tidal waves," but tides do not create them. Earthquake-generated tsunamis may sweep ashore and cause damage both locally and at places thousands of miles from the earthquake's epicenter.

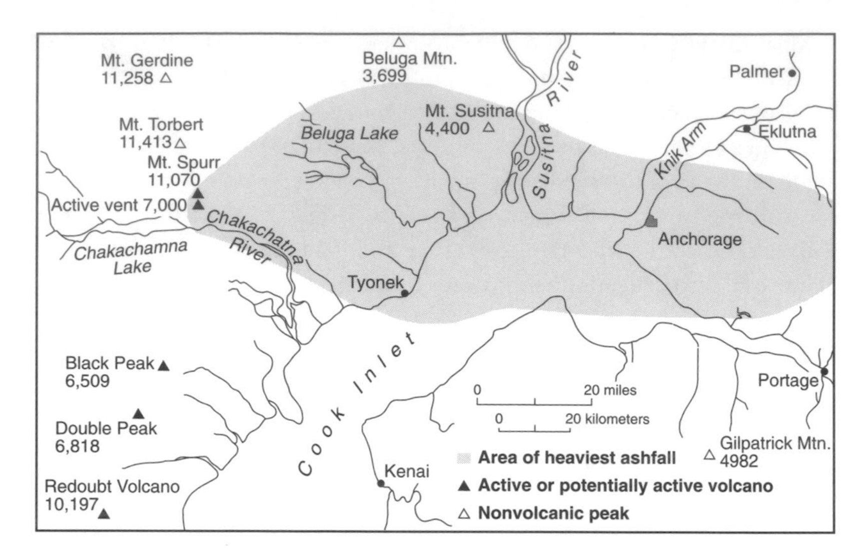

Figure 3.4 The potential impact of volcanic ash is illustrated in this figure, which shows the area of heaviest ashfall from the 1953 Mount Spurr eruption. Damaging ashfall from this volcano could affect the state's largest city, Anchorage. Figure modified from USGS Bulletin 1028N.

More than 90 percent of the deaths from the 1964 earthquake were the result of tsunamis—106 Alaskans died from these waves, and an additional 16 people died from tsunamis in California and Oregon. Some of the latter had gone to the beach to watch for the tsunami!

Large tsunamis are produced by regional uplift or subsidence of the seafloor during an earthquake. Tsunamis started in this way can travel long distances and cause destruction thousands of miles away. Submarine volcanic eruptions can generate regional tsunamis, and local tsunamis have resulted from pyroclastic flows plunging into the sea, as noted above.

Underwater landslides are another cause of tsunamis. Destruction in Seward, Whittier, Valdez (fig. 3.5), and other places in 1964 was the result of tsunamis generated by submarine landslides. Such tsunamis usually have only local effects. Abovewater landslides can also cause local tsunamis if they enter a body of water. On July 9, 1958, in Lituya Bay, Alaska, a large earthquake started a giant landslide that ran into the head of the bay and generated a tsunami. The wave ran more than 1,720 feet up a mountainside on the opposite side of the bay. Two fishing vessels anchored in the bay were sunk, and two people died.

Tsunamis started by volcanoes are uncommon, but they present a real threat to residents of the lower Cook Inlet region, the Alaska Peninsula, and the Aleutian Islands.

Fortunately, tsunami damage can be minimized through land-use planning, preparation, and evacuation. Tsunamis tend to strike the same localities again and again. If tsunamis have damaged an area before, they are likely to do so again. One choice is to avoid living in or using areas with significant tsunami hazard. At the very least, communities should review land use in these areas and ensure that no critical facilities (e.g., hospitals and police stations), high-occupancy buildings (e.g., auditoriums or schools), or petroleum-storage tank farms are located there.

Figure 3.5 Valdez waterfront destruction caused by the 1964 earthquake. The old waterfront of Valdez disappeared in a huge submarine slump. The present shoreline is at the top of a steep scarp. Dashed lines indicate destroyed buildings and docks. From "Submarine Landslides: An Introduction," by H. J. Lee, W. C. Schwab, and J.S. Booth.

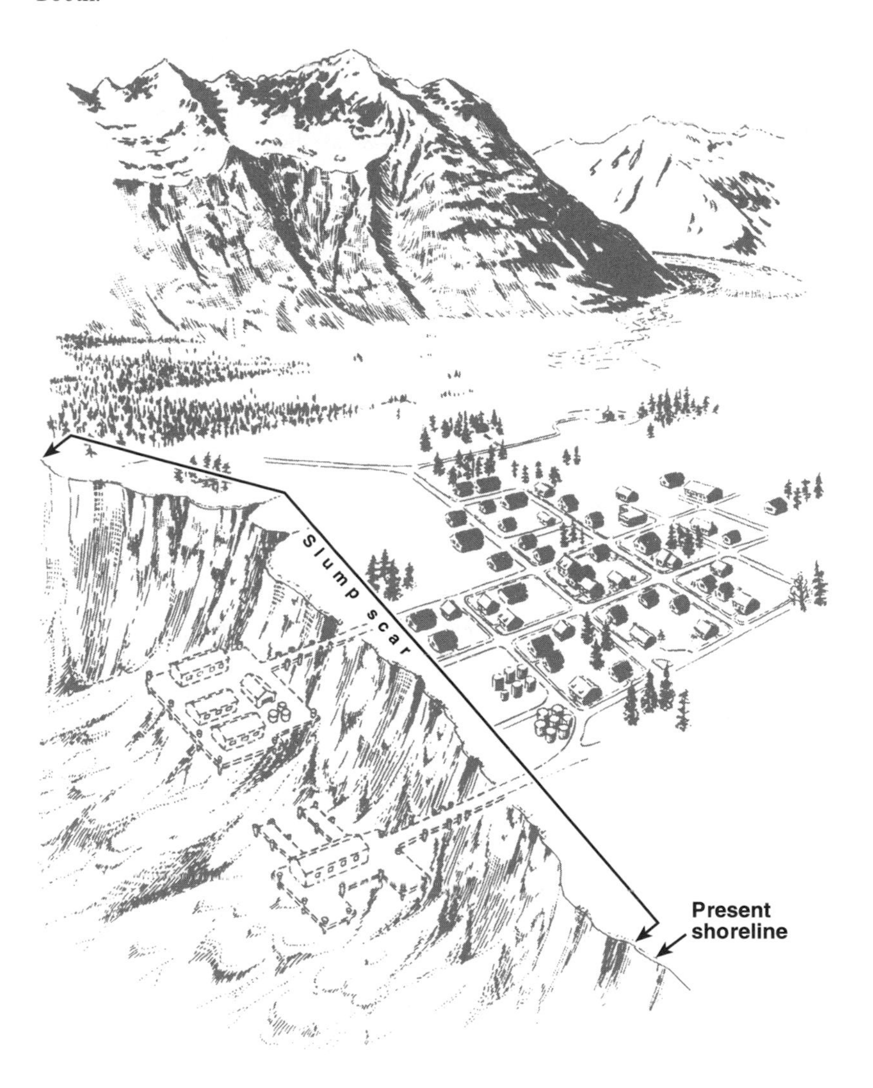

When the 1964 earthquake occurred, Alaskans in coastal areas who did not feel the quake had little or no warning that a tsunami was on its way. As a result, the Alaska Tsunami Warning Center (ATWC) was established.

Tsunami Warnings and Watches

When a large earthquake occurs near the coastline of the northern Pacific Ocean, an automated system at the ATWC rapidly determines its location (epicenter) and magnitude. If the earthquake is considered large enough to generate a tsunami, a tsunami warning is issued for a limited area near the epicenter of the earthquake and issued through the military, U.S. Coast Guard, National Weather Service, Alaska Division of Emergency Services, Federal Aviation Administration, and other federal agencies. A tsunami watch is issued to adjacent areas of Alaska, Canada, and West Coast states as appropriate, alerting them to a possible tsunami threat.

If a significant tsunami is detected by instruments that measure tides near the epicenter of the earthquake, the warning will be expanded to the entire coastline of the region. In case of a warning, people should immediately evacuate inland or to high ground. The ATWC will begin issuing its warning about 15 minutes after an earthquake occurs, but that is not fast enough if there is a local tsunami. People near the shore who feel an earthquake 30 seconds long or longer should heed nature's warning and move quickly to higher ground. People on boats when an earthquake occurs should understand that the safest place to be is in deep, wide water where wave energy is diffuse. If no wave was generated, the warning will be canceled. Although a warning is sometimes issued when no tsunami is present, the alternative—to issue a warning after a wave strikes a community—is unthinkable.

The ATWC works closely with tsunami warning centers elsewhere in the world because tsunamis generated in distant parts of the Pacific Ocean, such as Japan or Chile, have reached Alaska. A tsunami from northern Japan would take 4 hours to reach Adak Island and 8 hours to reach Kodiak, which allows Alaskans time to prepare if a watch has been issued. A tsunami travels from Peru or Chile to Kodiak in 16–18 hours.

Appendix A presents tsunami safety rules. If you live in an area prone to tsunamis, read them, learn them, and heed them.

4 The Problem of Unstable Slopes

Douglas N. Swanston

Much of coastal Alaska is characterized by steep slopes, shallow soils, and exceptionally high rainfall, all factors that contribute significantly to unstable conditions for construction on slopes. Intense scouring of the land by glaciers, active mountain building, uplift associated with rebound of the land after glacier retreat, and vigorous weathering of exposed bedrock and glacial till have led to the widespread occurrence of coarse-grained soils on slopes that frequently exceed the maximum angle of soil stability.

Soil mass movements, or "landslides" involving the downslope movement of soil primarily under the force of gravity, are the principal natural processes of erosion on steep mountain slopes, and they constitute a major hazard to the communities and highways that infringe on such unstable terrain. Soils on the mountain slopes are youthful (by geologic standards), postdating the last major continental glaciation approximately 10,000 years ago. Most hill-slope soils are shallow (generally less then 3 feet thick), coarse grained, and have very little clay. Because water readily soaks into these soils, drainage is primarily by water flow within the soil column.

During major storms, when soil moisture levels are high, temporary water tables are produced. These zones of water saturation tend to develop in concave linear depressions such as old revegetated landslide tracks and abandoned or intermittent drainages where the water flowing in the soil column concentrates. When the water table reaches the surface, surface flows develop, but the thick mat of forest humus and the understory plant cover are generally adequate to protect the soil from appreciable surface erosion. As a result, the principal effect of saturation and local water table development is to create an unstable soil cover at steep-slope sites, making them more vulnerable to downslope movement.

At the same time, on the lower slopes and along the valley floor, increased runoff produces stream flooding, with shifting channel beds and channel scour. Under these conditions, surface erosion is mostly restricted to undercutting streambanks and eroding areas of bare soil where natural processes and human activities have removed the vegetation.

Heavy snow accumulations at upper elevations coupled with extensive cornice development on lee slopes as a result of strong winds make snow avalanching a common annual process in many areas and a significant hazard to some communities.

Soil Mass Movements, or Landslides

Landslides are a widespread phenomenon in coastal Alaska. Some are the direct result of logging and logging road construction, and some result from urbanization and highway, pipeline, railway, and services expansion. Most landslides, however, are natural events caused by unstable topography and materials, the frequent occurrence of heavy rainstorms during the fall, and occasional warm rain-on-snow events during the winter. Soil mass movements range in scope from slow creep and earthflows to rapid high-energy rockfalls, but the majority are technically classified as debris avalanches or debris flows: the rapid downslope movement of a mixture of soil, rock, and forest debris mixed with a high water content.

Creep and Slump-Earthflow Processes

Creep occurs on almost all slopes in the coastal zone regardless of geology or soil type. For the most part, creep processes in rock and compacted glacial till are so slow as to be undetectable in the short term. Slow annual creep is detectable in steep-slope soils and in glacial lake and raised marine terrace deposits that are high in silt and clay and occur along valley floors and adjacent to the existing shoreline. In these materials, visual evidence in the form of catsteps, turfrolls, overturned soil profiles, and curved tree trunks is abundant.

In steep-slope soils, creep movement occurs primarily in the upper 6–12 inches. The movement is mainly by mechanical shifting of particles, but without well-defined shearing planes. The rate of movement in steep-slope soils is about 0.25 inch per year. Such movement produces a stress buildup over time and may increase landslide susceptibility on steep, unstable slopes.

Creep leads directly to landslides in glacial lake and raised marine terrace deposits. In these silt- and clay-rich materials, the movement is primarily a flow similar to molasses flowing down a sloping board (fig. 4.1), involving the mobilization of the soil mass by breakdown of silt and clay structures and the progressive failures of small blocks within the soil mass. Creep in these materials may reach or exceed several inches per day. In the mid-1980s, during the construction of the ore concentration mill at the Greens Creek silver mine on Admiralty Island, accelerated creep in the raised marine terrace deposits on which the mill was built posed a significant hazard to continuing development until it was halted by drainage diversion and pumping down of the water table. If it continues, such rapid deformation leads to failure at some point on the slope, resulting in the development of rotational slumps and slow earthflows (fig. 4.2).

Raised marine terrace and glacial lake deposits are widespread in coastal Alaska. They occur as distinct benches at various elevations inland from the present shoreline and as broad, bog-covered lowlands on the mainland and the islands of Cook Inlet, Prince William Sound, and the Alexander Archipelago of southeastern Alaska. Glacial lake deposits are produced by the formation of temporary lakes in valleys during deglaciation, and raised marine terraces are formed by widespread uplift of land above the present level of the sea associated with earthquake activity, rebound from removal of glacier weight, or a combination of the two.

Active postglacial rebound is still going on along portions of the coast between Yakutat and Ketchikan, where recent deglaciation has resulted in widely varying rates of crustal movement. For example, current uplift rates of glacial rebound in the Juneau area range from about 0.6 to 0.8 inches per year. Rates at Bartlett Cove in Glacier Bay National Monument exceed 1.5 inches per year. These rates decrease to the north and south, averaging about 0.3 inches per year at Yakutat and 0.06 inches per year at Ketchikan.

Figure 4.1 Creep and slump-earthflow processes. Creep is a slow downslope movement of the upper soil profile which under the right circumstances can lead to the more rapid processes of slump and earthflow.

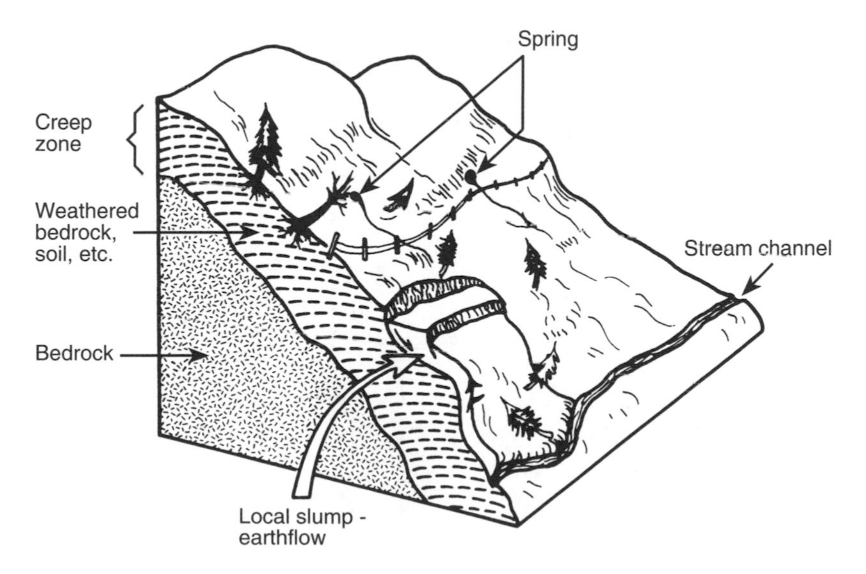

Such uplift rates have had an important effect on landownership along this portion of the coast. The state of Alaska claims ownership of all tidelands below a datum of mean lower low water, and property owners in both Juneau and Glacier Bay with boundaries originally surveyed at mean low low water have successfully claimed the additional land resulting from uplift since the original survey was made. So far, the courts have upheld the claims of the private landowners. Elsewhere along the coast, particularly in Prince William Sound and on the Copper River delta, glacial rebound is probably still occurring but is masked by the dominance of recent earthquake activity, which has caused local large-scale up and down shifts of the land surface relative to sea level.

Landslides in glacial lake and raised marine terrace deposits are occurring with increasing frequency in urban areas and in semiwilderness areas as road systems and construction expand onto previously undisturbed terrain. Very large landslides are rare, however; most landslides result in small slumps that close roads temporarily or damage structures.

Rock Slides

Rock slides are common along the north Pacific coast, where mechanical freeze-and-thaw processes dominate the weathering regime. Rock slides in-

Figure 4.2 Massive slump and earthflow developed in weathered bedrock and overlying glacial till along the southwest shore of Saginaw Bay, Kuiu Island, Alaska, during a major fall storm on October 26–28, 1988.

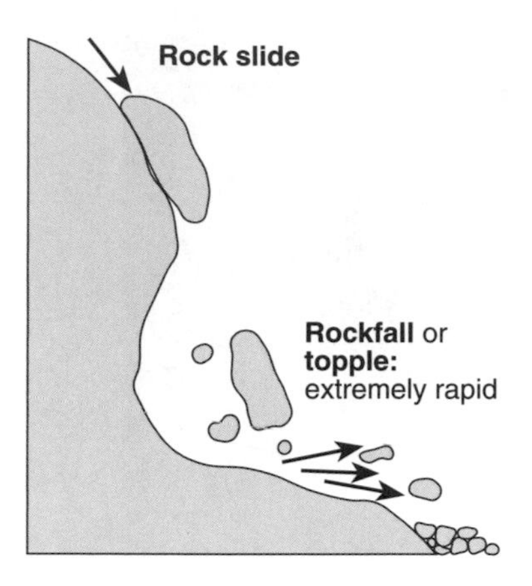

Figure 4.3 Rock slides are commonly caused by the expansion of water during the freeze-thaw cycle.

volve the collapse of rock cliffs and bluffs due to hydrostatic pressures generated by confined movement of water within and between rock units (along bedding planes and fractures), and by freezing of water in near-surface joints and fractures (fig. 4.3). The movement is very rapid, ranging from several feet per minute to several feet per second, and involves initial sliding of blocks of rock and overlying materials along zones of weakness such as joints and fractures in exposed cliff faces and along steeply dipping bedrock contacts. Excess water may be present, but water is not an important "lubricator" in transport of materials. Almost all of the transported material—a mixture of soil and woody debris dominated by large, coherent blocks of rock—is carried to the valley floor or adjacent stream channel by free fall, rolling, and bounding (fig. 4.4). Most rock slides occur in association with the collapse of cliffs and bluffs at remote wilderness sites or along highways outside urban areas, although a rock slide occurred in downtown Juneau in 1913 that seriously damaged a home and sent its occupants running downhill. A rock slide above Basin Road in Juneau in the 1970s resulted in one fatality (table 4.1).

Debris Avalanches, Debris Flows, and Debris Torrents

These movements typically occur on steep valley side slopes and within narrow V-notch tributary valleys. Debris movements begin as the sudden failure and sliding of a more or less intact mass of soil above an impermeable boundary such as bedrock or compact glacial till. Once movement occurs, the sliding mass quickly breaks up into an avalanche of soil, rocks, and forest debris. Downslope, such debris avalanches frequently become channeled into shallow, preexisting gullies and linear hollows formed by old

Figure 4.4 A rock slide developed in highly jointed diorite rock near the mouth of Robinson Creek, on Behm Canal east of Ketchikan, Alaska.

landslides and ephemeral stream flows, producing debris flows as the water content substantially increases (fig. 4.5). In a debris flow, boulders and logs are carried downslope "floating" in a matrix of mud, sand, and gravel. The materials in the flow are commonly deposited at the base of the slope as a debris fan or as a series of debris lobes extending onto the valley floor. These debris deposits are a jumble of rocks, trees, and soil susceptible to erosion by surface runoff, at least until vegetation cover is established.

Debris avalanching into canyons and deep, narrow gullies frequently results in the construction of temporary dams of rock, soil, and woody debris. Depending on the composition and volume of the deposit and the quantity of water flowing in the channel, these dams may persist from a few hours to many years. The failure of such debris dams during floods produces especially spectacular high-volume debris flows called debris torrents or debris floods. Flow velocities are high, ranging from several feet per minute to several feet per second. Debris torrents pick up large quantities of additional soil and organic debris along the channel that frequently doubles or triples the volume of material delivered to the valley floor and adjacent channel systems. The high speed and water volume of debris torrents carry the debris directly into a main stream, producing temporary heavy sediment and

Table 4.1 Landslides in the Juneau Urban Area, 1913–1954

Date	Time	Rainfall (inches)	Type	Location
Oct. 18, 1913	2100	3.5/24 hr	Rockfalls/ rock slide	Mt. Maria on Basin Road[1]
Sept. 25, 1918	—	6.32/24 hr	Debris avalanche	Slope behind Gastineau Hotel destroyed[2]
Sept. 25, 1918	—	6.32/24 hr	Debris slide	7th and Goldbelt into Evergreen Bowl[3]
Sept. 25, 1918	—	6.32/24 hr	Debris avalanche	Gastineau Hts.[4]
Jan. 2, 1920	1130	Warm, melting snow, heavy rain, 1.79/24 hr	Debris avalanche	Gastineau Hts.[5]
Sept. 27, 1935	—	2.89/24 hr	Debris avalanche	S. Franklin at AJ oil tanks[6]
Nov. 27, 1935	1530	3.35/48 hr	Debris avalanche	Third Avenue above Harris[7]
Nov. 27, 1935	—	3.35/48 hr	Slump	5th Street above Kennedy[8]
Nov. 27, 1935	—	3.35/48 hr	Debris slide	Evergreen Bowl[9]
Oct. 16, 1936	0800	1.43/3 hr	Debris avalanche	Gastineau Hts.[10]
Nov. 22, 1936	1930	3.89/24 hr	Debris avalanche	Gastineau Hts., above cold storage plant[11]
Nov 30, 1936	—	—	Debris avalanche	Thane Road near Standard Oil[12]
Oct. 31, 1949	—	2.36/24 hr	Debris avalanche	Gastineau Hts.[13]
Oct. 1, 1952	—	1.85/24 hr	Debris avalanche	S. Franklin by old Columbia Lumber Co. kiln[14]
Oct. 1, 1952	—	1.85/24 hr	Debris avalanche	Gastineau Hts., piled behind 475 S. Franklin[15]
Oct. 1, 1952	—	1.85/24 hr	Debris avalanche	Above Johnson Bldg., 263 Gastineau Avenue[16]
Dec. 16, 1954	—	Warm, melting snow, 2.0/24 hr	Debris avalanche	Irwin Street before Gold Street[17]

[1]Homes damaged [2]Apt. bldg., Gastineau Hotel damaged, $25,000 [3]Cabin destroyed [4]None [5]3 people killed, $50,000 damage [6]Road blocked [7]2 homes wrecked, one damaged [8]None [9]None [10]One woman injured, 2 houses damaged, Alaska Hotel damaged [11]14 died, 9 injured, apt. house, boarding house, 2 houses ruined [12]Road closed [13]House destroyed [14]Road closed [15]House destroyed [16]House destroyed [17]1 house badly damaged

Figure 4.5 Natural debris avalanche in the Marten Creek valley, southeast Alaska. The sliding surface is glacially polished grano-diorite. The difference between this and the rock slide in figure 4.4 is the large amount of matrix in which the boulders are "floating."

large organic debris loads that are deposited and redistributed downstream on the streambed and along the channel banks.

Occurrence and Distribution

While creep and slump-earthflow processes are active almost universally throughout the coastal zone, except for rare events such as the Saginaw Bay slump-earthflow (fig. 4.2), the events tend to be small and rarely cause damage in developed areas. Long-term creep may cause some localized damage (e.g., to basement walls, retaining walls, and sidewalks). Rock slides and topples, while more frequent, tend to be restricted to the alpine zone and to areas immediately below exposed bluffs and cliffs.

By far the most common and most hazardous types of landslide active along the Alaska coast are debris avalanches, debris flows, and debris torrents. Sites of recent debris avalanches and debris flows are clearly identified by linear bare strips on valley side slopes and deep V-notch channels where soil and vegetation have been removed (fig. 4.6). Increasingly older avalanche and flow scars are identified successively by strips of pioneering species such as willow and alder, and finally by even-aged stands of Sitka spruce that are younger than the surrounding timber (fig. 4.7). Such successional regrowth on older slide traces is easily recognized on aerial photographs and provides a convenient, if crude, means of estimating a slope's present stability and past sliding history.

Snow avalanches, which produce similar linear tracks on timbered slopes, can be distinguished from debris avalanches by the following characteristics. If the track starts below the timberline, it is probably a debris avalanche track. If the track extends above the timber boundary, it probably originated as a snow avalanche. Also, a snow avalanche produces little soil deposition at the slope base, and there is little or no removal of soil along the slide track.

Assessing the Risk of Landslides

The balance between slope stability and slope failure is delicate on steep slopes. The slope gradient controls the stability at a particular location because the primary force acting to pull an unstable soil mass downslope is gravity. As the slope angle increases, so does the pull of gravity acting on the

Figure 4.6 Aerial view of a slide near Walker Cove, Behm Canal. A debris avalanche was apparently initiated by a rockfall from a bluff directly above the head of the slide trace. The resultant debris flow lower on the slope carried debris directly to the sea.

Figure 4.7 Older avalanche and debris flow traces show successional regrowth of willow, alder, and Sitka spruce in this view of a slide trace on the south side of Blake Channel Narrows. Note successional regrowth starting with willow and alder in the foreground, with spruce moving in behind.

soil mass. This gravitational stress is counteracted by the soil strength, which is composed of three distinct elements: cohesion, root strength, and frictional resistance.

Cohesion is the capacity of individual particles to stick or adhere together due to cementation, capillary tension, or weak electrical bonding of clay particles. Cohesion is absent or minimal in the shallow, coarse-grained soils characteristically developed on steep slopes in coastal Alaska and is generally disregarded when making an assessment at this level.

Root strength is the resistance to failure generated by the anchoring and reinforcing effects of tree roots. The anchoring effect of root growth through the thin soil and into joints and fractures in the bedrock, and the lateral reinforcing of unstable soil masses across the slope by intertwining root systems, increases the strength component by as much as 10 percent. On steep, unstable slopes, roots may be critical in resisting the development of a landslide.

The dominant component of soil strength is the frictional resistance developed between the soil mass and the underlying failure, or sliding surface. Frictional resistance is controlled by the weight of the soil mass, its angle of internal friction, and the level of water in the soil. The angle of internal friction of the soil is a measure of the interlocking of individual soil grains and is a distinct property defined by laboratory analysis. For shallow, coarse-grained soils, this angle approximates the maximum gradient at which these soils will remain in place against the pull of gravity.

In coastal Alaska, this maximum gradient is about 34 degrees (67 percent). Slopes above this gradient are highly unstable even in a dry, undisturbed condition. Water saturation of the soil and a rising water table at the site produce an active pore water pressure exerted against the base of the soil column. Increasing pore water pressure reduces the weight of the soil mass, thereby reducing the frictional resistance between the soil mass and the potential sliding surface. In effect, the soil mass is "floated" above the failure surface if pore water pressures become high enough.

Slope gradients approaching or exceeding the maximum angle at which the soil mass will remain in place are common on mountain slopes within the coastal zone. Under normal, low-rainfall conditions (nonstorm), however, the various forces acting on the soil mass have adjusted to one another, and the mass will remain in place indefinitely unless disturbed. This semistable state may be weakened or destroyed during periods of high rainfall by development of pore water pressures associated with a rising temporary water table. With the soils in such an unstable condition, only a small triggering force is required to cause total failure and rapid downslope movement by landslide processes. The trigger can be produced by rapid additional increases in pore water due to a big rainfall, by ground motion caused by earthquakes and blasting, by rapid increase in weight of the soil mass caused by rockfalls and by trees blown down during storms, and by direct destruction of the stabilizing root mass by tree blowdown or by logging and clearing. The weight of the trees growing in the soil is not a factor in the soil mass because most of the trees' weight is carried on the bedrock surface by roots that pass through the soil and lie on the bedrock.

Influence of Precipitation

The distribution pattern of debris avalanches and debris flows along the Alaska coast exhibits a distinct relationship between occurrence and rainfall. Two areas that exhibit a clear correspondence between zones of maximum rainfall in excess of 6 inches in 24 hours and zones of high landslide occurrence are on Revillagigedo and Prince of Wales Islands and around Peril Strait on Baranof and Chichagof Islands.

During the fall, when these giant rainstorms occur, the shallow soils reach their full water capacity rapidly and quickly attain saturation and maximum instability. Groundwater fluctuations generally show a close correlation between temporary water table levels and variations in rainfall. A short lag in temporary water table development following storm onset usually occurs at the beginning of the rainy season and reflects the time necessary for soils to reach saturation. Thereafter, fluctuations in the temporary water table closely follow rainfall variations.

Snow accumulation and the timing of snowmelt on the upper slopes can substantially affect the amount of water in the soil over short periods, primarily through retention of rainfall in the snowpack and delayed release of large quantities of water during warm rain-on-snow events in winter and spring melt.

Influences of Parent Material and Rock Structure

Debris avalanches and debris flows typically develop in soils constructed by weathering of shallow materials derived either from the accumulation of landslide-transported bedrock fragments or in glacial till plastered on valley side walls. Such soils have coarse grains, are very permeable, and have a high proportion of angular rock fragments and some organic material. They are basically without cohesion.

Fractures in the bedrock are also an important factor determining the stability of many potentially unstable slopes. Highly jointed bedrock slopes with principal joint or fracture planes parallel to the slope provide little

Figure 4.8 Headwall area of a debris avalanche in colluvial soil. Note the extensive development of tension-release fractures parallel to the slope in this diorite bedrock. The fractures occurred when the rock expanded following removal of the weight of the overlying glaciers that once covered this area.

support for overlying surficial materials. The joint and bedding planes also create avenues for concentrated subsurface flow and active spring development as well as functioning as ready-made failure surfaces for overlying material. Surface-parallel fracturing is widespread in the coastal region and is an important factor in the occurrence of debris avalanches and flows on shallow colluvial soils overlying bedrock. Surface-parallel fracturing in igneous and metamorphic rocks is due to stress release near the surface produced by the removal of the weight of overlying glacial ice (fig. 4.8).

Sedimentary rocks with bedding planes parallel to the slope function essentially the same way. The uppermost bedding plane within the rock provides a smooth surface that serves as (1) an impermeable boundary to subsurface water movement, (2) a layer that restricts the penetration of tree roots, and (3) a water-charged potential failure surface with reduced resistance to movement of the overlying soil.

Influence of Gradient

Slope gradient is a principal factor in slope stability (fig. 4.9). More than 75 percent of the debris avalanches and debris flows in coastal Alaska begin on slopes with gradients in excess of 34 degrees (67 percent). An additional 15 percent are initiated on slope gradients between 26 and 33 degrees (49–66 percent). The remainder occur on lower-gradient slopes and usually begin on local bluffs and cliffs.

As a general rule, slope gradients exceeding 25 degrees are susceptible to debris avalanche and debris flow development, especially where soil overburden is thin or poorly bonded to the underlying surface of bedrock or compact glacial till. Slopes with similar characteristics and gradients greater than 35 degrees are *highly* susceptible to these processes. These controlling gradient characteristics are used widely in the coastal region to identify broad areas of landslide hazard, and, coupled with local terrain features such as the presence of shallow linear depressions and gullies where ground-

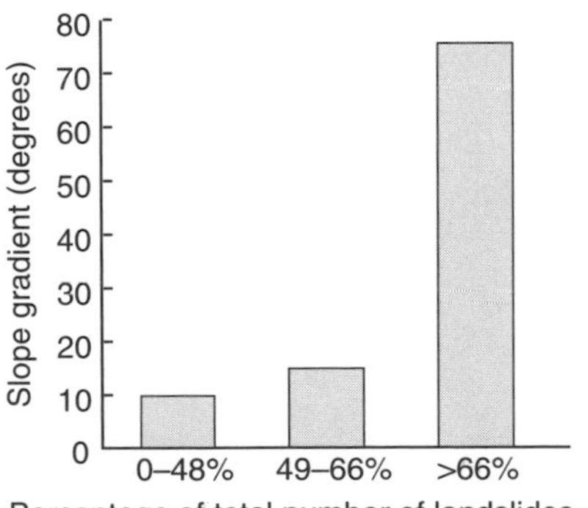

Figure 4.9 Distribution of landslides in coastal Alaska by slope gradient. Clearly, steeper slopes are the most susceptible to landslides.

water flows converge, they provide a ready means for identifying potentially hazardous sites.

Influence of Vegetation

Slide-prone soils may be partially stabilized by the anchoring effect of roots through the soil and into the bedrock and unweathered glacial till deposits, and by the binding of intertwining root masses across local zones of weakness on the slope (fig. 4.10). Stump excavations at various sites show that the tree root patterns on steep slopes consist of widely spread lateral roots lying along the unweathered till and bedrock surfaces, with small sinker roots penetrating through the weathering profile and lodging in cracks and crevices in the underlying material.

Destruction of these roots by windthrow or by decay following insect infestations or timber cutting may substantially increase the susceptibility of extremely steep slopes to sliding. Windthrows, which occur frequently in southeast Alaska forests, are directly correlated with increased debris avalanche and flow occurrence. A growing body of evidence suggests that the 4- to 5-year time lag in increased landslide activity following logging is related to loss of the anchoring effect of roots.

The Degree of Hazard

Debris avalanches, debris flows, and debris torrents pose a significant hazard to communities and highways in coastal Alaska when structures and roads are placed within or adjacent to deposition and run-out zones. No buildings in place, no hazard! Poor site choice results from a combination of ignorance of the potential hazard and lack of effective land-use planning by local governments. The rapid expansion of the population outside traditional community boundaries within the coastal zone brings the public and its associated support systems into direct conflict with natural processes that have always existed but previously could be ignored.

The greatest losses of lives and property as a direct result of landslide processes have occurred in Juneau, the capital city (table 4.1). One of the most picturesque cities in the Alaska coastal zone, Juneau was established in 1881 at the foot of a steep slope, primarily to serve gold miners in the area. Apparently, little thought was given to long-term planning or to potential hazards from the clearly identifiable landslide and snow avalanche tracks that course downslope through the townsite.

More than 15 catastrophic debris flow events have occurred since the town was founded, resulting in at least 17 deaths and extensive property losses. One of the most tragic events occurred at 7:30 P.M. on November 22, 1936. A debris flow that started above Gastineau Avenue during extended heavy rains smashed across Gastineau Avenue and into the Juneau Cold

Figure 4.10 Excavated stump at Hollis showing root distribution. Note the large lateral and buttressing roots which penetrate the soil column and spread out along an underlying impermeable surface of bedrock or compact glacial till. These large roots support the weight of the tree and provide resistance to overturning during windstorms. The root masses are anchored to the underlying substrate by numerous small sinker roots which penetrate the substrate through joints and fractures in the rock and along cracks and zones of advanced weathering in the compact till.

Storage Plant, destroying an apartment house, a boarding house, and two homes; 14 people died and 9 were injured. Similar events, luckily with no fatalities, have continued over the years, and a large part of the city is considered to be in a high-hazard zone for debris avalanches and debris flows.

Figure 4.11 shows the numerous debris flow and snow avalanche channels entering the Juneau urban area. Gastineau Avenue, the uppermost street on the right side of the photo, has undergone de facto zoning as most of the houses on the upslope side have been removed by repeated debris flows. The most recent event in the area was unusual because it occurred in the summer—at approximately 7:00 A.M. on July 16, 1984—during a heavy rainstorm on already wet soils. Rainfall at the Juneau Airport was recorded at 1.06 inches for the 6-hour period during which the event occurred, and a high-intensity burst was observed during a 30-minute period about an hour before the landslide. The combined debris avalanche and debris flow swept down the mountainside behind the Fred Meyers store, destroying a small

Figure 4.11 Aerial view of downtown Juneau showing the many debris flow gullies and snow avalanche paths entering the urban area. Debris flow paths are identified by gullies and by linear strips of lighter-colored alder within the darker old-growth spruce-hemlock forest covering the slopes. Snow avalanche paths begin above timberline and are funneled to the base of the slope along well-defined gullies. Several gullies still contain snow. The light-colored, linear vegetated feature on the right side of the photo above Gastineau Avenue, the uppermost street, is the trace of the November 22, 1936, debris avalanche.

hydroelectric dam; badly damaged two homes, destroying a garage; and piled soil, rock, and organic debris across Glacier Highway and into several local businesses.

Elsewhere in the coastal zone, debris avalanches and debris flows continue to affect property and public safety. Probably the most spectacular event affecting the city of Ketchikan occurred sometime during the early morning of November 29, 1969, when a debris flow initiated by overflow of an emergency spillway completely destroyed the Upper Lake Silvis powerhouse, throwing the city into partial darkness. In Wrangell on October 9, 1979, a debris flow that began during an intense rainstorm swept downslope, striking an apartment house, destroying several trailers, and blocking the main highway south of the community. Rainfall recorded at Wrangell for the 24-hour period during which the event occurred was 4.49 inches.

While such large landslides are spectacular and worthy of note, they are uncommon. Smaller events occur in most Alaska coastal communities on a regular basis, linked directly to high-intensity storms, poor drainage controls, and disturbance of the hill slopes by clear-cutting and construction.

The best mitigation against the hazards of creep, slumping, debris flows, landslides, and rock avalanches is to avoid areas where they occur. Do not build on steep slopes (gradients greater than 25 degrees) or the toes or bases of such slopes and associated run-out areas, especially where avalanche chutes are present, where scars of past events are reflected in missing vegetation or patches of pioneer vegetation, or where timber clear-cuts are visible. Muskegs, evidence of poor drainage on hillsides, hummocky topography, leaning trees or structures, and thin soils are all indicators of instability. Permafrost underlying a thaw zone provides perfect conditions for ground failure and soil flow. If construction in such areas is an absolute necessity, consult with a construction engineer and apply stabilizing techniques (e.g., grade slopes to stable angles, install drainage systems, maintain vegetation cover, anchor foundations in bedrock, and insulate foundations to prevent thawing of permafrost), even if these measures are costly.

Snow Avalanches

The combination of steep slopes, a moist maritime climate, high levels of snowfall and mixed rain and snow during fall and winter, and strong winds off snowfields and ice fields leading to extensive cornice development on lee slopes make the Alaska coastal zone particularly prone to snow avalanches. Such events are widespread within the glacial valleys and along the steep mountain fronts that skirt the numerous fjords, channels, and straits that comprise most of the coastal waterways.

Assessing the Risk

On heavily forested slopes of any gradient, the accumulated snow is well anchored by trees and brush, and few avalanches of any size are initiated. On open alpine slopes and unforested gullies, however, there is a general correlation between slope gradient and frequency of snow avalanches. Typically, large avalanches are likely to occur on slopes with gradients between 25 and 50 degrees. Avalanches do occur on slopes with gradients less than 25 degrees, but they are unusual and small. On very steep slopes (greater than 45 degrees), snow tends to slough off soon after falling, and loose snow avalanches are common. The steepest slopes (80–90 degrees) usually do not accumulate enough snow to feed avalanches, although failure of cornices constructed at such sites by drifting snow can be significant avalanche triggers for the slopes below.

Snow avalanches are also influenced by slope profile. Convex slopes are particularly dangerous because they favor slab-type avalanches. Concave slopes are less hazardous and probably provide better peripheral anchorage for areas of unstable snow.

Because of the transient nature of snowfall in the coastal zone, with its frequent rain and mixed rain and snow, a normal snow cover consists of a number of distinct strata, each representing one snowfall or snowdrifting period separated by ice lenses formed during an episode of rain, sleet, or thaw between snowfalls. The position of the weakest stratum within the snow cover determines the location and the size of any developing avalanche. If the weakest stratum happens to be near the ground surface, the avalanche will involve the entire snowpack. If the weakest stratum lies near the snow surface, then the avalanche will likely involve only that stratum and those above it. Clearly, the degree of hazard of avalanche development is greatest when the weakest stratum lies at or near the ground surface.

When snow breaks free from the slope, it may run from the surface in localized "sloughs" if it is fresh, loose snow; or it may begin to slide in depth over an extended area as a coherent slab if it is packed and bonded. Generally speaking, loose snow avalanches occur most frequently during and immediately after snowfall and in the springtime when snow becomes wet and loose. They are very common and often quite small. Slab avalanches are initiated by rupture of a coherent mass of snow over an extended area and are frequently large and destructive events.

The moisture content of the snow is also important because it affects the physical properties of the avalanche mass. Once in motion, dry snow pulverizes and fine fragments diffuse into the air like a dust cloud. The turbulent mass of airborne snow moves rapidly downslope, is not greatly impeded by minor obstacles in its path, and pushes ahead of it a strong and often destructive blast of air. In contrast, wet snow remains coherent after disruption, moves at a slower rate that is controlled by local topography, and runs near the surface as a dense flood capable of inundating and destroying objects and structures in its path.

With normal levels of snow on the steep terrain characteristic of the Alaska coastal zone, much of the newly deposited snow slides as surface avalanches and sloughs during or shortly after falling, greatly reducing the possibility that snow avalanches of major proportions will occur. Even these shallow avalanches can be destructive, however, because heavy snowfalls are relatively common in this northern temperate, maritime environment. Single snowfalls of greater than 24 inches occur frequently at elevations above 2,000 feet, and the sheer weight of the snow mass coupled with a major rain can result in a catastrophic avalanche.

Avalanches are caused by factors that reduce the strength of the snow or increase the stress applied to it. The snow's strength is directly reduced by the destruction of the interlocking spikes and branches of newly fallen snow crystals that occurs during melting and refreezing. This loss of strength is commonly accomplished by temperature rise, especially when the snow temperature is near the freezing point, and by rain, a warming agent that is not only an effective destroyer of crystal bonding but also serves as a lubricant. The formation and growth of fragile new crystals by melting and refreezing at the base of the snowpack can lead to deep avalanches due to collapse of structure along this weak layer. Gradual increases in the weight of the snowpack by repeated snowfall, snowdrifting, and rain are the most common means of increasing the stress on the snowpack. In addition to these controlling influences, avalanches are commonly "triggered" by external influences such as falling snow cornices, avalanches from upslope, rockfalls, animals, and humans. This is the primary theory behind the use of explosives to bring down avalanches before they get too big.

Snow avalanche paths are easily recognized linear tracks through the forest extending from above timberline to the valley floor. At least a qualitative measure of the path's age and the frequency of avalanche occurrence can be obtained by inspecting the avalanche path for degree and type of disturbance, extent of vegetation cover, and recentness of damage to vegetation within and adjacent to the avalanche path. If the track is denuded of vegetation and bare soil, if trees and brush along the edge of the track show evidence of recent damage, and if rock and broken woody debris are deposited in the run-out zone, the avalanche probably occurred within the last year. If the lower end of the track is covered with low willow brush, huckleberry, and devils club, avalanches probably occur at 1- to 2-year intervals. A covering of low alder suggests a return interval of 5–10 years.

The Degree of Hazard

Probably the most tragic snow avalanche event in the history of Alaska occurred on April 3, 1898, in Chilkoot Pass east of Skagway when at least 67 gold stampeders died. Such catastrophic events are rare, and until recently snow avalanches have been of little concern to the general public because they occur mostly in uninhabited areas. This is changing rapidly, however, as populations expand beyond existing towns and villages, and transportation infrastructure, property, and public safety are more often directly affected. Since 1952, extensive but poorly documented damage to public and private property has occurred repeatedly, and at least 46 people have been killed or injured in snow avalanches within the coastal zone (table 4.2). A car driver was killed below Bird Hill along the Seward Highway in May 1952. Two skiers died on Harbor Mountain near Sitka in February 1952. Between 1971 and 1975, two skiers and a road maintenance worker were killed in the Juneau area at three separate avalanche sites adjacent to the Juneau road system. Five died within the popularly used Chugach State Park directly adjacent to the city of Anchorage.

Table 4.2 Documented Avalanche Fatalities in Alaska, 1898–1979

Date	Location	Activity	Fatalities
April 3, 1898	Chilkoot Pass, Dyea	Gold stampeders	67 known
1899	Lynx Creek, Kenai Peninsula	Gold miners	4
May 9, 1952	Bird Hill, Mile 91 Seward Highway	Car driver	1
Feb. 27, 1954	Harbor Mountain, Sitka	Lift skiers	2
Jan. 10, 1971	Juneau area, Tongass National Forest	Climber	1
April 12, 1971	Eklutna Glacier (3,500 ft.), Chugach State Park	Climbers	2
July 1972	Mt. McKinley (19,600 ft.), McKinley National Park	Mountaineers	3
Dec. 30, 1972	Flattop Mountain, Chugach State Park	Hiker	1
Jan. 1, 1974	Taniana Peak, Chugach State Park	Climber	1
Feb. 7, 1974	Thane Road, Juneau area	Highway dept. snowplow driver	1
Jan. 16, 1975	South Fork, Campbell Creek, Chugach State Park	Tour skier	1
Mar. 21, 1975	McGinnis Glacier, Central Alaska Range	Mountaineer	1
May 10, 1975	20 Mile River, Portage area	Hunter	1
Nov. 15, 1975	Granite Creek, Juneau area	Hiker	1
Feb. 19, 1976	Mt. Marathon, Seward area, Chugach National Forest	Climber	1
Aug. 11, 1976	Mt. Foraker (7,900 ft.), McKinley National Park	Climbers	3
Dec. 12, 1976	Sheep Mountain, mile 113 Glenn Highway	Hiker	1
Feb. 1, 1977	TAPS storage tank, Valdez	Worker	1
Jan 21, 1978	Taylor Creek, Chugach National Forest	Ski mountaineers	4
May 16, 1978	Mt. Foraker (11,000 ft.), McKinley National Park	Mountaineers	2
May 9, 1979	Mt. Hunter (10,000 ft.), McKinley National Park	Mountaineer	1

Source: Compiled by Alaska Division of Parks, Chugach State Park.

The potentially most damaging snow avalanche in an urban area occurred at 5:05 A.M. on Tuesday, March 22, 1962, above Behrends Avenue in downtown Juneau when a fast-moving, largely airborne dry-snow avalanche slammed into the western portion of Juneau's Highland district, inflicting varying amounts of damage on more than two dozen homes (fig. 4.12). Because of the early hour, few people were up and about, and only one person was injured. Avalanches along this particular track have a return interval of 14 years, and the development at its base is considered one of the most dangerous in North America in terms of future damage and potential

Figure 4.12 The Behrends Avenue snow avalanche track into downtown Juneau. During winter storms, heavy snow drifting on the lee side of the ridge causes massive cornice development above timberline. Small avalanches are common along this path during major rain-on-snow events and spring thaw. Large avalanches such as the destructive dry avalanche of 1962 have a recurrence interval of about 13 years.

for fatalities. One of the most spectacular snow avalanches ever also occurred in Juneau, in 1972, when a slab avalanche off Mount Juneau crashed into Gold Creek Valley, immediately behind the city. The resulting cloud of snow boiled over the ridge behind the town, blocked out the sun, and engulfed the entire downtown area. Tree limbs and twigs were scattered widely, but no one was injured and little damage was done.

Like rock avalanches and landslides, the best mitigation against snow avalanches is to avoid areas subject to the process (i.e., on and below steep slopes with unstable snow cover). As discussed above, snow avalanche scars are readily visible, and avalanche areas are likely to see repeat events. But all slopes are vulnerable. Sometimes the hazard can be reduced by generating small avalanches or using explosives to collapse cornices. Adding catchment basins in the run-out area may help keep damage in check in areas where avalanches recur. Where housing developments, schools, medical, and other service facilities are threatened by repeated snow avalanches, a systematic program of relocation should be developed.

5 Wind, Ice, and Sea

Ted Fathauer

Seismic hazards represent the biggest and potentially most destructive events likely to affect the coastal zone, but hazards driven by climatological processes are the most frequent, most pervasive, and most likely to be experienced by coastal zone residents. The collective impacts of storm-driven winds, currents, surges, waves, and ice on the entire coastal zone are similar in magnitude to a big, quick seismic event in terms of economic losses due to property damage. The difference is that localized flooding, coastal erosion, and ice override damage don't get the public's attention—or that of planners and regulators—like a big earthquake, spectacular volcanic eruption, or tsunami does. Nevertheless, everyone living in the coastal zone should understand and prepare for natural hazards associated with climatic events and processes.

Storm Surges

Storm surges, commonly known as coastal floods, occur when the sea is driven above the high-tide level onto normally dry land. The high water that occurs during coastal floods is not the primary concern, however. The factor that makes coastal floods one of the leading causes of property damage in Alaska is the powerful and destructive surf that occurs in conjunction with the high water (fig. 5.1). This surf can do direct damage to structures. In addition, the shoreline erosion that accompanies most Alaskan coastal floods can remove roads and undercut structures.

Many coastal communities, particularly those on the Bering and Arctic coasts, are situated on low-lying land. Particularly vulnerable to coastal floods are villages built on sand spits (e.g., Homer Spit), barrier islands (e.g., Shishmaref), and river deltas (e.g., Kwigillingok on the Kuskokwim Delta). Such land has the advantage of being level and close to the sea, but its low elevation makes it susceptible to shoreline retreat and coastal flooding. For example, the November 12, 1974, coastal flood on the Bering

Figure 5.1 High winds blow spray from breaking storm waves into downtown Kodiak. Photo courtesy of the Kodiak Historical Society.

Sea coast of Alaska inundated areas as much as a mile inland between Cape Nome and Topkok (see appendix C, *Forecast Procedure for Coastal Floods in Alaska*). The debris from this extraordinary storm is still distinctly discernible as a line of deposits marking the high-water line. Local residents refer to this as the driftwood line or, more commonly, the trash line.

Erosion and change are part of the life cycle of low-lying coastal land. Normal wave action and currents cause gradual and continuous change in the shape and location of sand spits, barrier islands, and river deltas. In a sense, any structure placed in such an area has a limited life expectancy. During major coastal floods, the rate of topographic change increases enormously. A sand spit may change more during one day in a major coastal flood than it would normally change in several years.

Damage during large coastal floods in Alaska is not limited to the effects of high water and surf. If sea ice is present, as it was during the 1963 Barrow storm, blocks of ice carried by high water and driven by waves and wind can wreak considerable destruction on structures. In addition, Alaska coastal floods are normally accompanied by high wind, rough seas, and extensive precipitation.

Although large-scale coastal floods in Alaska are limited to the Bering Sea and Arctic coasts, small sections of the shoreline on the Gulf of Alaska coast and even in southeast Alaska also experience coastal flooding. These are enumerated later in this chapter.

Causes of Coastal Floods

Coastal floods in Alaska are caused by wind-driven seawater affected by tide levels and atmospheric pressure.

Wind-Driven Transport of Seawater

This is the most important factor in coastal floods. During the 1974 Nome flood, for example (see fig. 2.5), the total rise in static water level was almost 12 feet. Tide levels and atmospheric pressure accounted for about 2 feet of this total; the other 10 feet were attributable to wind-driven transport of the sea. Coastal flooding in Nome used to be a much greater problem before the seawall was constructed (fig. 5.2).

It is the friction between air and water at the air-sea interface that drives the water movement. The rotation of the earth (which causes the Coriolis force) causes moving objects, such as a mass of water, in the Northern Hemisphere to be deflected to the right of their direction of motion. Thus, surface ocean currents driven by the wind are at about a 20-degree angle to the right of the wind.

Figure 5.2 A 1902 storm in Nome. Boat and building owners are busy rescuing their property. Photo courtesy of Ethel Becker; reprinted with permission from *Alaska Geographic,* vol. 11, no. 1 (1984).

Each successive layer of the ocean acts on the next lower layer just as the wind acts on the surface of the sea. As a result, wind-driven currents in the Northern Hemisphere turn steadily to the right (clockwise) with increasing depth, an effect commonly known as the Ekman spiral. This explains the movement of icebergs in the Arctic noted by Fridtjof Nansen during the historic voyage of the *Fram* in 1893–96. Nansen observed icebergs in the Arctic moving in a direction about 30 degrees to the right of the wind. Such movement is the net result of surface and subsurface ocean currents and the force of the wind acting on the portion of the iceberg above the waterline.

In the deep ocean, net mass transport of water is at a 90-degree angle to the right of the wind. In the comparatively shallow depths of the Bering Sea and Alaska's Arctic coastal shelf waters, net transport is roughly 45 degrees to the right of the wind.

The exact direction of water transport is not particularly important. The point is that wind flow need not be directly onshore to cause shoreward movement of seawater. Indeed, the wind direction that will cause the greatest onshore transport of seawater is *not* perpendicular to the coast, as one might think, but at an angle about 45 degrees clockwise from the coastline. Thus, the winds over the Bering Sea that are most dangerous to Nome come not from the south but from the southeast. Residents of Unalakleet, in eastern Norton Sound, become concerned when there is a strong wind from the south through southwest; it often presages high sea levels during early winter storms.

The amount of wind-driven transport increases as the wind speed and the duration of a storm increase, as do the wave height and surf. Thus, potential coastal flood damage increases with increasing strength of the wind and with the duration of strong winds.

Wind-driven water transport and waves also increase as the length of the fetch increases. *Fetch* is the distance across open water over which a wind field is blowing. For instance, if the wind in Norton Sound is from the southeast, the fetch upwind of Nome is about 120 miles. If the wind in Norton Sound is from the north, the fetch at Nome is zero. One factor to consider with respect to fetch is the extent of sea ice, which reduces the effect of the wind on the sea—both on wind-driven water transport and on wave buildup.

Friction at the air-sea interface increases as the air becomes colder than the sea surface. When the air is colder than the surface of the sea, the first few meters of air above the water become turbulent due to updrafts from the warm sea (warm air rises) and compensating downdrafts of cold air from above. This overturning increases the transfer of momentum and energy from the wind to the sea. Thus, as the temperature of the air drops below that of the sea surface, wind-driven transport of seawater and wave heights increase. On the other hand, when the air is warmer than the sea, there is no tendency for air to rise and the low-level temperature profile is stable and resistant to overturning. This reduces transfer of energy from the wind to the sea.

Thus, the rapid cooling of the air above the relatively warm (after the summer), largely ice-free ocean greatly favors coastal flooding in Alaska during the autumn and early winter months. During winter, sea ice may or may not cover the outer coast of the Kuskokwim Delta and Bristol Bay; this varies from year to year according to the winter temperatures. Sea ice is not a factor in the few locations on the Gulf of Alaska coast and in southeast Alaska where coastal flooding is possible.

Tide Levels

Tide levels can be computed for many coastal locations in Alaska using the tide tables published annually by the National Ocean Survey. Typical ranges between high and low tides vary considerably from location to location (table 5.1).

The greatest tides occur at the heads of large bays, such as Cook Inlet and Bristol Bay. The tide ranges listed in table 5.1 are typical ranges. During spring tides the ranges are larger; at Anchorage, for instance, the difference between high and low tides can be more than 35 feet, 6 feet higher than usual. In evaluating the shoreline, it is very useful to read the ocean's writing. Physical signs such as erosion of the shore, a high-water mark of drift-

Table 5.1 Tidal Ranges in Alaska

Location	Range between high and low tide (feet)
Ketchikan	15.4
Juneau	16.3
Yakutat	10.1
Valdez	12.1
Homer	18.1
Anchorage	28.8
Kodiak	8.5
Dutch Harbor	3.7
Naknek River entrance, Bristol Bay	22.6
Nushagak Bay	19.5
Goodnews Bay	8.9
Saint Michael	3.9
Nome	1.6
Kiwalik	2.7
Barrow	0.4

wood and debris, and changes in life-forms marking the extent of high and low water (which together indicate the magnitude of tidal ranges) are all obvious to the experienced observer, and are valuable information to a potential purchaser or builder.

During the 29.5-day lunar month, the difference between high and low tides reaches its maximum when the sun, earth, and moon are all aligned. This happens at new moon and full moon, and produces the high spring tides. The height of spring tides varies from one month to the next. A coastal flood that comes on a particularly high spring tide can inflict considerably more damage than it would if it came during a neap tide (when the moon is at first quarter or last quarter). At locations such as Nome and points along the Arctic coast where the magnitude of astronomical tides is small, the phase of the moon is far less important than it is at the head of Bristol Bay. Nonetheless, past experience shows that the worst coastal floods at Nome have all occurred during a new moon or a full moon.

Atmospheric Pressure

The atmospheric pressure influences the seawater level through the so-called inverse barometer effect. As atmospheric pressure drops over a region, the sea level in that region can rise in response. The common unit of pressure measurement is the millibar, a metric unit. The average sea level pressure is just over 1,000 millibars. A 10-millibar drop in atmospheric pressure at sea level will allow the seawater level to rise about 4 inches. A drop of 30 millibars below average barometric pressure can bring up the sea level a whole foot. On the other hand, barometric pressure well above normal presses down on the sea and can lower the sea level about 4 inches for every 10 millibars of pressure above average.

Like air in the atmosphere, seawater flows from areas of high or rising pressure to areas where pressure is low or falling. This flow takes time—on the order of hours; if the pressure change is very fast, the sea level will adjust more rapidly.

Areas Susceptible to Coastal Flooding

Figure 5.3 shows areas of Alaska's coast that are susceptible to coastal flooding and their relative vulnerability to coastal floods. The relative vulnerability is necessarily a subjective rating, not an objective quantity. Further, in areas where there are neither people nor development, there is, strictly speaking, no danger of damage or loss due to coastal flooding. Since it is not possible to project future shoreline development and coastal population, figures 5.3 and 5.4 are drawn to indicate the potential severity of coastal flooding along the Bering and Arctic coasts. Unmarked locations are

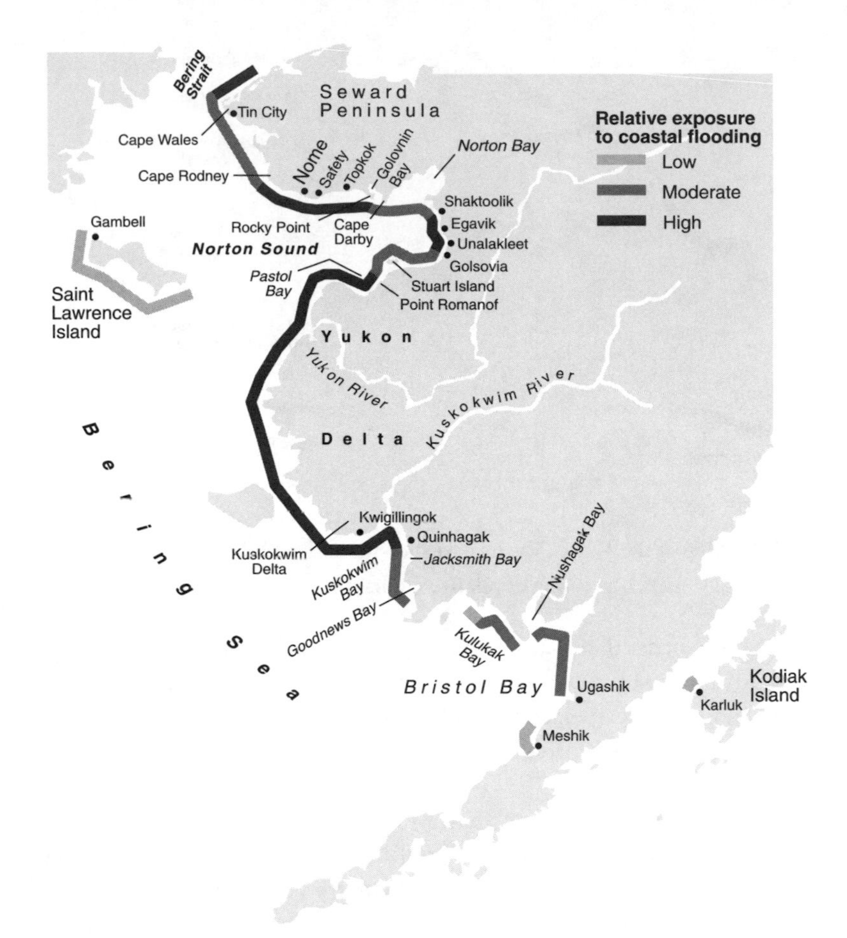

Figure 5.3 Map of the Bering Sea coast showing susceptibility to flooding during coastal storms. The rating of relative vulnerability is explained in the text.

presumed to have only minor exposure to coastal flooding. Low danger means that flooding is infrequent and generally is not severe when it does occur. On stretches of coast with moderate exposures, flooding may be expected to occur about once every 3–5 years, and the magnitude of floods may be sufficient to cause beach erosion several yards or more inland and damage to structures as much as 6 feet above normal high-tide and surf levels. High-danger portions of the coast can expect significant coastal flooding every 2 years or so, and the flooding may cause significant beach erosion and damage to structures more than 100 yards inland if these are situated within 10 feet or less of the normal high-tide and surf elevations. Most coastal floods do not produce the maximum damage possible. The rating system used in preparing this map is based on studies of past storms (see, e.g., appendix C, *Forecast Procedure for Coastal Floods in Alaska*), the

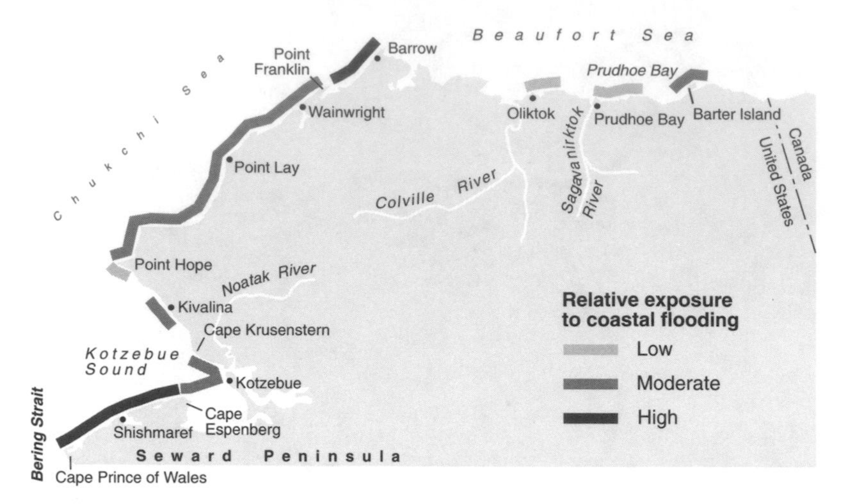

Figure 5.4 Map of the Arctic Ocean coast showing susceptibility to flooding during coastal storms. The rating of relative vulnerability is explained in the text.

frequency of storms that can produce coastal flooding, and the potential damage in major storms. High-exposure areas can expect minor coastal floods in summer interspersed with a few major storms, while areas marked low exposure can expect infrequent events that produce minor or moderate damage.

Alaska's meteorology is such that, in theory, any part of the coastline could be susceptible to dangerous coastal floods. The factors determining risk are primarily the topography, the bathymetry near shore, and seasonal sea ice cover. Practically the entire coast of southeast Alaska and Kodiak Island, for example, is steep and rocky, and thus very resistant to flooding and erosion. The high bluffs of the Kenai Peninsula north of Kachemak Bay make that area similarly resistant to flooding. Offshore sandbars absorb much of the energy of incoming surf and sometimes stall floes of sea ice offshore. One such protected area is Kotzebue. The Noatak River empties into Kotzebue Sound just north of Kotzebue, and the discharge of the Noatak moves southward just offshore of the beach of Kotzebue. This constant movement of water keeps a relatively deep cut parallel to the beach of Kotzebue dredged naturally year-round and maintains an offshore sandbar to the west of the core of this current.

The impact of storms on a given section of coastline can change as the years pass. For instance, the construction of a seawall at Nome before the November 1974 storm no doubt greatly mitigated the effects of this storm. On the other hand, efforts to restrain beach erosion at Kotzebue and Shishmaref during the past 10 years have not succeeded. There, each successive storm escalates the threat to existing structures and the shoreline as the ocean cuts progressively inland, and in some cases as buildings are constructed closer and closer to the shore.

To evaluate the danger of coastal flooding at a particular point on Alaska's coast, consult this book and take advantage of the expertise of longtime local residents. Sheltered coasts with high, firm ground are best, but hundreds of miles of the Alaskan coastline do not offer such ground. In areas with significant exposure to coastal flooding, set structures well back from the waterfront. If possible, bury utility lines and tanks at a secure depth.

Examples of Storm Surge Problems in Alaska

Auke Bay Harbor is a popular marine recreation area located just north of the city of Juneau. During a high spring tide, a strong weather front passing through from the west followed immediately by a large, rapid rise in atmospheric pressure will bring large swells, 6–8 feet high, into Auke Bay from the west in spite of the very short fetch upwind. During the past 25 years, such swells have hit Auke Bay exclusively during the autumn, with October and November being the favored months. Past events include one in October 1972 and the big Thanksgiving Day storm in 1984. Auke Bay rarely has high waves or strong winds, so a sudden and unusual blast of gales and rough seas from the west can be expected to do damage to docks and may pull boats from their moorings. Significant coastal floods appear to occur about once every 10 years in Auke Bay. Coastal flood exposure is rated as low.

The Thanksgiving Day storm of 1984 did a great deal of damage in *Tenakee Springs.* One nonoceanographic, nonmeteorologic factor involved in this rare case was the age and considerable weathering of the existing structures on and along the docks at Tenakee. Many of the buildings standing at Tenakee in 1984 were built on very old, weathered pilings, as were the wharfs. Buildings set back from the shore and situated on foundations on the ground did not sustain any damage from the sea. Coastal flood exposure here is rated as low.

Karluk is a small town on the north shore of Kodiak Island. The shoreline at Karluk is east-west, so a north wind blows directly onshore. On January 8, 1978, a north wind of 40–57.5 mph blew over an ice-free fetch of 150 miles for nearly 24 hours, one day before the maximum monthly tides. Heavy surf caused considerable damage and shore erosion. This is the only known case of coastal flooding at Karluk. The weather maps at the time of the storm showed nothing unusual, so minor coastal flooding can be expected again; exposure to coastal flood is rated low.

Homer Spit is something of a topographic anomaly—a low-lying sand spit surrounded by a bay with fairly high and steep shorelines. Like Shishmaref and Kotzebue, Homer Spit is subject to an ever-growing exposure to coastal flood damage with the passing of each storm. The road on Homer Spit has been eroded by storms several times, and each time reconstructed. Coastal flood exposure here is rated moderate.

As noted in table 5.1, the tidal range in *upper Cook Inlet,* north of Homer to Anchorage, is the largest in Alaska. This poses two problems. First, boats that anchor near shore during a high spring tide may be aground for some time after the high tide has passed. Consult the tide tables carefully in this area: the larger the magnitude of the tides, the larger the possible error in the tide predictions. Wind and atmospheric pressure may alleviate or aggravate problems posed to mariners by abnormally high or low tides. In Turnagain Arm, upper Cook Inlet (just south of Anchorage), a tidal bore is sometimes observed. This is most likely to happen when a very low tide is followed by a very high tide. The height of the incoming wave, the tidal bore, is often as much as 6 feet. This is reason for caution on the part of those present on the tidal flats at the onset of high tide.

Meshik is a village on the south shore of Bristol Bay, near Port Heiden, located on low-lying land. On October 27–28, 1976, a fairly typical midwinter storm produced northwest gales of 35–52 mph blowing persistently over a 300-mile fetch. Maximum monthly tides occurred the day before the storm, which lasted 36 hours and brought heavy surf into the town. The exposure to coastal flooding in Meshik is rated relatively low, like Karluk, Tenakee Springs, and Auke Bay.

The Bristol Bay coast from Ugashik up along the east shore of Nushagak Bay and along the east shore of Kulukak Bay is wide open to a fetch of several hundred miles to the southwest. The east shore of Kuskokwim Bay from Goodnews Bay to Jacksmith Bay is similarly situated. These coastlines are moderately exposed to coastal flooding. Recent notable cases include the November 1974 Bering Sea storm and the floods along the Yukon-Kuskokwim Delta of November 8–10, 1979.

The Yukon-Kuskokwim coast from Quinhagak to Pastol Bay slopes very gently into the sea and is totally exposed to storms in the Bering Sea. The 1974 and 1979 storms brought extensive damage to these coasts, which have a high exposure to coastal floods.

The southern Norton Sound coast from Point Romanof to Golsovia is somewhat less exposed due to the protection afforded by the Yukon Delta and Stuart Island to the west.

North of Golsovia, through Unalakleet and up to Egavik, exposure to the west is totally unprotected. Exposure to coastal flood danger in this region is high.

Shaktoolik, Norton Bay, and Golovin Bay are slightly more protected by the shorelines of the southern Seward Peninsula, Rocky Point, and Cape Darby. Exposure in these areas is moderate.

From west of Rocky Point to Cape Rodney, coastal flood exposure is high. There is an essentially unlimited fetch to the south-southwest. The November 1974 coastal flood inundated areas up to a mile inland from the normal shoreline.

West of Cape Rodney to Bering Strait the shoreline is conducive to coastal flooding, but the proximity of Saint Lawrence Island to the southwest affords some protection, and exposure is moderate.

The south and west shores of Saint Lawrence Island are subject to some coastal flooding. Much of the south shore is protected by a barrier island. The village of Gambell is on a high beach ridge plain, but the airstrip is built on flat, low-lying ground that is exposed to the west and southwest. The airstrip and village at Gambell have some exposure to coastal flooding.

On October 3, 1960, westerly winds of 55–85 mph blasted across Norton Sound, building seas to 20 feet. Wave action at Unalakleet was described as violent. A 55-year resident said it was the worst storm in memory at Unalakleet. The most recent large coastal flood recorded in Norton Sound occurred during the storm of October 5, 1992, which affected much of the south coast of the Seward Peninsula, including Nome and the road east to Topkok.

The northwest coast of the Seward Peninsula, including Shishmaref, is wide open to surf and wind from the west and northwest. Its exposure to coastal flooding is high, though the ice cover lessens the duration of the coastal flood season in the early winter. Coastal floods have occurred as early in the year as August (1975) and as late as November (1974 and other years).

One factor pertinent to the severity of beach erosion on most of the Bering Sea coast and along all of the Alaska coast north and east of the Bering Strait is the presence of permafrost in nearly all beaches. Once a storm brings seawater to a coast underlaid by permafrost, thawing of the permafrost inevitably proceeds, and the bluff collapses into the sea. This has been a factor in erosion at Shishmaref, Kotzebue, and Barrow.

Kotzebue Sound is partly protected by the arms of Cape Espenberg and Cape Krusenstern. The region is moderately exposed to coastal flooding in autumn and early winter. Once Kotzebue Sound becomes covered with land-fast sea ice, the danger of coastal flooding is past. Indeed, the ice cover in Kotzebue Sound during winter is so robust that snow machine travel across the ice is routine. In midwinter, the passage of a storm that would cause a coastal flood during the open-water season may cause overflow of seawater through cracks in the ice sheet. This has caused problems for travelers because it occurs beneath the snow covering the ice and is thus not easily discerned until the traveler drives into it—and risks becoming frozen in if the weather is cold and windy. Significant overflow occurred at Kotzebue during the November 12, 1974, storm in the Bering Sea. The partial ice cover over Norton Sound was blown ashore along the coast from Nome to Rocky Point.

North of Kotzebue Sound, the low-lying villages of *Point Hope* and *Kivalina* have suffered repeatedly from coastal floods. The new village at Point Hope is on a less threatened area of the spit not subject to flooding for the last century, in contrast to the erosion-prone old town at the cape. Coastal flooding has threatened the barrier island community of Kivalina for the past 35 years. Erosion may occur on both the ocean side of Kivalina and on the lagoon side. Exposure to coastal flooding at Kivalina is rated as moderate.

The Arctic coast from north of Point Hope to Point Barrow is completely open to the west and southwest. Past coastal floods from north of Point Hope up to Point Franklin have been limited, and exposure is moderate. Beyond Point Franklin to Point Barrow, the nearshore topography is lower and exposure is high.

Wainwright is built on a high permafrost bank. Ongoing thawing of the exposed permafrost that constitutes the entire foundation of this high ground has caused the bank to recede each year. Erosion of the beach below the bank was accelerated during the September 20, 1986, storm, which left four homes hanging precariously over the edge of the bank.

Significant erosion of the bluffs southwest of Barrow—on the order of 50 feet inland over the past 50 years—has been reported. Major coastal floods on this coastline are not frequent but can be severe. On October 3, 1963, westerly winds of 50–75 mph blasted across an abnormally ice-free Chukchi Sea. At the town of *Barrow,* the sea was driven inland about 400 feet. In the closing phases of the storm, large chunks of sea ice, some up to 12 feet thick, were driven about 15 feet inland. Several houses were lifted up by the high water and destroyed. No storm since has equaled it. The most recent significant coastal flood in Barrow was on September 20, 1986.

The most vulnerable area in Barrow is the road going north from Barrow through Browerville, and on to the old Naval Arctic Research Lab. Years of sand removal have greatly increased the danger. Unless this road is built up and fortified, it will continue to experience at least minor coastal flooding about once every 5–10 years.

East of Barrow, ice cover is a deciding factor in limiting exposure. The coast near the old Distant Early Warning line site at *Oliktok* has low exposure to coastal flooding, while the coast near *Barter Island*—closer to the waters of Mackenzie Bay, just east of the Canada border and ice free for much of the summer—has moderate exposure. The only real damage likely to occur during coastal floods on these two stretches of coast is to the airstrips, which are built low down along the beach.

The oil complexes at and around Prudhoe Bay are built so high and so strongly that they are beyond the reach of the sea. This region does have idiosyncrasies, however. First, the ice bridges built for wintertime construction work may, in unusual cases, be affected by overflow just as the land-fast ice in Kotzebue Sound is affected. On December 10–11, 1984, a 35-millibar drop in sea level pressure and a shift of the wind from east to west opened cracks in the land-fast pack ice along the shore at Prudhoe Bay and delayed construction of an ice road. Of course, just as the wind can bring in the sea, it can take it away. On September 25–26, 1974, the Arctic coast of Alaska had a near record 200 miles of ice-free waters from the Beaufort Sea out into the Chukchi. East winds of 25–50 mph, accompanied by blowing sand, blew September 24 and 25, and let up on the 26th. A long, vigorous, large-scale easterly gale on the Arctic coast produced a drop in sea level in Prudhoe Bay of several feet on September 25–26. Marine operations were brought to a halt during this time.

Ice Override

Ice override occurs when floating sea ice is driven ashore by wind during coastal floods (fig. 5.5). The Inupiat call this *ivu* (see appendix C, particularly "Onshore Ice Pile-up and Ride-up," and "Fast Ice Sheet Deformation during Ice-Push and Shore Ice Ride-up").

Ivu occurs much less frequently and affects a much smaller area than coastal floods do. Very specific weather, oceanographic conditions, and shoreline topography are required for ivu to occur. Ivus have been reported in autumn and early winter, but there are no reports of them in late winter or spring. By late winter, the shorefast ice has become too strong a barrier and the intensity of storms has dropped significantly.

During an ivu, either shorefast or mobile pack ice may be floated free and driven ashore. A typical ivu lasts less than an hour. They may occur during stormy weather—which increases their likelihood—or, in rare instances, during benign weather (see appendix C, "Another Ivu at Nome").

The winter pack ice in Kotzebue Sound and along the Arctic coast east of Point Barrow is shorefast. Weather satellite pictures consistently show the so-called shear line delineating the boundary between mobile pack ice and shorefast ice in Kotzebue Sound and in the Beaufort Sea. Both shear lines are marked by large arcs, hundreds of miles long, constituted by open or recently refrozen leads.

Eastern Norton Sound has shorefast ice in most years, although the extent varies. The really mobile ice is driven by the ocean current that enters the Arctic Sea and flows vigorously northward through the Bering Strait,

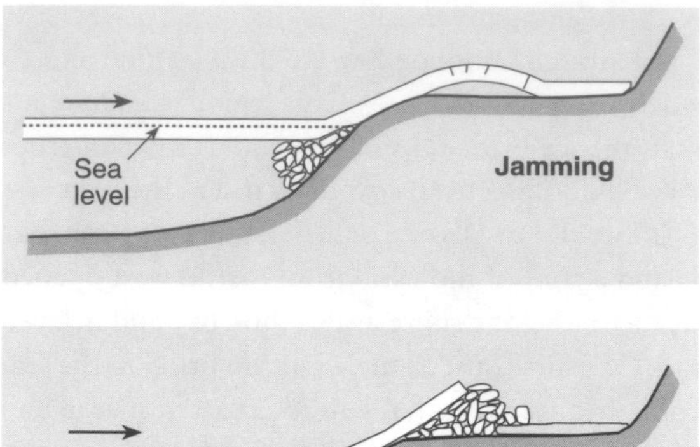

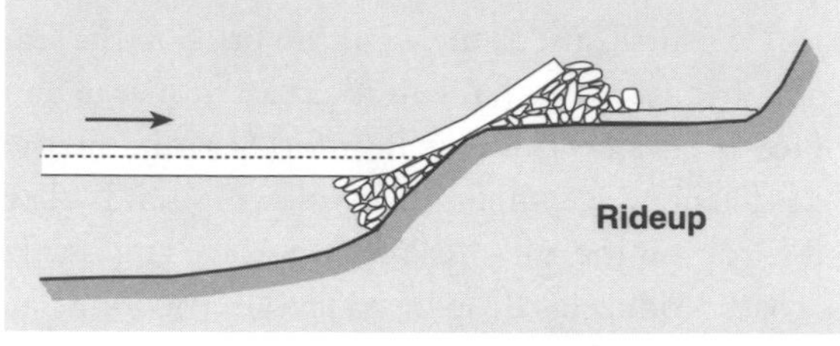

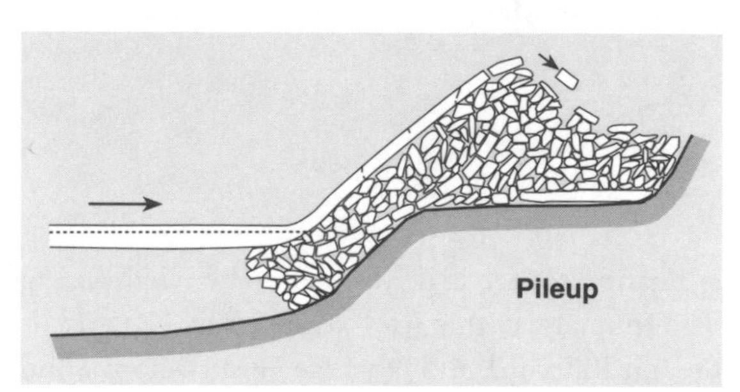

Figure 5.5 Types of onshore ice pileup. In extreme cases, ice penetrates more than half a mile into the mainland beyond the beach. Communities often battle small-scale ice rideups with bulldozers.

north to Point Hope, thence northeast to Point Barrow, and northwest off into the clockwise-moving Arctic gyre. Ivus have been reported on the Seward Peninsula coast from west of Rocky Point to Cape Rodney, the village of Gambell on Saint Lawrence Island, the northwest coast of the Seward Peninsula (including Shishmaref), and the Arctic coast from Point Hope to Point Barrow, particularly the beach north of Point Franklin.

A 300–500-year-old sod house was excavated several years ago on Alaska's Arctic coast. The roof had been forcefully collapsed, apparently by ivu, and the residents had died in this sudden catastrophe (see appendix C, "Sealed in Time").

An unusual form of ivu occurred at Nome on the afternoon of December 9, 1987. The ice cover was less than 2 weeks old and had formed during abnormally calm weather. It was therefore probably quite flat, uniform, and free of fractures and rafts. Tides at Nome are mixed: during part of the lunar month there are two high tides and two low tides per day, and during the rest of the lunar month there is only one high tide and one low tide per day. From December 7 to December 11, 1987, Nome had only one tide cycle per day, which reduced the frequency of disturbance to the ice. On December 11, light south winds set in over Norton Sound in conjunction with semimonthly maximum tides, and a nearly uniform sheet of ice, approximately 2 feet thick, began riding up on the beach at Nome and spilling over the seawall onto parts of Front Street. No damage was done, and by midafternoon the ivu had ceased just as mysteriously as it had begun. Local residents could not remember anything like it.

February 1989 was vastly different. On the western boundary of an enormous cold air mass, which set numerous low-temperature records in interior Alaska, a southeast storm began blowing over Norton Sound on February 2. Had there been open water, the wind alone would have posed a serious threat of coastal flooding. Instead, after 2 days of powerful wind, the sea level on the southern Seward Peninsula coast came up 6–8 feet, and a classic ivu took out the Safety Bridge east of Nome on February 3–4. This event occurred at the time of the semimonthly maximum tides—a coincidence that is often repeated in the timing of coastal floods and ice overrides in Alaska.

The Seward Peninsula coast near Nome commonly has large ice ridges along the shore that are typically built up in a series of onshore thrusts of the sea ice. These normally pose no problem; however, when a storm elevates the sea level along the shore, then an ivu is possible if the wind is onshore.

In most cases, the ivu penetrates only a few tens of feet. A 450-foot onshore movement was reported at Barrow in December 1978, however, and there is evidence that other ice sheet movements have gone ashore as far as 100 yards. Multiyear pack ice in the Arctic Ocean is typically 7 feet thick, but the ice in an ivu must be 4 feet thick or less. This is why ivus are limited to relatively new ice in autumn or early winter.

Ivus occur only on very gently sloping beaches in the limited portion of northwest Alaska described above. Since the frequency of these events is quite low, it may be difficult to determine if a certain stretch of beach is susceptible to them. Deposits of beach gravel on the tundra near the shore may be evidence of a past ivu. Residents who have lived in the area a long time may be useful sources of information. Structures built in areas that appear or are known to be susceptible to ivus are best set well back from the shoreline; utility lines and tanks should be buried securely if possible.

Although it is difficult to predict ivus, they normally occur in stormy weather that favors coastal floods. The meteorology and oceanography of coastal floods is well understood, and warnings of the storms that cause them are normally broadcast 12 hours or more in advance. Coastal residents can therefore be better prepared by routinely following the broadcasts of National Weather Service coastal forecasts and warnings.

High Winds

Compared with most of the state's interior, Alaska's coasts are often windy places. Strong winds are most frequent from September through January, although calm periods can occur at any time.

Nature's handwriting is an excellent guide to identifying the areas that experience the strongest winds. Coasts with old-growth forests, common in southeast Alaska, are better protected than bare, rocky capes. Indeed, elevated capes that jut out from the coast are the windiest places known to Alaska's meteorologists. The most notorious example is Cape Prince of Wales, on the east side of Bering Strait. The weather station at Tin City, just south of the elevated terrain of the cape, often records the strongest winds in the state. A protracted windstorm during the winter of 1974–75 brought peak gusts of more than 115 mph to Tin City.

Coastlines that lie beneath high mountains, as at Valdez and Douglas (just west of Juneau), are susceptible to strong and very gusty winds when cold Arctic air is pulled strongly offshore by ocean storms in winter. Cold air is dense, and downhill flow along steep, high slopes accelerates it. Such windstorms are usually characterized by strong gusts. Strong south winds from warmer maritime areas do not have an effect anywhere near as intense as cold winter winds from the north, out of the Arctic. Strong gusts are a threat to structures; a steady 70 mph wind is much more easily borne than a wind that fluctuates rapidly between 45 and 90 mph. Wind with a high gust factor hammers all that lies in its path. In general, the steeper and more elevated the surrounding terrain along a shoreline, the greater the likelihood of particularly strong winds. The strength of peak winds is normally amplified at sharp capes along the coast.

Construction design for the Alaska coast must take high winds into account. Existing communities normally have some information on the strongest winds that may be expected. Shorelines distant from mountainous terrain rarely have winds above 75 mph, while coastal areas close to high, steep mountains or high, sharp capes may well expect peak winds of 140 mph. If construction, harbor design, and the like take these factors into account, the danger from Alaska's ferocious windstorms may be mitigated. In established communities and harbors, public safety officials can minimize hazards by keeping up-to-date on coastal weather forecasts.

6 Shoreline Erosion

One of the major problems facing Alaska's shoreline communities (and beachfront communities everywhere) is the retreating shoreline. Intimately related to the erosion process are other hazards such as storm surges and tsunamis (see chapters 3 and 5). As the shoreline retreats, the surf zone moves closer to shorefront buildings, and the potential for damage from ice, storm surge, and waves increases. Shorelines and beaches exist in a dynamic equilibrium involving four factors: wave and tidal energy, sea level change, sediment supply, and location in space (shoreline position and profile) (fig. 6.1). These four parameters control the behavior of most shorelines.

Wave and Tidal Energy

Waves and, to a lesser extent, tides carry sand and gravel to and from beaches. One of the principal processes involved in surf-zone sediment transport, especially on sandy beaches, is longshore transportation. This shore-parallel transportation occurs when wave crests strike the beach at some angle other than head-on (fig. 6.2). When this happens, a portion of the energy of the breaking waves is directed along the beach, forming a current in the surf zone. The breaking waves put sediment into suspension, and the longshore current carries it along the coast. Although beach sediment is usually carried in both directions along a beach, most beaches have a dominant direction of longshore sediment transportation. Corresponding to the terminology used in rivers, the direction of net annual sediment transport on a beach is termed downdrift (downstream), and the reverse direction is referred to as updrift (upstream).

Storms tend to move beach sediment offshore. In this case, the impact of waves on sediment transport may be enhanced by strong seaward-directed bottom currents capable of carrying large volumes of sand-sized material far away from the beach.

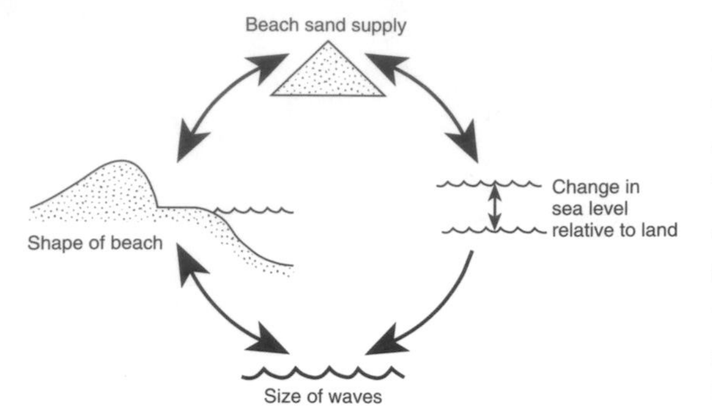

Figure 6.1 Factors controlling the dynamic equilibrium of beaches. When one factor changes, the others adjust, in effect making the beach a giant, flexible buffer zone between land and sea.

In Alaska, ice plays a significant role in shoreline evolution. Ice may serve a protective role by reducing or eliminating wave impact when the nearshore water is frozen. In addition, freezing of the sediment making up beaches and bluffs during the winter greatly reduces the impact of waves. In contrast, ice can be destructive during pileup and ride-over, and is probably responsible for moving a lot of material close to and onto the beach during the spring ice breakup.

Sea Level Change

The level of the sea may be changed temporarily by tidal and climatic effects, or "permanently" by tectonic forces or global water budget changes. A rise in sea level allows waves to attack the back of the beach and causes the shoreline to shift its position landward. The combination of a rise in sea level and high wave energy, as during a storm, may flatten the beach profile. One of the most common causes of short-term sea level rise is storm surge, a rise induced by the combined effects of an atmospheric low-pressure center and the wind forcing water ashore.

The long-term sea level is changing worldwide, and this is an important factor in the shoreline retreat equation. In most places the sea level is rising due to the increased water volume from melting alpine glaciers and to thermal expansion of the upper surface layers of the sea caused by global warming. Tectonics, or earth movements, are having a large effect as well. For instance, the land in the vicinity of Juneau is rising, and as a consequence, the sea level is dropping and the shoreline erosion problem is reduced somewhat. On the other hand, the continental shelf is sinking slightly along the U.S. East Coast due to the weight of the water over the shelf, and the rate of sea level rise is increasing as a consequence. Along many coasts of the world unaffected by recent glaciation, the sea level is rising at a rate of about 1 foot per century. In Alaska, ongoing mountain building and the isostatic rebound that accompanies glacial retreat make the importance of sea level rise as a shoreline erosion factor highly variable. In areas where the shoreline skirts a delta, sediment subsidence due to compaction may cause a relative rise in sea level.

The higher the level of the sea, the farther inland the shoreline will be. The faster the rate of sea level rise, the greater the rate of shoreline retreat—other factors such as sand supply and wave energy being equal. There is a strong possibility that the rate of global sea level rise will increase in future decades due to the greenhouse effect.

Sediment Supply

Both the volume and the grain size of beach sediments are critical variables in the shoreline erosion equilibrium. In general, the larger the supply of sediment, the lower the rate of erosion. If the sediment input is high enough, shorelines will accrete—that is, build out. Accretion is common along some of Alaska's coast near river mouths. Grain size also helps to determine erosion rate. For example, other things being equal, a beach covered by boulders will retreat much more slowly than one made up of sand. A boulder beach such as Seward will be affected only during storms, while a fine sand beach such as the one at Shishmaref may change its shape daily. Grain size even determines the beach shape. Boulder beaches tend to be very steep, while fine-grained sand beaches (in which the grains are barely

Figure 6.2 Longshore currents are formed by waves striking the shoreline at an angle and are responsible for lateral transportation of sand and gravel along beaches. Halting the flow of beach sediment through engineering structures, beach mining, or inlet dredging results in increased erosion of adjacent beaches (e.g., see fig. 11.8).

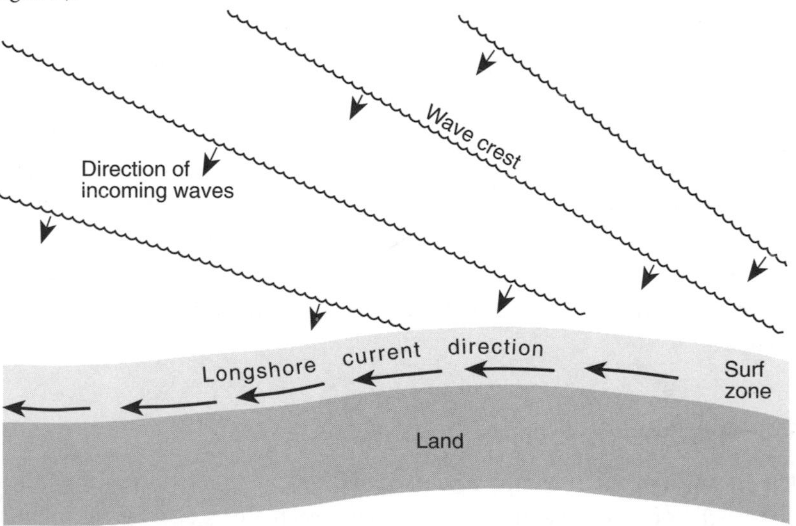

Figure 6.3 Abandoned house at Kividlo on the Seward Peninsula. Older settlements located on barrier islands were often protected from erosion by high dunes.

visible to the naked eye) are very flat between the high- and low-tide lines.

Beaches obtain their sediment from a number of sources; the most important are rivers, bluff erosion, the continental shelf, and nearby beaches (from longshore currents). Shorelines also have sediment sinks, places where sand is lost from the beach. These include sand dunes and washover fans on adjacent uplands where material derived from the beach is deposited by wind or wave action. Adjacent beaches, spits, bay-mouth bars, the offshore continental shelf, and flood- and ebb-tidal deltas associated with inlets are also sediment sinks.

The relative importance of sources and sinks is often difficult to determine, yet it is important to know this information in order to protect community beaches (fig. 6.3). For example, damming a local stream or river might cause beach erosion by trapping sand upriver that would normally supply the adjacent beaches. Building a seawall or groin updrift from a beach may cause a deficit in sand supply and increase the rate of beach retreat. Halting bluff erosion by seawalling the base of a bluff may cut off the sediment supply to the beach and lead to accelerated erosion. Bluff stabilization is a common practice in Alaska (e.g., in Barrow, Kenai, Homer, and Kodiak).

Shoreline Position

The location of the shoreline is determined by the interaction of the other three factors in the dynamic equilibrium of beaches (fig. 6.1). Events in 1964 on the Kenai Peninsula along Cook Inlet illustrate how the dynamic equilibrium works. During the 1964 earthquake, the shoreline sank, typically about 3 feet or so. In other words, the sea level rose. What was once a broad, expansive beach became a much narrower strip of sand as the shoreline retreated in response to the subsidence. The bluff that was once protected by a wide beach suddenly became vulnerable to wave erosion (fig. 6.4). Today the eroding bluff is slowly causing the widening of the beach by adding bluff sand to the beach sand supply. The position of the shoreline is being slowly pushed seaward in response to a healthy supply of sediment, but the bluff line is retreating back, to the real distress of some property owners with buildings too close to the bluff edge.

As a beach's profile changes, the other equilibrium parameters are altered in turn. For example, as a beach widens and its profile flattens, waves dissipate their energy over a broader strip of beach surface, which has the effect of reducing the waves' impact on the land. Often the bluff or first row of dunes is attacked during storms, in part because the rise in sea level pushes the surf zone landward. This facilitates movement of sediment seaward to flatten the beach (fig. 6.5), and at the same time adds material eroded by the waves to the beach sediment supply. When the storm is over, the shoreline is usually left with a wide, flat beach with a high-tide line farther inland than it was before the storm. Part or all of the beach sediment that was carried sea-

Figure 6.4 Trees about to topple over an eroding bluff along the Kenai Peninsula, Cook Inlet shoreline. The bluff was protected by a wide beach before the 1964 earthquake caused the land to sink. The accelerated erosion of the bluff caused by the subsidence is slowly furnishing new sand to a gradually widening beach. In a few decades, the beach will widen enough to protect the bluff and reduce the erosion rate.

Figure 6.5 A steep gravel upper beach along Sakie Bay on Dall Island. Gravel beaches tend to be steeper than sand beaches, but all types of beaches flatten during storms. This flattening causes storm wave energy to be dissipated over a widened zone, thus reducing the amount of beach erosion.

ward may be moved back in from the continental shelf in the months and years following the storm and added to the beach. However, sand lost too far out to sea will not be recovered and is no longer part of the beach's sediment supply. When sand is lost to sediment sinks or as the result of human activities (e.g., mining and construction), the dynamic equilibrium is permanently altered. Perhaps the most spectacular example of beach mining increasing storm hazards is at Barrow (described in chapter 9).

Two important principles must be understood by anyone dealing with the management of an eroding shoreline.

1. *There is no shoreline erosion problem until someone builds next to the shore.* Humans are responsible for the shoreline erosion problem because we build too close to retreating shorelines. Thousands of miles of remote Alaskan shorelines are retreating, but only a relatively few miles have erosion "problems."

2. *The beach is not threatened by shoreline erosion.* Under natural conditions, the beach simply changes its position in space while retaining its general shape in response to storms and sea level changes. When we stabilize a shoreline, it is done to save buildings, not beaches. Beaches don't need saving under natural conditions. Only when we interfere with the system, especially through shoreline engineering, are beaches threatened.

Erosion Solutions

A community with a serious erosion problem can employ one of three "solutions": (1) shoreline armoring, (2) soft stabilization, or (3) relocation. *Hard stabilization,* or shoreline armoring, involves the use of hard, immovable objects such as seawalls to halt erosion. *Soft stabilization* refers to the emplacement of additional beach material in order to push the beach seaward. *Relocation* refers to any means by which buildings are moved out of harm's way, allowing the shoreline to march on and the beach to exist in its natural form.

Armoring the Shoreline

Hard Stabilization

Advantages

—Holds shoreline in place (the land in back of beach)

—Provides limited protection to property in back of the beach

Disadvantages

—Usually results in destruction of the beach (cuts off sediment supply, changes beach profile, changes wave dynamics)

—Costly

—Temporary

—Unsightly

—May limit access

—May promote overdevelopment in storm-susceptible areas

There are two basic kinds of hard stabilization: groins and seawalls.

Groins are walls built perpendicular to the shoreline extending across the breaking waves, usually into the water. Their purpose is to trap sand moving in the longshore transport system and thus cause the beach to widen by building seaward. The problem is that the trapped beach sediment was going somewhere, and wherever that somewhere was is now suffering a deficit in its sand supply (fig. 6.6). Deficits translate into erosion; therefore groins cause downdrift erosion. Although not a common approach in Alaska, groins are a widely used engineering approach to halt shoreline erosion elsewhere in the world.

Groins can be made of a variety of materials, including concrete, stone, steel, gabions, and even lines of construction debris. In Kotzebue, shore-perpendicular lines of concrete-filled 55-gallon oil drums are used as groins. Except under special circumstances where downdrift erosion won't occur (updrift from a rocky shoreline or on a short pocket beach), groins

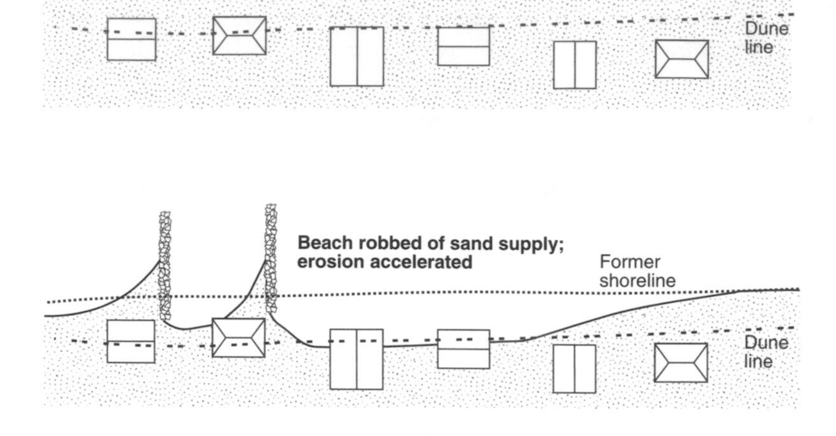

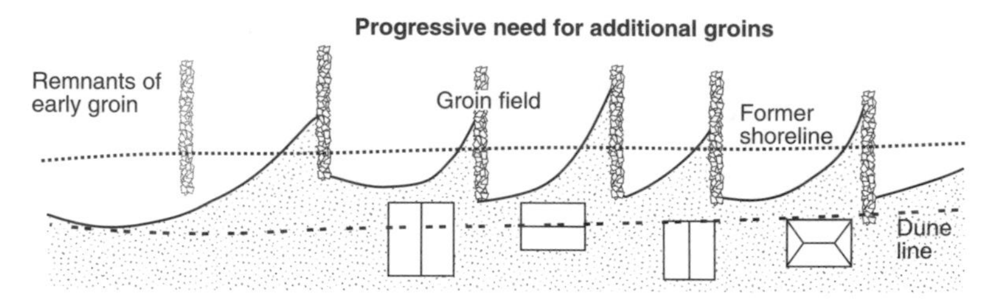

Figure 6.6 The groin saga. Groins are walls built perpendicular to the shoreline for the purpose of trapping sand and widening the beach. This causes a sand deficit, leading to shoreline retreat at downdrift locations.

probably cause too much long-term damage to be considered a useful approach to halting shoreline erosion. Jetties, shore-perpendicular structures at river or inlet mouths, are designed to keep sediment out of channels. These structures are not designed to protect beaches, but their effect is to cause updrift accretion and downdrift erosion, as at Nome.

A wide variety of structures fall within the seawall family. The two most important are seawalls and revetments. Seawalls are freestanding structures constructed parallel to the shore and designed to withstand the impact of storm waves (fig. 6.7). The Nome seawall is perhaps the most famous in all of Alaska. It is triangular in cross section: 35 feet wide at the base and 18 feet high above mean low tide.

Revetments often have the same external appearance as seawalls, but they are not freestanding structures. Instead, revetments consist of some kind of armor covering a slope of land facing the sea. The rocks protecting the landward terminus of the Homer Spit are a revetment. The railroad track at the base of the sea bluff overlooking Knik Arm in downtown Anchorage is revetted, as is much of the shoreline in front of Kotzebue.

The revetments at Homer and elsewhere are generally constructed of boulder-sized stones or blocks of broken rock called rip-rap. One advantage of using such materials is their wave-absorbing qualities. When a wave breaks, much of its water is absorbed in the interstices between the rocks. Thus there is little backwash to carry beach sediment seaward. In contrast, on a vertical, impervious seawall, almost all of the water that comes forward to strike the wall is reflected seaward or refracted toward the end of the wall. Alaska revetments need to be able to withstand the impact of sea ice jamming up against them. That is why the revetments and seawalls at Barrow, Shishmaref, and Kotzebue are faced with concrete blocks connected by wire. This design produces a flexible surface able to respond to ice override.

Important lessons can be learned from the seawall experience in Alaska. The revetment protecting Kotzebue has greatly diminished the time-honored practice of using the beach as a community workplace where fishermen and others can land and store small boats, load and unload supplies and catches, and mend nets. The Kotzebue armor has been in place for several years and has collapsed in a few areas (fig. 6.8). In nearby Shishmaref, a revetment of similar design failed within a few weeks of its

Figure 6.7 The boulder and cobble beach at Seward. In the foreground is a vertical sheet steel seawall. This type of seawall makes some recreational use of the shoreline difficult. The recreational vehicles in the background are on a site that is highly susceptible to inundation by tsunamis in future earthquakes. An alarm system warns those in RVs to head to higher ground when a tsunami threatens.

Figure 6.8 The cement blocks in this revetment at Kotzebue are connected by cables. The system is designed to be flexible in case of ice push. The revetment has suffered a partial failure because gravel has washed out from behind it. In Shishmaref, where the beach is fine sand, total washout occurred (fig. 9.4) and the revetment collapsed within a few weeks after construction. Ironically, low-cost gabions on an adjacent portion of the Shishmaref shoreline have successfully held the shoreline in place for more than 20 years.

emplacement. The difference is that the Kotzebue revetment has gravel behind it, while the Shishmaref revetment was backed by fine sand, which simply flowed away under the revetment during the first minor storm.

The great irony of the Shishmaref "seawall," which failed at great cost to the state and the village, is that the community already had, at a different location, a successful revetment made with old-fashioned gabions—wire baskets filled with stone (fig. 6.9) which are then piled on top of one another to form a stepped revetment. The gabions of Shishmaref had been in place for 12 years when the fancy, high-tech concrete block wall captured the fancy of the village elders. When in doubt, use the approach that has worked elsewhere. New technology is not always better technology.

Specific details of seawall and revetment design are well beyond the scope of this chapter; however, a few principles are worth noting. Most seawalls and revetments should be backed by filter cloth and should be well anchored. Filter cloth allows water that ponds behind walls to pass through but prevents sand and gravel from doing so. In Alaska, winter ice is often a problem for seawalls. Typical ice-resistant revetments consist of concrete blocks tied together with cables, which give the structure some flexibility in case of ice override.

If a seawall is intended to protect buildings behind it, the height of the wall must be greater than the height of the expected storm surge. If a wall is intended only to halt shoreline erosion and not necessarily to prevent a storm surge from entering developed areas, height is not a factor as long as the wall covers the entire eroding surface. Seawall footings must be deep and well anchored, and the structure must be stabilized by tie-backs into the land behind the wall in order to prevent the wall from failing under the onslaught of storm waves. A seawall or revetment at the base of a bluff such as those along the Kenai Peninsula and in Homer will not halt bluff erosion, because much of the erosion is caused by rain runoff, frost action, chemical and physical weathering, groundwater seepage, and other processes unrelated to the sea.

The following generalizations about seawalls (and similar shore-hardening structures) should be considered by any community planning to install them.

Seawalls destroy beaches. You can see this for yourself. Negligibly small beaches or none at all exist in front of most of the seawalls and revetments in Alaska. In most cases the beach was lost because the wall or revetment was constructed at the low-tide line. But in any situation where a shoreline is eroding and a wall is installed, the shoreline will continue to retreat; and

Figure 6.9 This rapidly eroding bluff along the East Homer shoreline is protected by gabions and boulders. The gabions are the rectangular wire containers filled with gravel in the foreground. Although gabions have been successful in slowing shoreline retreat along some Alaska shorelines, here the bluff is continuing its gradual landward march.

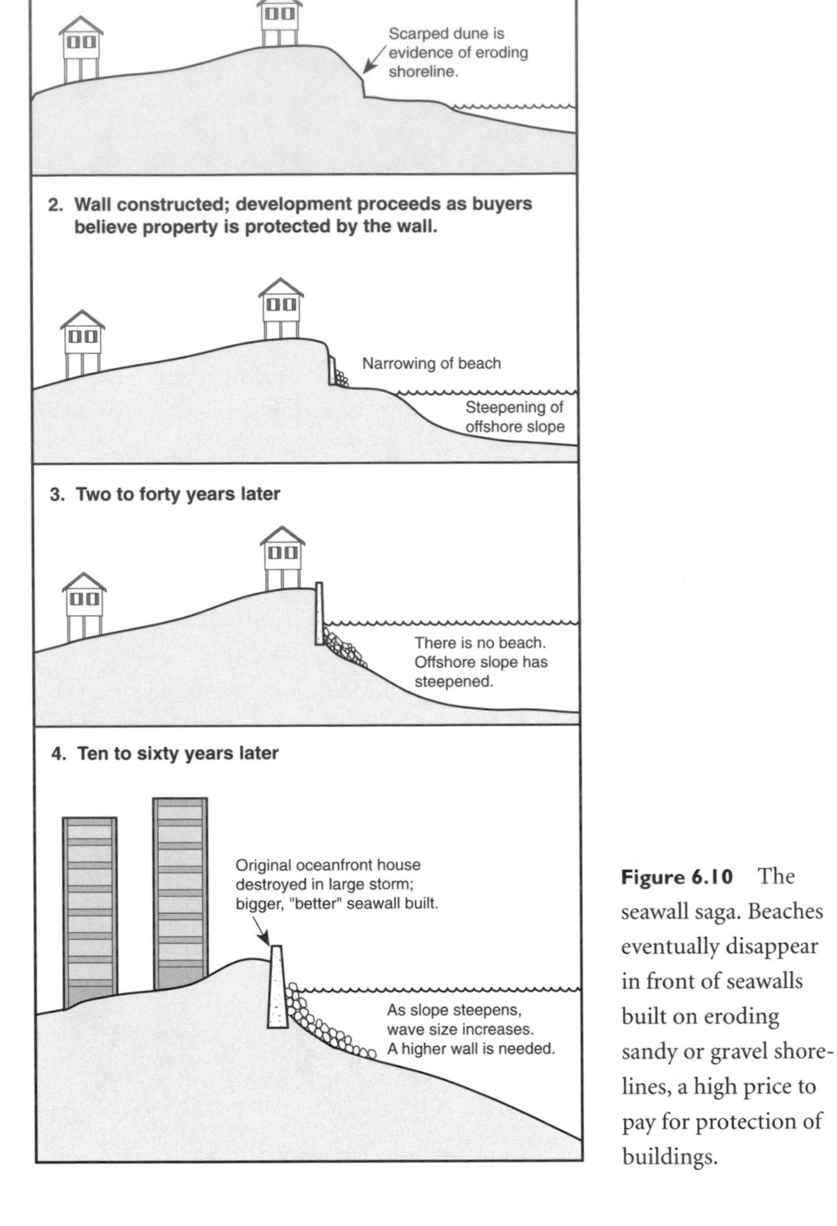

Figure 6.10 The seawall saga. Beaches eventually disappear in front of seawalls built on eroding sandy or gravel shorelines, a high price to pay for protection of buildings.

as it backs up against the wall, the beach will get narrower and narrower (fig. 6.10). In other words, the passive wall is responsible for beach loss.

Seawalls get longer. Seawalls reduce the sand transport in front of them, increasing the erosion rate on adjacent beaches. The lack of sand combined with the refraction of wave energy to the end of the wall produces an area of particularly high erosion, a sort of "end-around effect," at the immediate ends of seawalls. Downdrift erosion can be halted only by lengthening the seawall. This principle can be seen in Nome, where the shoreline armoring is growing, as the town grows, in a westerly direction. As a result of the big October 1992 storm, the revetment was extended to the east.

Seawalls get bigger. There is a strong tendency for seawalls to grow in bulk and height (on a time scale of decades) as the beach in front of them gets narrower and steeper. The loss of the beach fronting a wall means that less of a breaking wave's energy is expended on the beach sand and more is expended on the wall. Hence the need for bigger walls with time. The massive Nome seawall started out as a few tiny scattered wooden bulkheads.

Take the long-term view. Remember, as seawalls grow in size and length, they cost more to replace and repair, and they damage a greater length of beach. Unlike their counterparts to the south, Alaska's beaches aren't used much by swimmers. But beaches are a part of communities' recreational, tourist, and working lives, and sometimes play an important role in subsistence hunting and fishing. Degradation of beaches by seawalls may take decades, so community leaders must take a long view of the future (fig. 6.11).

Take the long-distance view. Hard stabilization of any kind eventually reduces the transportation of sand in the downdrift direction, and sometimes in both directions. Thus, a large stone barge-loading facility built perpendicular to the shoreline at Cape Nome is causing erosion on Safety Spit to the east. Stabilization of bluffs to the west and east of Homer Spit poses a long-term threat to the spit's sediment supply. The seawall in Barrow threatens the town's beaches to the northeast. Examples of threats to downdrift beach sediment supplies can be found in almost every Alaskan shorefront community.

Look at examples. If an engineering firm suggests a particular design for a shore protection structure, don't just take the firm's word for it. Ask to see examples where the design has worked. Visit nearby communities with similar erosion problems in similar oceanographic settings. Find out what worked and what didn't work for them. The troubles with the Shishmaref revetment could have been avoided if this rule had been followed.

Don't be the first. Never be the first to try out a new device to halt shoreline erosion. There are a thousand designs on the market, and most of them will not work for your community. Let someone else be the pioneer. When it comes to shoreline stabilization, there are advantages to being old-fashioned and out of date.

Seek a solution for the whole community. Don't solve the erosion problem for single buildings. Protection of single buildings, as is being attempted in Homer by the Elks Club (see fig. 11.8), rarely works because sooner or later the building will end up projecting seaward like a minicape and will be vulnerable to erosion. More important, as has happened around the Homer Elks Club, the erosion rate on neighboring properties immediately in-

Figure 6.11 Sandbags and various forms of debris are fighting a losing battle against the shoreline erosion problem on the lagoon shoreline of Kivalina. Before long the small oil tanks will be at the edge and will have to be moved or abandoned.

creases. If you stabilize one portion of a developed shoreline, you must stabilize all of it. If it was entirely stabilized, however, the Elks Club bluff would begin to choke off the supply of sediment to Homer Spit and would probably cause the spit's erosion rates to increase.

Think soft. In Seward, consideration is being given to putting in a revetment on the only remaining portion of the beach that is not armored. The eroding section of beach is rocky and not easy to walk on, but it is used much more by people who simply stroll the shoreline than the adjacent boulder revetments, which are dangerous to walk on. A few dump-truck loads of coarse gravel each year will help preserve an important community asset for future generations. A revetment will destroy it.

Soft Stabilization

In many Alaska coastal communities, intertidal zones have been narrowed by a combination of shorefront construction, seawall emplacement, sand mining, and shoreline erosion. Barrow is an example of extreme beach loss due to sand mining.

Beach replenishment is being used increasingly for erosion control and storm protection on the East and Gulf Coasts of the United States. Very little beach replenishment has been carried out in Alaska up to this point, although plans are now under way for replenishment in Barrow and Wainwright. Alaska tends to go for hard solutions such as the construction of seawalls or rock revetments.

The advantage of beach replenishment is that it preserves the intertidal zone for future generations (as opposed to hard stabilization, which destroys it). There are disadvantages, too.

Beach Replenishment

Advantages

—Improves the beach

—Provides storm and erosion protection while the beach lasts

Disadvantages

—Costly

—Temporary

—May promote overdevelopment in storm-susceptible areas

Beach replenishment is any procedure that replaces lost beach materials and builds the beach out in a seaward direction (fig. 6.12). In the lower 48 states, more than 450 miles of beaches have been replenished on all three ocean coasts, mostly by dredging sand and pumping it ashore. In a few cases, beaches have been replenished from inland sand sources. In Virginia Beach, Virginia, more than 100,000 dump-truck loads of sand per year have been emplaced in years when erosion losses were severe.

Costs on the East Coast are typically on the order of $1 million or more per mile. Typically, replenished beaches on the East Coast last 2–4 years and then have to be re-replenished. Miami Beach is an exception; the replenished beach there has lasted quite well for 15 years. Some beaches have been replenished 10 times since 1965. Clearly, this is a costly alternative.

If gravel-sized or larger material can be obtained, however, artificial beaches will probably have much longer life spans than the sandy beaches of the East Coast. Many Alaska beaches are very coarse grained to begin with, and coarse material abounds along large areas of the shoreline thanks to the past actions of glaciers. This means that using land sources for beach material may be very economical in Alaska.

Since there is essentially no experience with replenishment on the Alaska coastline, there is no basis for estimating costs. Very likely costs will be highly variable depending on the quality, quantity, and location of available sand and gravel.

Experience with beach replenishment elsewhere provides some useful principles. These include the following:

—At present, we cannot accurately predict the costs and required volumes and the lifespans of replenished beaches. Use of gravel instead of sand will greatly increase life spans.

—The best basis for estimating the life spans of a replenished beach is previous experience on the beach in question or on adjacent beaches. (This won't help Alaskans.)

—Storms are the primary process responsible for beach loss. Other factors such as beach length and even beach size seem to play a secondary role.

—Replenishment projects can be quite small (e.g., 10 dump-truck loads to patch a storm-cut notch). The U.S. Army Corps of Engineers generally insists that projects be very large if they are to be involved, but often it is not necessary to build large beaches.

One possible source of low-cost beach material is inlet, channel, and harbor maintenance dredging, although sometimes such material is polluted. Nearby gravel and sand pits can probably provide relatively low-cost beach replenishment materials for most communities in Alaska.

The Do-Nothing and Relocation Alternatives

Seawalls, groin fields, and breakwaters are costly and damaging to the environment. Beach replenishment is costly and temporary. The sea level is rising, and the rate of the rise may accelerate along with the rate of erosion. These facts argue for moving buildings back from the retreating shoreline or letting them fall into the sea. Moving buildings to higher elevations also reduces the risk to them from tsunamis and ice override.

Increasingly the do-nothing alternative is being taken quite seriously, particularly in states where preservation of the recreational beach is essen-

Figure 6.12 The beach replenishment alternative for retreating shorelines. Beach material may be obtained from sand or gravel pits on land (Kotzebue) or by pumping dredged material from lagoons or the open ocean (Wainwright).

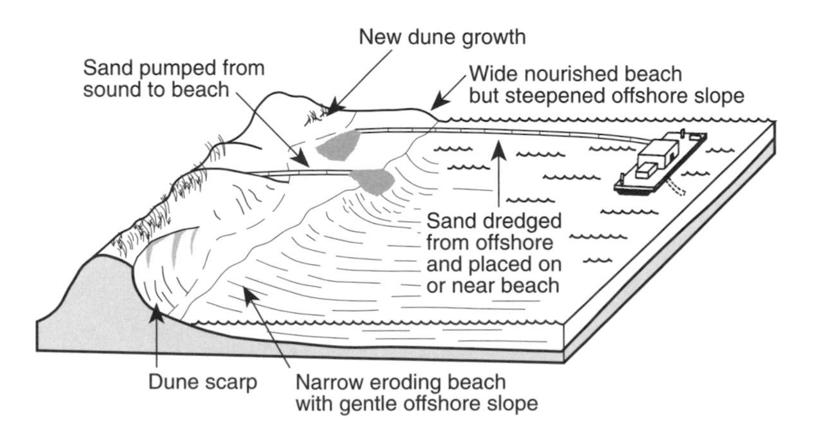

tial for the economy. Looking at it quite broadly, the advantages and disadvantages of relocating buildings are as follows.

Relocation

Advantages
- —Usually the best way to preserve beaches
- —May save money in the long run
- —Moves buildings away from hazards

Disadvantages
- —May be costly
- —Politically difficult
- —Depends on available space

Alaskans won't find the relocation alternative quite as startling as it was to some residents of the Lower 48 when the idea was first introduced a few years ago. In part, this is because most coastal towns and villages in Alaska are very small, the cost of other management alternatives is beyond most communities' means, and Alaska has been a leader in the relocation alternative. Furthermore, the shorefront buildings in many villages are small and easily moved, unlike the situation in highly developed beachfront communities such as Fort Lauderdale, where 25-story condominiums reside cheek-by-jowl next to the restless sea. Alaskans have already accepted the relocation approach. Pioneering efforts in this regard include two northern villages, Point Lay and Point Hope, which were moved farther inland. Valdez is a classic example, and other small villages have been relocated as well (e.g., on Kodiak Island). Shishmaref and Kivalina are contemplating the possibility.

In a November 1946 town meeting attended by 200 of Nome's citizens, a critical vote was held. By a large margin, the citizens voted to move the whole town back to the World War II airfield at higher elevations behind Nome to escape the pounding Bering Sea storms. But dollar signs clouded Nome's window of opportunity. Leading businessmen in the town objected to the move and started the ball rolling to get the large Corps of Engineers seawall completed by 1951. That was the end of the famous Nome Beach, but not the end of the storms. In October 1992 a storm flooded the streets of Nome just like in the good old days. The town still faces the relocation option, but with the increased growth and costs added by half a century.

Some federal government policies are promoting the relocation alternative. For example, the Upton-Jones amendment to the National Flood Insurance Program was intended to provide communities covered by federal flood insurance with financial incentives to move erosion-threatened buildings or demolish them in place. In addition, the amendment required all federal shoreline stabilization projects, such as breakwaters or replenished beaches, to consider the relocation alternative. If relocation was found to be feasible or cheaper, it was to be the management approach recommended to the community. The Upton-Jones program was discontinued in September 1995.

Shoreline erosion is not the only hazard that threatens economic loss, and coastal hazards such as storm surge, ice override, and tsunamis are not the only hazards Alaskans must consider and plan for. Relocation is also a viable mitigation alternative for buildings located on riverine floodplains, in zones of slope failures, and in areas subject to failure during seismic events. Communities within the coastal zone that are also on the floodplains of rivers are usually at high risk from several hazards and must plan and act accordingly.

7 River Flooding in Coastal Alaska

Riverine flooding is not usually a hazard that people associate with coastal communities, but it can be a significant and recurrent problem. As with other natural hazards, flooding is not a problem until humans build on the floodplain.

The response of a river to rainfall is related to the size of the drainage basin, the intensity and duration of the rainfall, and the amount of ground cover. In general, vegetation provides a cushion for absorbing rain, as does a low saturation level within the ground. Once the ground is saturated, water flows directly into streams and rivers. When slopes fail, debris flows may produce small dams, impounding floodwaters that may be suddenly released when the temporary dams rupture. Glacier-fed streams may pose an additional flood hazard.

A graph plotting the flow of most northern Alaska and glacial rivers can be compared to an electrocardiogram. Usually there are very large pulses of runoff water in the summer, from June to August. These pulses reflect the annual ice breakup flood, which defines the major distributaries of the river system but usually does not produce catastrophic flooding in deltas (fig. 7.1). However, more rapid than usual melting or high amounts of snow may overload streams and cause flooding. Rainfall on snow creates even higher flood levels because the warmer rain melts the snow, creating a combination that can be doubly catastrophic.

In southeast Alaska, as much as 90 percent of the annual rainfall falls during several weeks in the autumn. Because drainage basins are so close to the river deltas, the lag time between rain and discharge into the sea may be only days or less. Flooding is a common threat to lower delta communities, which face high tides and storm surges in addition to riverine floods (fig. 7.1).

Geologists use probability theory to predict how often streams will flood. A 100-year flood has a 1 in 100 (1 percent) chance of occurring in any given

Figure 7.1 Flooding during the summer of 1986 at Tuntutuliak along the Yukon River. Here, on the lowermost delta, close to the sea, river floods combine with high tides and storm surges to increase the frequency and severity of coastal flooding.

year; one or more 100-year events have a 40 percent chance of occurring in 50 years. It is well to recall that, as in flipping a coin, although a 100-year flood may occur *on average* once in 100 years, there is always a chance that two 100-year storms will occur in rapid succession.

Our notions of flood levels are based on records of past events. In North America, written flood documentation extends back only about 150 years. Geologists can extend the record by studying and dating the succession of flood layers in riverbanks and deltas, but this type of data is not available for Alaska coastal communities.

Records of twentieth-century flooding in Alaska are limited but instructive for predicting future flooding. We know that floods are likely to strike suddenly in areas with small catchment basins and high rainfall, a setting typical of many south-central and southeast Alaska communities. Skagway, situated on a river delta, is the community most repeatedly threatened by large-scale flooding, with 10 major floods in this century alone (table 7.1). The 1967 Skagway flood resulted when 4 inches of rain fell in a single day. The town of Seward, located on an alluvial fan, suffered large floods in 1936, and the Resurrection River, north of town, was the focus of flooding in October 1986. The 1986 flood in Seward followed 18 inches of rain over a 3-day period. The outbreak of glacially impounded Russell Fjord occurred several days before the flood but was probably not related to the same rainfall.

Creek flooding near Anchorage and the Mat-Su Valley is an increasing menace as developers choose to build along creek banks. While the high water in June 1949 hit undeveloped areas, flooding on the same small creeks in August 1989 caused $10 million in property damage. A flood in August 1971 affected Palmer and the Mat-Su Valley after 3–9 inches of rain fell over the course of a week. The Matanuska Valley area has the greatest continuing problem with flooding and river erosion (*Anchorage Daily News*, May 7, 1995). Homeowners along the lower part of the river are continually asking for state assistance in stopping the damage. Prospective buyers should reflect on the problems of flooding and erosion before buying property. Too often, homeowners avoid responsibility and expect society to solve problems generated by their own refusal to accept the course of nature.

Landforms and Risk Analysis

Some landforms come with built-in risks. It is important to know the characteristics of the land surface, and their associated risks, before you buy or rent.

Floodplains

Floodplains are the relatively flat floors of river valleys that are formed by the deposition of river flood sediments. They exist *because* they are flood areas. If you are living on a floodplain, you should know the likely scale and magnitude of the inevitable future floods. Find out the past flood stage heights and build or buy well above them. This information is available on Flood Insurance Rate Maps (or FIRMS), which should be available from the Federal Emergency Management Agency (FEMA; see appendix B) and in city halls and other government offices. If FIRMS are not available, the natural environment offers clues that can help you evaluate your site. Ask yourself the following questions to assess the likely risk of flooding:

1. How stable are the banks? (i.e., Are they steep and collapsing?)
2. Is there evidence of fresh (annual) erosion?
3. How old is the vegetation cover on the banks? (Are there 300-year-old trees or annual shrubs?)
4. Is there evidence of ice scour on the trees atop the riverbanks?
5. Is there a thin cover of sediment from last year's flood?
6. How far does the likely flood zone extend?

Flooding can affect areas located a considerable distance from the riverbank itself. Examine topographic maps carefully and think about the behavior of the floodwater once the banks have been overtopped. Look for inactive river channels—linear depressions that look like canals—and imagine the possible course of the river in flood. In some cases, the flooding

Table 7.1 Chronology of Major and Other Memorable Floods and Droughts in Alaska, 1943–1990

Flood or Drought	Date	Area affected	Recurrence Interval[*] (years)
Flood	Oct. 1943	Skagway[1]	50
Flood	Oct. 1944	Skagway[2]	<50
Flood	Oct. 1949	Skagway	
Flood	June 21, 1949	Anchorage area[3]	10 to >100
Drought	1950–52	Southeastern Alaska (mainland southeast of Juneau and islands near Ketchikan)[4]	10 to >35
Flood	Aug. 12–13, 1961	Juneau area[5]	10 to >100
Flood	Oct. 2–15, 1961	Ketchikan area[6]	10 to >100
Flood	Mar. 27, 1964	Gulf of Alaska coastal area[7]	Unknown
Drought	Spring 1965–Spring 1966	Southeastern Alaska (southern mainland and islands)[8]	Unknown
Flood	Sept. 1967	Skagway	30 or less
Drought	1968–71	South-central (Copper, Matanuska, and Susitna River basins, Anchorage area), southwestern Alaska (Kuskokwim River), and Yukon River in Alaska (middle Yulkon, Tanana, Nenana, and Koyukuk River basins)[9]	10 to >35
Flood	Aug. 8–11, 1971	South-central (lower Matanuska and upper and middle Susitna River basins, Palmer area) and southwestern Alaska (Kvichak River basin)[10]	10 to >100
Drought	1972–76	South-central and southwestern Alaska (southern one-half of these regions, excluding lower Kenai and Alaska Peninsulas)[11]	10 to >35
Drought	1981–86	Southeastern Alaska (islands near Ketchikan and on mainland north to Juneau)[12]	10 to >30
Flood	Oct. 8, 1986	Russell Fjord Lake[13]	Unknown
Flood	Oct. 10–12, 1986	South-central Alaska (Seward area, lower Susitna River basin, and streams west of Susitna River mouth[14]	
Flood	Aug. 25–29, 1989	Anchorage area[15]	10 to >100
Flood	Sept. 1990	Skagway	25 to >100

Sources: Modified from the National Water Summary 1988–1989. Recurrence intervals calculated from U.S. Geological Survey data; other information from U.S. Geological Survey, state and local reports, and newspapers.

*The average interval within which streamflow will be greater than a particular value for floods or less than a particular value for droughts.

Table 7.1 Continued

[1]Flood of record [2]Nearly as high as 1943 [3]Largest known peak discharge in some small streams in Anchorage [4]Deficit flows from Dec. 1949 to May 1952 in small streams; less severe near Ketchikan [5]High water in Juneau and mainland streams as much as 40 miles southeast [6]High water in island streams and mainland streams near Ketchikan [7]Tsunami; local and seismic sea waves from Good Friday earthquake extensively damaged coastal communities; lives lost, 115 [8]Short, extreme low-flow period of about 1 year in many streams [9]Severe streamflow deficits on southern side of Alaska Range; less severe in other areas; drought began in Aug. 1968 and ended in July and Aug. 1971 [10]Mostly result of widespread rainstorms; antecedent rain and high water were also significant; damage, $10 million [11]Deficit flows from Cordova to the Kenai River, and Anchorage areas; less severe on Kodiak Island and Bristol Bay drainages; started in fall 1971 and ended in fall 1976 [12]Deficient runoff in some island streams near Ketchikan; lesser deficits elsewhere; extended from May 1981 to Oct. 1986 [13]Breakout of lake formed when advancing Hubbard glacier dammed a fjord [14]More than 15 inches of rain near Seward and 8 inches near Talkeetna; damage, $20 million, mostly to roads and the railroad [15]Local floods on two urban streams in Anchorage and, to a lesser extent, on other streams in Anchorage area; damage, $10 million, mostly from inundation of residences

might be comparatively slow (over a period of hours), allowing time for a planned response.

Those wishing to purchase riverside property should consider comparing topographic maps or even aerial photographs from different time periods in order to determine how rapidly the river is changing. Aerial photographs are available from commercial firms in Anchorage and from the Geo-Data Center of the Geophysical Institute in Fairbanks. If you compare photos, remember that you must locate the same point and measure in the same direction, accounting for any differences in scale between the maps or photos. Homeowners should consider measuring distances annually to record the erosion rate. Once again, use a fixed point and direction (bearing) to maximize the accuracy of the measurement. Living in a flood zone also requires listening attentively to weather reports during the flood season, watching the river, and developing a personal and family disaster plan (see appendix A).

Glacial Outburst Floods

The extensive ice cover of south-central Alaska poses a unique threat for down-valley residents and port facilities. More than 750 temporary ice-dammed lakes are mapped in the coastal mountains of south-central Alaska. Very few present problems for humans, but the few that do could wreak sudden havoc. The steady onslaught of ice down-valley can sometimes block the drainage in a tributary valley, preventing the release of water and creating a lake. Ice dams, as one might expect, are not stable structures and may rupture in a number of ways ranging from overtopping as water levels in the lake rise, to the weakening of the dam by earthquakes, floating of the ice dam, subglacial melting by volcanic heat, or drainage through cracks in the ice.

Outburst floods are spectacularly short-term events; many of the existing lakes fill and empty each year. Although an ice dam is the ultimate cause, the sudden addition of water from storms can push the lake toward a breakout flood within a matter of days or hours. Catastrophic flooding is often produced by a combination of meteorological factors and can occur even in the winter. For example, in January 1969, a release of water into Skilak Lake near Kenai led to the rupture of the ice cover and produced ice jams that flooded downriver residents.

The hazard of outburst flooding is more common than Alaska residents may appreciate. Outburst floods are a possibility in the Mat-Su valley and along the proposed Taku River road connection to Juneau, and flooding in Juneau from the Mendenhall glacier can be exacerbated by fall storms. Flooding from the Scott glacier could raise water levels in Eyak Lake east of Cordova and threaten lakeside properties. The well-documented blockage of Russell Fjord by the Hubbard glacier in 1986 was predicted at least 15 years earlier. Few people were affected by this event, but fishing boats could easily have been caught unaware. Oil storage facilities on the Drift River (west shore of Cook Inlet) are threatened by outburst floods when the Redoubt volcano melts glacial ice. Volcanically influenced floods can lead to mudflows that fill entire drainages.

Pacific Typhoons

Alaskans usually don't spend much time worrying about typhoons, but residents of south-central Alaska occasionally suffer nearly torrential downpours associated with the big Pacific storms. In mid to late September 1995, intense rain persisted for a whole week after a low-pressure zone that formed 2,000 miles south-southwest of Anchorage hit the last gasping winds of Typhoon Oscar, which had already raked the Tropics of the western Pacific. The reinvigorated system hit the Kenai Peninsula with full fury, unleashing massive rains. Kenai, Seward, and a host of other communities along Turnagain Arm suffered flooding, highway washouts, mudslides, and millions of dollars in property losses.

Staying out of Harm's Way

The best way to mitigate natural hazards is to avoid them. Do not build in the risk zone. For example, pipelines, highways, and port facilities should not be located in areas subject to outburst flooding. When this rule is ignored or cannot be followed realistically, then designing with nature is the next option (e.g., elevating structures or otherwise making them flood resistant through construction techniques). Alternatively, individuals and communities may try to out-engineer the natural processes. To reduce flood risk, levees, dikes, and flood-control lakes behind artificial dams are often the structures of choice. These are analogous to seawalls (discussed in chapter 6) in that they require maintenance, often fail to produce the promised protection, and may change the natural response (e.g., channel reorientation), sometimes enhancing the danger in adjacent locations. The best way to mitigate damage from natural hazards is to stay out of harm's way by intelligent site selection.

8 Human-Induced Hazards and Health Risks

Although natural hazards create the greatest risk to life and property in Alaska, humans themselves are responsible for some hazards, predominantly through the technology associated with modern industrial society. Human technology is clearly the immediate cause of oil spills and catastrophic fires. Several other hazards arise from human activities in the coastal zone as well, ranging from pollution to the harmful effects of dredging and offshore mining. These hazards and others pose potential health risks to Alaskans.

The potential for the most costly harm to Alaska's shores arises with the source of its greatest wealth, petroleum, and the decision to transport it in single-hulled ships through narrow fjords during the hazardous winter storm season.

Oil Spills

The psyche of Alaskans was irrevocably altered on March 24, 1989, the day of the infamous *Exxon Valdez* oil spill (fig. 8.1). The spill defiled pristine shores over a stretch as long as the coastline from Massachusetts to North Carolina. Despite two and a half centuries of European intrusion, the spill marked Alaska's coming of age into the modern world. Not large by global standards, the *Valdez* spill spewed more than 10.8 million gallons of crude, a mere drop compared with the catastrophic 275 million gallons unleashed into the Persian Gulf during the 1991 Gulf War. Fortunately, only one-sixth of the ship's total cargo was released, and more than 40 percent of the discharged oil either evaporated or was recovered during the costly cleanup in the months after the spill. Despite this measure of luck, 1,200 miles of shoreline were fouled with oil of varying consistency, generally decreasing in severity from thick sludge on Knight Island in Prince William Sound to

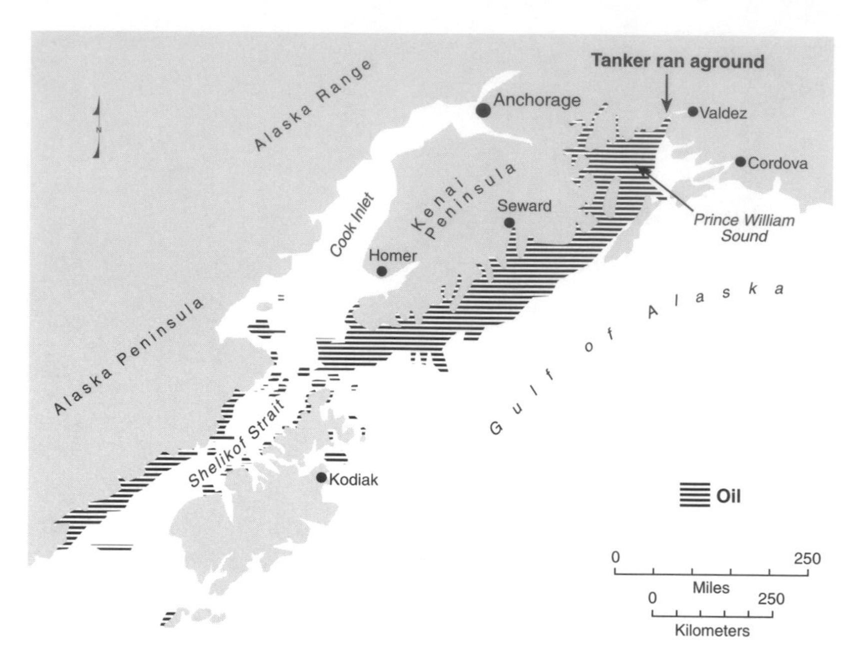

Figure 8.1 Area covered by the oil spill caused when the tanker *Exxon Valdez* ran aground. This was one of the most environmentally damaging oil spills in North American history, and in spite of all precautions to the contrary, it could happen again.

a wave-stirred froth ("mousse") on the beaches of Kodiak and the Alaska Peninsula (fig. 8.2).

Government oversight combined with poor planning and inadequate response by Alyeska Pipeline employees contributed to the catastrophe. Overly optimistic low estimates of spill probability in south-central Alaska doubtless led to a certain complacency on the part of regulators. In the previous 10 years, Alyeska had actively fought contingency planning (see appendix C, *Exxon Valdez Oil Spill Information Packet*). Statisticians fed this complacency, stating that the likelihood of a spill was low, about once in 240 years. The first 72 hours following the spill were relatively calm, a circumstance that should have aided in containment of the oil. Nearly 18 hours passed before booms and skimmers arrived at the beached tanker, however—three times longer than promised by Alyeska Pipeline Service Company. Exxon Corporation blames the state of Alaska for preventing the use of chemical dispersants, a fact disputed by the state, which counters that the several attempts made by Exxon were ineffectual because of a lack of equipment and adequate chemicals.

In the following weeks, the spilled oil flowed in coastal currents down the coast of Alaska, ensnaring innocent biological bystanders in its wake. The oil moved rather languidly due to the sluggish currents of the season. The prevailing current flow in the sound and the gulf is westward and southwestward. The effect of currents on the spill in the first days indicated that oil would probably not move to the eastern or southeastern parts of the sound, toward Cordova. Oil did not reach Cook Inlet until April 12, and it entered the head of Resurrection Bay near Seward on April 17, three weeks after the spill. The shores of Kodiak Island and the Alaska Peninsula weren't soiled until after mid-May. Some oil was deflected offshore by localized eddies. In some places, oil transport was complicated by the convoluted topography of the steep, rocky cliffs. As waves converged due to offshore obstructions, oil became trapped and subject to tidal and wind effects that led to delayed transport onto beaches and embayments. Comparatively little oil reached beyond Prince William Sound; only 10 percent of it rafted beyond southern Kenai Peninsula's Gore Point; and only 2 percent entered Shelikof Strait.

Three lessons should be learned from the *Exxon Valdez* spill.

1. Much worse spills are possible and should be expected.
2. Probability estimates of an event's likelihood are more political rhetoric than policy statements. Constant surveillance of the regulators is necessary.
3. Complacency is the greatest threat to preventing spills.

The impact of crude oil on the biota of the coasts is readily apparent, especially on the larger mammal and bird species that were coated with oil and virtually drowned or smothered (see appendix C, *Five Years Later*). As many as 500,000 birds belonging to 90 species were killed by the spill, along with 4,500 otters—30 percent of the population (fig. 8.2). Estimates of the number of bald eagles killed are uncertain and range from 200 to 900, as high as 10 percent of the total population in the spill-affected area. More subtle—and still largely unknown—are the effects on spawning salmon, tidal invertebrates, and bottom feeders. All in all, an oil spill represents a dire threat to the animal life in the coastal zone. Worse still, the effects linger for untold years in subsurface layers or as metabolized chemicals that may retard individual survivability. Native subsistence fishers and marine mammal hunters believe that the edibility of shellfish and marine life has been severely compromised, despite repeated claims to the contrary by marine scientists.

Oil spills are a potential hazard on much of the Alaska coast. Offshore platforms in Cook Inlet and the Beaufort Sea could generate small spills, as could planned development in the Chukchi Sea off northwest Alaska. Any cleanup operation would be totally ineffectual in the frigid Arctic, especially if heavy pack ice were present. Areas of potential oil lease sales include

Norton Sound and even Bristol Bay, the site of bountiful fisheries; no part of the coast is totally immune from the threat of oil pollution. Small spills from fuel oil–laden barges, freighters, and fishing boats are a continuing problem in southeast Alaska and elsewhere.

Figure 8.2 Oil spill cleanup workers head home across an oil-blackened beach at the end of a long workday. Photo by Jacinto Trías for USGS.

Oil and the Coastal Environment

Experience gained from spills elsewhere in the world can offer some instructive rules for Alaska communities. First, remember that oil concentrates in calm-water areas such as tidal flats and marshes; these vital environments should be protected from oil as quickly as possible. They play an enormous role in carbon cycling and as bird and fish nurseries. Oil is rapidly dissipated on high-energy coastlines (e.g., rocky headlands exposed to the full fury of the sea). Don't worry about oil in such areas. Sand and gravel beaches pose special problems. The wide spaces between cobbles on gravel beaches allow oil to penetrate to a considerable depth while the upper layers are often washed free of oil during winter storms. The lower beach may be sandy, but oil may be sealed in if there is gravel on the upper beach. Oil may persist for a long time in such a setting. On very fine sand beaches, oil forms a solid coating and will need to be removed, but little oil will penetrate below the surface. The response of beaches with coarser sand is more complex because this grain size is extremely mobile. Oil on the surface will be covered quickly by sand, forming a complex structure of thin oil layers and thicker sand deposits.

Health Effects of Oil Spills

Although contact with petroleum products is a part of daily life, people generally ignore their toxicity, even to complacently inhaling the cancer-causing benzene fumes from gas pumps. The crude in the Alyeska Pipeline is loaded with benzene and toluene, among the most toxic of all hydrocarbon compounds. Fortunately, these compounds are light, and they evaporated from the spilled oil. Cleanup workers were unequipped to work in proximity to the oil and lacked proper inhalation equipment; a few workers faced aggravated respiratory problems. Some workers reported nausea and dizziness, especially on warm days. The worst effects of the benzene should have been reduced by evaporation during the first days, and beached oil should not have posed a considerable health risk—or so health experts claim. Carcinogenic hazards are, of course, cumulative, so ongoing exposure to a compound may become critical decades later.

Concerns with the toxic effects of the spill continue to affect the Sound's subsistence consumers, primarily Chugach and Koniag natives. Reports of brain-damaged or cancerous seals and oily-tasting mussels and salmon were soon commonplace throughout the spill zone in 1989. Professional tasters, samplers, and laboratory data all downplay the hazards of fish or shellfish that do not overtly smell of petroleum products. In fact, one study shows that pre-spill Kodiak harbor petroleum levels (which can be attributed to harbor-related discharge) in salmon approached that in oil-contaminated fish from the sound after the spill. Subsistence harvesters continue to view resources from the spill area with concern (although they ignore carcinogens uncritically ingested from outside-derived agricultural and processed foods). Time will tell if these concerns are justified.

Physiological Hazards

It is with good reason that Alaska's waters are not known for recreational swimming. Even in the recent past, many Alaskan natives did not even know how to swim. The 38–45°F summer waters off Alaska are so cold that survival time is measured in minutes, although some lucky people have survived for 45 minutes or longer. In addition to heat loss associated with cold water, several other factors are important. Loss of breathing control may lead to hyperventilation, reduced carbon dioxide levels, and diminished blood flow to the brain. In addition, most people are disoriented in turbulent water and are less able to cope with the emergency. The best remedy or prevention for cold shock is proper dress, which includes a personal flotation device or survival suit.

Fire

Fires in Alaska originate from both meteorological and human activities. Wildfires can and do occur in southern and southeast Alaska, despite the region's well-deserved reputation for high precipitation. Fires often follow extended dry spells, or droughts (see table 7.1), and are coupled with high-pressure systems from the interior that generate thunderstorms. Fires in small towns can be devastating, especially if wood is the prevalent construction medium. Most towns do not maintain sufficient water reserves or possess the fire-fighting equipment to counter large fires. Massive fires devastated Nome in 1934 and Wrangell in 1906 and 1952, and hit the small village of Tenakee Springs in 1993. The possibility of a marine fire puts fear into the hearts of firefighters throughout southeast Alaska, although units train regularly for such an event. Few resources are available to contain an out-of-control fire aboard a tanker or cruise ship.

Beach and Offshore Mining, and Dredging

Alaska's shorelines are occasionally affected by society's urgent need for construction material, but the removal of beach sediments may ultimately have catastrophic consequences for communities down the coast. Beaches serve as the first bulwark against storms, and the removal of sand and gravel can expose developed areas to considerable harm. Beach mining at Barrow has exacerbated the area's coastal erosion problem. Most of the bluffs in town have only narrow beaches to seaward, the result of beach mining in the 1950s and 1960s coupled with the effects of storms in the following years. The Point Barrow spit was once buffered by a gravel storm ridge more than 10 feet high; the removal of this gravel for roads in the 1950s left the subsistence camps at the Shooting Station in harm's way. Beach mining is also a particularly severe problem at Kotzebue. Two areas are mined at Kotzebue: the updrift bluffs and the updrift beach ridges south of town. Mining on the beach ridges has proceeded to the point where a powerful storm surge could create a new inlet south of town and alter the character of the lagoon behind the town as well as affect the sediment supplied to the beaches in town. If community planners or construction companies are forced to obtain gravel and sand from the shore, their prime concern should be to minimize its removal from updrift locations. Remember that sand and gravel are carried in the nearshore zone, and their removal in the updrift direction will leave the beaches downdrift in short supply.

Dredging for navigational purposes, such as at Dillingham, Bethel, and at the port of Nome, is often conducted by the U.S. Army Corps of Engineers for the best of intentions: to provide access for vessels that carry supplies or engage in commercial fishing. But dredging can have adverse consequences for homeowners in the area. The maintenance of the inlet near Nome is a case in point. The early efforts at dredging during the 1930s probably led to the dramatic decrease in sand supplied to the beach at Nome and contributed to its severe erosion by storms in the 1940s. Dredging in harbors sometimes remobilizes contaminants, and improper disposal of dredge spoil may alter water quality at other locations.

Offshore mining is comparatively rare in Alaska, but it does occur on a significant scale near Nome. As everyone knows, the sands of the beaches of Nome provided the gold that looms so large in its history. The low-gradient continental shelf off Nome also contains considerable gold, and large dredges sift through this material as economic circumstances allow. The mining process appears to involve little actual removal of sand and gravel from the shelf; however, the rearrangement of sand and gravel on the ocean floor may alter the shoaling of waves during storms. If dredges create channels in offshore sandbars, storms may hit Nome with more punch. Citizens should examine environmental impact statements regarding potential adverse effects of mining. One environmental concern with mining is that hazardous chemicals may be put into suspension by wave action associating with the mining process. Chemical analyses collected by the Corps of Engineers assure us that the mercury concentration is well below hazardous levels (see appendix C, *Nome Harbor, Alaska: Dredged Material Disposal Environmental Assessment*).

The limited building space in mountainous coastal areas of southeast Alaska often means that builders are tempted to add gravel and sand to the shore to serve as foundations. This artificial fill is a double-edged sword if the land in question lies in earthquake country. New land is provided, but liquefaction resulting from lengthy seismic shaking turns such sites back into a watery milieu that may be incapable of supporting buildings or other structures. Gold-mining tailings were often used as fill; this is the case in Juneau, where much of downtown is built on artificial fill. In general, risk increases with the percentage of fine-grained materials such as clay or silt at a site. One should always worry about land settling and liquefaction in areas with artificial fill.

Getting Down to the Nitty-Gritty

Anyone armed with a general knowledge of both natural and human-induced hazards can evaluate his or her own region, community, or specific homesite with respect to its level of risk. The following four chapters are arranged by region to provide Alaska residents with information and case histories that are specific in terms of geological, geomorphic, and human-made hazards. Although you may live in only one of the regions, planning decisions made in other regions may involve large public outlays of funds. All tax-paying citizens of Alaska—residents, planners, and elected officials—should understand the problems and financial costs of living with the Alaska coast.

Part III
Risk Evaluation for Alaska's Coastal Communities

9 Arctic Alaska

Introduction (RM 9.1)

The coast of Alaska north of the Bering Strait is a schizophrenic creature. For eight to nine months of the year, its two seas, the Chukchi and the Beaufort, are nearly sealed in ice several feet thick. Much of the ice shifts with currents and breaks, jostling under the pressure to form a complex corrugation of upturned ridges and open canal-like passages known as leads. Leads open and refreeze many times during the winter. Fierce winds force the snow into drifts and dunes that cover much of the ice surface. Polar bears and ringed seals thrive in snow dens. Around June, the reverse polarity, summer warmth, usually 50–60°F, takes hold and the Arctic coast resembles most other temperate shorelines. Birds return from the south, and the polar bears migrate farther north to the perennial ice pack. Waves and tides are the dominant forces on most of the northern Alaska coast in summer. Storms erode the beaches and bluffs, and fair-weather waves allow recovery, as everywhere. At Barrow and in the Beaufort Sea, however, the sea ice never melts in some years and is sometimes forced onshore by wind.

The northern coast is not easy to typecast. While more than half of it is fringed by sandy or gravelly spits or barrier islands, long stretches consist of steep bedrock cliffs, especially from southern Kotzebue Sound to Cape Lisburne. Silty bluffs, 20–100 feet high, comprise much of the coast 100 miles in either direction of Point Barrow, the northernmost tip of Alaska at 71 degrees north latitude. The bluffs have a high ice content but are susceptible to catastrophic collapse after storm erosion, especially after some summer melting has occurred. On all of the northern coast, the open-water ocean is still the most forceful geomorphic agent—excluding the fearsome spectacle of sea ice overriding low bluffs or barrier islands. Storm surges occur with surprising regularity during the fall months and can cause considerable damage to coastal communities. For all the fury of occasional storms and the winter hardships it faces, however, the Arctic coast is not seismically active and is spared the earthquake and tsunami hazards of southern Alaska.

Village Centers and Subsistence Camps

Native settlements scattered along the Bering and Chukchi Seas present a unique case of shoreline development. Nearly all villages (Prudhoe Bay is the exception) lack road access, and many receive supplies through regular barge service. Although villages of several hundred people serve as primary residences, the typical Native American family also has several camps for harvesting specific resources: a fall camp for berry picking, a spring sealing camp, and so on. Camps are usually temporary shelters of plywood that contain barrel stoves and various supplies. The number of camps along the coast is surprisingly high. Visitors often mistakenly believe the camps are abandoned and are tempted to scavenge.

Even in the prehistoric past, when sod houses were the rule, storm surge was a problem. Sod-block houses required a substantial time investment in quarrying and in construction, and could not be moved when rising waters threatened. Summer pole structures were, of course, considerably more mobile. The demands of modern technological equipment make settlements even less flexible and less movable. Each village has a multi-million-dollar high school and airstrip that anchors it to that location despite annual storms or surges. Some coastal villages lack indoor plumbing and rely on "honey buckets" and a sludge wagon for waste disposal into a local pond, tidal flat, or other site. Larger towns, like Barrow and Kotzebue, are slowly converting to plumbing, at great expense and further decreasing the people's mobility.

Barrow (RM 9.2)

A cosmopolitan village at the top of Arctic North America (latitude 70° N), Barrow is the seat of the nation's largest county-level government body (in land area), the North Slope Borough. Barrow, the largest town (pop. 3,465) in the borough, is situated at the northern apex of Alaska, at the junction between the Arctic Ocean's Chukchi and Beaufort Seas. The seas off Barrow are ice locked 9–10 months a year or longer. In summer, the perpetually shifting Arctic Ocean pack ice is only miles away, and onshore winds sometimes bring ice to the shore and prevent ships from reaching the oil stations at Prudhoe Bay, 200 miles southeast of Barrow (fig. 9.1). Wind is a persistent feature of winter here, chilling the bones and threatening structures. As winds shift around the compass during the icy seasons, large leads open in the ice. Leads provide avenues of travel for air-breathing whales in

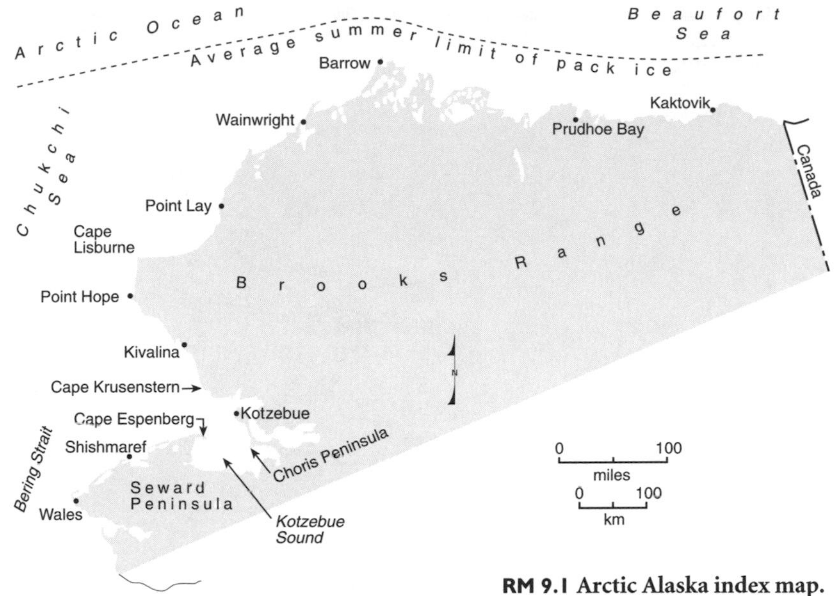

RM 9.1 Arctic Alaska index map.

their northward journey to feed on the summer bounty of small crustaceans along the ice edge.

Promontories such as that near Barrow allow access to migrating sea mammals and explain the long-term persistence of settlement at Barrow. People have lived here for nearly 4,000 years. Although archaeologists have not been able to prove it, Barrow's earliest settlers may have been the ancestors of the modern-day Inupiat Eskimos.

Although considered the capital of the Iñupiat Eskimo nation, Barrow is one of the most ethnically diverse communities in Alaska, with sizable Filipino and European communities and residents from nearly every U.S. state. Built on permafrost and without direct sun for 4 months in winter, the town boasts an Olympic-sized swimming pool, ethnic restaurants (Italian, Chinese, and Mexican), and four hotels. Despite its penetration by the Presbyterian church in 1892 and subsequent establishment of schools and a trader's store, Barrow remained a traditional settlement until after World War II. In 1947, the U.S. government established a research station just north of town and a Distant Early Warning station soon afterward to buffer Alaska during the cold war. The North Slope oil boom of the 1970s irrevocably transformed Barrow and its inhabitants by providing a large tax base.

Today, nearly all the town is hooked up to indoor plumbing and has a government committed to frequent bus service, its own scientific research on marine mammals, a community college, and locally financed health and social services. Several multistory buildings, including a state court building, are clustered within the central business district, which is just yards

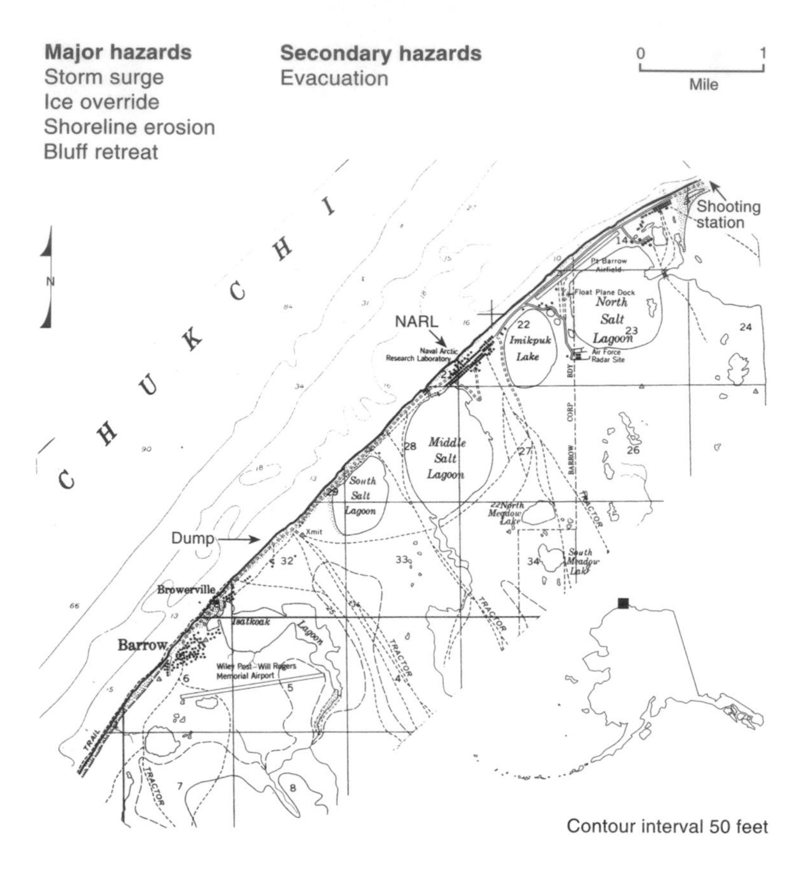

RM 9.2 Barrow

away from the ocean. In 1988, Barrow's moment of fame arrived during an international rescue mounted to save two gray whales trapped under an early buildup of sea ice.

Barrow includes four different neighborhoods, or districts, which are either on low bluffs or on a gravel spit to the north. Two are predominantly residential: Barrow proper, the ancient town of Utqiagvik ("where one hunts snowy owls"), and, across a lagoon, Browerville, originally the site of the first scientific mission to Barrow in 1881 and the store constructed by the nineteenth-century trader Charlie Brower. North of Browerville is NARL, formerly the Naval Arctic Research Laboratory, presently a borough-run educational facility and research center. The Shooting Station, at the base of the spit, is a popular bird-hunting encampment with numerous plywood cabins.

The 25-foot bluffs that underlie Barrow proper are a mixture of interbedded sand and silt deposited on the continental shelf during higher sea levels 125,000 years ago. Differences in the percentage of sand and silt influence how the bluffs erode. Water within the sediments is permanently frozen (permafrost) and is bonded into a solid mass resembling concrete. However, the heat and kinetic energy of waves, with the help of summer temperatures, can carve erosional cavities several yards deep at the base of the frozen sediments. When such a cavity reaches a critical distance, the bluff becomes unstable and collapses in huge blocks. Bluffs with a higher percentage of silt contain more ice and are more likely to collapse in large blocks than sand-rich bluffs. Under the archaeological site of Utqiagvik (at least 1,200 years old), at the southwest margin of Barrow, the bluffs are sandier, and thus less erodible, than they are just 100 feet to the north at the mouth of Kugok ravine.

Bluff erosion and collapse at Barrow are generated by a combination of storm activity and wind operating across hundreds of miles of open water to create high waves—a situation not uncommon in the fall at intervals of every few years. Storms from the south or southwest cause the greatest damage, as in 1963, 1968, 1969, 1972, 1986, 1990, and 1992. The amount of erosion during a single storm can be significant, upward of 5 yards, as large blocks are undercut (see fig. 1.8). This makes it essential for homeowners and planners to take the long view. Erosion is a perennial problem on the southern margins of Barrow. Browerville has lower bluffs and less erosion.

Only a small stretch of the Barrow shoreline is protected by artificial structures. An area several hundred feet long from south of the library to

Figure 9.1 Onshore winds pushing small ice blocks ashore on a Barrow beach during the summer. Although swimming in the ocean will never become a major form of recreation in Barrow, the beach is nonetheless very important to the social life of the community.

the location of the former veterinary clinic (fig. 9.2) is protected by a revetment formed of concrete blocks interconnected by wire with polyvinyl bags of sand and gravel at the top. An area south of this revetment has already started to experience heightened erosion. Seawalls cause erosion, and residents at either end of the revetment are faced with decisions and pressures to expand the seawall. This situation is complicated by the presence of archaeological materials, including burials, in the bluff. As yet, Barrow has no consistent plan for dealing with erosion, and the appearance of an erosion-exhumed body—as occurred in 1981, 1987, 1990, and again in August 1994—places the city and borough in a crisis mode. The lack of residential development on the old Utqiagvik site itself provides a buffer against erosion to some extent. Relocation of buildings from the bluff margin would be the most expedient solution for Barrow, especially since potentially usable land is abundant farther inland.

A complicating factor in bluff erosion is the repeated mining of the Barrow beach and bluff for sand and gravel construction material. Without a buffer of beach below the bluff, any wave energy is free to erode the bluffs. The U.S. Navy started mining Barrow beaches for road and runway gravel in the 1950s by removing a 10-foot-high beach ridge at the Shooting Station on the Barrow spit, north of town. This removed an important storm protection feature. The community has continued the mining to the point that sand is often in short supply on Barrow's beaches. During storms, the North Slope Borough mobilizes a small army of heavy equipment to bulldoze beach material into high embankments with the short-term gain of protection, but at the expense of long-term sand loss on the lower part of the beach. Current plans in Barrow involve beach nourishment, a solution that will require constant vigilance and repeated efforts to resurrect the high gravel ridges and comparatively wide beaches reported by the first Europeans and remembered by elders. The selection of a source for nourishment materials is critical. One proposed source is a sandbar only three quarters of a mile offshore. Considerable study should be undertaken before mining this area. The offshore bar may be actually providing sand to the beach or acting as a breakwater against storm wave energy. Scientists need to establish the direction of sand transport and the wave climate that maintains the bar. Failure to do this could compromise the beach nourishment.

Storm surges could affect much of Barrow, especially its commercial and administrative center, including the Top of the World Hotel, the public safety center, the main Stukpuq (store), and the state court building. All are within 250 feet of the sea and not far above sea level. To the north of Browerville, the Barrow dump is located within 200 yards of the coast and may be at risk from storms. Fortunately, the prevailing direction of longshore transport is away from downtown Barrow.

Figure 9.2 Erosion has increased in Barrow, in part because of the "seawall." Several buildings and a major archaeological site are threatened by future storms. Compare the narrower beach in front of the seawall with the wider beach in the foreground.

Ice overrides, termed *ivu* by the Iñupiat (see chapter 5), occur frequently and penetrate inland more than 10–20 yards on the Barrow spit near the Shooting Station. Usually these are not catastrophic, and bird hunters are no doubt aware of their likelihood. Outside the spit area, the threat of ice override is greatest in areas close to the margin of bluffs where piled-up ice can provide a ramp for further ice push. The full force of ocean currents can drive large blocks up such ramps and right onto the bluff top. In the late 1870s a bluff-top ice override trapped and killed a man in a subterranean entryway to a traditional-style sod block house. Ivu is also the cause attributed to the prehistoric collapse of a house that resulted in the preservation of the 500-year-old "Frozen Family" excavated by State University of New York archaeologists in 1981–82 and written up in *National Geographic* magazine.

All in all, while Barrow presents several hazards for homeowners, much of the settlement is distant from the sea and above the level of maximum storm surge. The three principal hazards here threaten those close to the sea: storm surge, bluff erosion, and ice override. The good news for Barrow residents is that virtually no earthquake activity occurs on the North Slope. Volcanic hazards are also virtually nonexistent. Given the absence of steep slopes, debris flows (outside of bluff-related events), avalanches, and landslides are also out of the question. High winds and blowing snow are threats to structures, but topographic channeling is absent. Tsunamis are also very negligible threats in Barrow, given the absence of recorded earthquakes offshore. By Alaskan standards, much of the Barrow area is comparatively hazard free, although a midwinter visitor might question that statement!

Kaktovik (RM 9.1)

Located on Barter Island, an eroding barrier island with bluffs up to 58 feet high, the village of Kaktovik is subject to ice override and storm surge erosion. Barter Island is one of many barrier islands in the Beaufort Sea. The closeness of the pack ice offshore limits the fetch distance and precludes major wave generation by southwesterly storms. Storms with easterly winds are most hazardous.

Prudhoe Bay (RM 9.1)

Prudhoe Bay is a unique coastal community; it exists only to extract oil. Shore protection and hazard mitigation needs here are unique and generally short term. Oil exploitation at this outpost on the Beaufort Sea has resulted in some experimental engineering in the coastal zone. The oil companies have constructed a number of artificial islands since 1980 by dumping several hundred thousand cubic meters of gravel fill and armoring the margins with plastic fabric sandbags. They claim that the method works reasonably well, although even small ice fragments gouge the sandbags. Coastal engineering in Prudhoe Bay has influenced cities and towns across northern Alaska. Engineers adapted a concrete mat revetment used in California during the 1970s; the first such revetment was laid at Resolution Island in 1980. A project at man-made Northstar Island in 1985 provides some insight about concrete mat revetments. The concrete mats extend 4 feet above mean low water and 18 feet below mean low water. Sandbags were placed both above and below the mats, and the entire island was mantled with mats. It is probable that the island surface is concrete as well. Clearly, any concrete mat revetment must be protected against ice scour at its base, overtopping by storm waves, and leakage. Further, engineers recommend anchoring mat revetments. None of the community revetments in northern Alaska measures up to the standards of the Northstar Island structure. Any Arctic community seriously considering a revetment should examine *Arctic Coastal Processes and Slope Protection Design*, published by the American Society of Civil Engineers in 1988 (see appendix C). Such communities should also consider the negative impact such structures have on beaches.

Wainwright (RM 9.1)

Wainwright faces north on the Chukchi Sea, and its 500 residents reside on a low mainland bluff that is extremely susceptible to storm-driven erosion. The hazards are similar to those at Barrow. Storm intensity is influenced by the extent of open water versus ice cover. Low ice years, such as 1986, witness extensive erosion. Fortunately, few houses are close to the bluff margin. Ice override is a minor threat in the Wainwright area. Compared with the rest of Alaska, Wainwright is comparatively hazard free, given the levelness of the surrounding topography and the generally quiet seismic environment. Camps upriver from Wainwright on the Kuk River and its tributaries can face flooding.

Point Lay (RM 9.1)

The former barrier island community of Point Lay (pop. 145) resettled on the mainland in the 1970s to avoid the storm erosion problems at the older site abandoned in the early 1960s. In a sense, this approach involved going back to their roots, because elders confirm the archaeological record of villages located on the mainland margins of lagoons. In the old days, barrier islands were wisely used as temporary camps for sealing or other activities.

Point Hope (RM 9.1)

Situated on a long spit jabbing into the Chukchi Sea, called the "index finger" in Iñupiat, the village of Point Hope (pop. 629) has an illustrious past. Some of the most spectacular archaeological remains in North America are from the Ipiutak site, which dates to A.D. 500–700. Despite the location, the Ipiutak people appear to have specialized in hunting caribou and seal rather than whales. Whale hunting did not take hold at Point Hope until about 1400. Point Hopers traditionally located their subterranean whalebone-and-sod-walled houses near the tip of the point. This area faced incredible erosion during the early twentieth century; more than two-thirds of the old village had been lost to the sea by the 1960s. In 1974, the entire town was relocated several miles to the east, but still on the gravel spit. The present town consists largely of modular (prefabricated) houses on widely spaced lots, although several frame buildings from the old town, including the Presbyterian church, were moved here. The town's decision to move in the face of imminent danger should be commended and perhaps emulated as an example of the retreat option in action.

Kivalina (RMs 9.1, 9.3)

Kivalina is a small subsistence village (pop. 317) located on a northwest-to-southeast-trending barrier island facing the open waters of the Chukchi Sea. Access to the mainland is by water in warm weather and across the frozen lagoon in cold weather. The villagers perceive that the island is eroding on both sides and is rapidly becoming narrower. At present, the open-ocean beach is quite wide and has a healthy, well-developed berm. There is no evidence of hard shoreline stabilization on the ocean side, past or present. In 1990, a sandbag revetment was emplaced on the lagoon side of the island near the inlet at the south end of town (see fig. 6.11). It has probably done little to halt the erosion because the bags don't cover the upper part of the unconsolidated bluff, which continues to be undercut by waves at high water levels. At the north end of the revetment there is a severe end-around effect and erosion is occurring at a rapid rate, probably in excess of 3–4 feet per year.

This island is indeed rapidly narrowing. No houses are threatened immediately by erosion on the ocean side, but several are close to trouble on the lagoon side. The tidal range is less than a foot. An engineering study indicates that the community is only 1–2 feet above the normal high-tide line. No ridges or dunes are present to protect it. In spite of the low elevation and the flat surface, the community has never been flooded completely by a storm, although village elders recall a number of close calls. The perception of the local people is that no single past storm stands out in memory, but each fall, particularly in October, storms arrive relentlessly on the island's shores. Fall storms cause erosion, produce freezing spray, and threaten the community with waves and ice override. A major problem is that these storms occur just when the lagoon is beginning to ice over. Storm surge waters raise the level of the lagoon and break up the ice, which means that escape is not possible by boat, snowmobile, or dogsled. Residents could be trapped during a fall storm. A plus side of storms occurring in the early freezing stage is that the surf on the open ocean side is often considerably dampened by the slush ice in the water column. The airstrip is immediately adjacent to the beach, and this location, combined with its low elevation, makes it vulnerable to erosion and overwash during storms, completing the isolation of the community. At least one recent storm in February broke up the lagoon ice and temporarily trapped people on the island. Some sled dogs drowned when they fell into storm-opened crevices in the lagoon ice.

The community is seriously considering a move to a mainland location 5 miles to the north. In this they would be following the lead of Point Hope and Point Lay, two native communities that have moved back from eroding shorelines in recent years. Kivalina considered moving back more than 30 years ago, but all the elements did not fall into place. Now the community is considering the move once again, this time under more urgent conditions. The Kivalina inhabitants' reasons for moving are numerous, but they see five main problems:

1. The island is eroding on both sides and loss of land is occurring rapidly.
2. Land on the island is already at a premium. There is no room to build

new housing for young couples, and a number of families are doubled up in single-family houses.

3. The community has neither running water nor sewers. Septic disposal, because of lack of space, is currently quite marginal and probably is creating a health hazard.
4. The town garbage dumps are full to overflowing and may be creating a health hazard.
5. The mouth of the Wulik River is directly across the lagoon from Kivalina. A major overflow from the waste disposal system of the large Red Dog Mine upstream has polluted the river. The community was not notified of this spill, and there is considerable fear that the Wulik can no longer be depended on as a water supply. At the proposed relocation site, water can be obtained from the unpolluted Kivalina River.

The experience of Kivalina in planning relocation provides a useful les-

RM 9.3 Kivalina

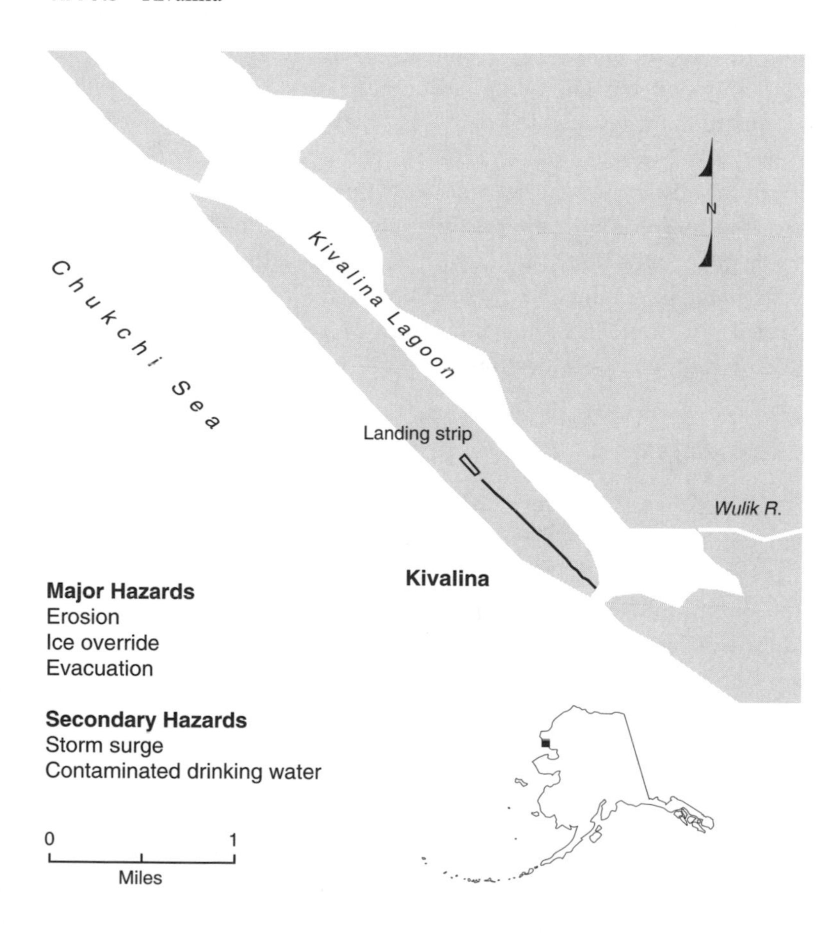

son for other Alaska communities. Consultants contracted by Kivalina residents considered eight locations on the adjacent mainland. Three are within several miles of the Wulik River mouth, and two are more than 5 miles away on the bedrock hillsides of the lower reaches of the Kivalina River. The 18 criteria used to assess the sites emphasize safety from storms and floods, accessibility to critical subsistence resources, as well as a good water supply, good soil conditions, use of the existing airport, and barge access. Residents ranked each location according to each variable. The resulting tabulation provided a more or less objective analysis of the sites. The highest score went to a Wulik River parcel that is not far from the present village and would allow residents continued access to maritime resources. Unfortunately, the parcel, named Kuugruaq, presents two geomorphic hazards: its location within the Wulik River delta renders it subject to river flooding, and storm surge flooding is possible at high tide. Data from a single year (1993) show that half the Kuugruaq parcel floods during ice breakup. The cost of moving Kivalina would not be insubstantial. The most feasible alternative involves building 90 new houses ($120,000 each), a new school ($15 million), a store, and possibly a new airport ($10 million). The total cost, without modern water and sewer, city buildings, fire service, or a community center, is $36 million. Adding the cost of community buildings and sanitation brings the total to at least $50 million—more than $150,000 for every man, woman, and child in the community. The remediation for hazardous substances required on the existing settlement would add to the costs. It is uncertain how the Kivalina move will be funded, but it is a safe bet that state and federal monies will be used. For this reason, further research should be conducted, and safer, even if more distant, sites should not be ruled out. Residents should be wary of repeating their mistakes by choosing another unsafe location.

Kotzebue (RMs 9.1, 9.4)

Kotzebue (pop. 2,751) serves northwest Alaska as a transportation, government, and retailing hub. The town is home to the Northwest Alaska Native Corporation (NANA), the Chukchi campus of the University of Alaska, and is the administrative center of the National Parks Service (Northwest areas) and the Northwest Arctic Borough. Located just above the Arctic Circle, the town lies on a wide beach ridge spit built by sand and gravel eroded from the Baldwin Peninsula by southwesterly winds and waves and carried northward by waves and currents (fig. 9.3). Kotzebue is a relatively new town, in that long-term settlement here dates only from about A.D. 1000. A large native trade fair held in the early nineteenth century (and perhaps earlier) gave the site prominence as a regional center. To the west, Kotzebue faces the sheltering embayment of Kotzebue Sound; low mountains to the northwest and north reduce fetch from those directions. The most intense storm energies reach Kotzebue from the west and southwest, where fetch can reach upward of 500 miles. Severe storms do hit Kotzebue, but with less severity than on more exposed shorelines, such as at Shishmaref, Point Hope, and Barrow. A lengthy sandbar about half a mile offshore serves as a natural breakwater that damps wave energy. The fall 1974 storm so omnipresent in northwest Alaska sent floodwaters 3–5 feet deep into the heart of Kotzebue. In late winter, it is not unusual for ice piles to reach impressive heights. Ice towers tens of feet high form at the edge of the beach when the pack ice shifts and breaks, forcing smaller fragments upward. Sometimes ice overrides onto the waterfront at Kotzebue. So far, at least in historic times, no lives have been lost. Evacuation could present severe problems for Kotzebue residents, especially during a sustained or sudden storm surge that cut off avenues of escape to high ground.

Ice is a major hazard in Kotzebue. Fall storms can send ice onshore, but the major danger is from spring breakup ice from Hotham Inlet north of town. The severity of ice override varies with the amount of snowfall and drifting atop the ice. It is conceivable that 100-yard-long blocks of ice could simultaneously affect much of the Kotzebue waterfront. So far, bulldozers have fended off such threats.

Kotzebue residents have tried several times to contain storm erosion and storm surge. Shoreline stabilization proceeded in a more or less typical evolutionary sequence: 55-gallon steel barrels used along most of the waterfront in the 1960s, gabions and groins emplaced in the 1970s, and a concrete brick revetment in the late 1980s. Now community officials are considering future options. Material for rock revetments and beach nourishment is in short supply. Most recently the town has employed the Armortex system of interlocked, wire-tied cement bricks that was so unsuccessful at Shishmaref. The cement brick revetment at Kotzebue covers one quarter of the waterfront area (see fig. 6.8). The revetment has largely succeeded at Kotzebue, perhaps because, unlike the Shishmaref revetment, it is backed by gravel, not sand.

The revetment largely protects private residences. A major gap extends seaward of a series of commercial establishments, including two restaurants, the Nulugvik Hotel, and two stores and a gas pump. These gaps originally were less eroded portions of the beach. Other gaps in the revetment provide critical work space for mooring boats, storing nets, and for fish and meat drying racks (fig. 9.3). Several low-tech methods of coast protection are also employed along the Kotzebue waterfront. The margin of the upper beach is often bolstered by gunnysack sandbags, which are surprisingly resilient, despite being split open for the most part. Traces of decayed gabions can be seen on the lower beach, especially near Hanson's store. The acute angle of the revetment apparently offers protection from

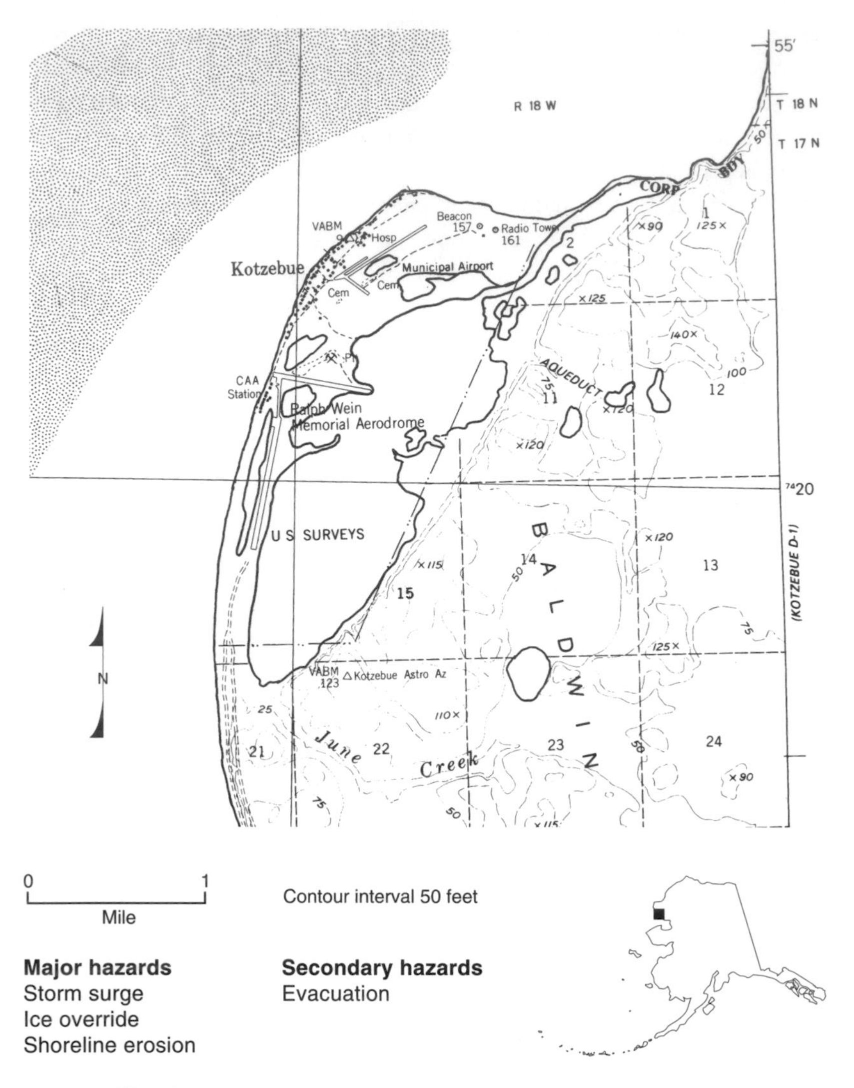

RM 9.4 Kotzebue

low-level storms despite its tendency to channel waves upward into the town.

Bulkheads protect several properties in Kotzebue, notably the corrugated form of Crowley Marine and the NANA office, which uses a low-tech system of concrete-filled 55-gallon steel drums. This 2-year-old structure is already failing and contributing to downdrift erosion. In one stretch of beach, a series of groins constructed of barrels or gabions in the late 1960s have trapped a fair amount of gravel on their updrift (south) sides.

Several other engineering problems confront Kotzebue residents as a result of the long-standing practice of mining the older updrift (south) ridges and beach for gravel. This area is also used as the town's dump. Removal of sediment is making the dump less safe from storm overwash and attack, and is probably severely impoverishing the beach farther downdrift in town. Any future efforts to mine gravel updrift should be resisted. Currently, the hunger for gravel to feed the town's roadways is leading Kotzebue contractors more than 100 miles to the south, to the Choris Peninsula beach ridges. No gravel should be extracted from this source as it may endanger several 3,000-year-old archaeological sites.

The principal hazards at Kotzebue are storm surge, erosion, and ice override. Although attempts to halt erosion are in place, the revetment solution used sporadically so far should be rethought. First, the gravel beach is used extensively by Kotzebue subsistence hunters and fishers. Revetted areas are less useful for net storage, drying racks, and boat launching. Beach nourishment, easily accomplished by the municipality's truck fleet, would be preferable. Gravel should not be taken from the beach.

Shishmaref (RMs 9.1, 9.5)

The ivory-carving, subsistence-hunting village of Shishmaref (pop. 433) is located on a small barrier island on the Chukchi Sea, just below the Arctic

Figure 9.3 The unarmored section of the Kotzebue waterfront provides a well-used beach area for drying fish and mooring boats. Where revetments have been built along most of the Kotzebue shorefront, the beach is much less useful to the community.

Circle. Like most barrier islands worldwide, Sarichef Island is slowing eroding on the seaward side and rolling over toward the mainland. In the last 10–20 years, according to the local inhabitants, more than 30 feet of erosion has occurred on the dune scarp in town. Erosion here can be a spectacular event, with 5-foot-long blocks of grass-bound sand detached in a single storm. The sand is carried northeast by the prevailing longshore current; hence, Sarichef Island is lengthening to the northeast. Tidal currents produce some progradation on the southwest end. Oblong Sarichef Island, 4 miles long and 0.5 mile wide, consists of a stable older core of higher dunes, up to 25 feet above sea level, and two tidal flats adjacent to the inlets to the southwest and northeast. Littoral currents mainly from the southwest are driving the island to the northeast. Dunes are building along the inlet margin north of town. The settlement of Old Shishmaref, occupied in the sixteenth to nineteenth centuries and then abandoned, was on the dunes along the inlet in a location less threatened by erosion and storms than the present village site. Perhaps the wise choices of the Shishmaref forebears should be emulated.

Shishmaref comprises a diverse assortment of structures, mostly plywood cabins, some more than 40 years old. The largest structures are the school complex, built in 1978; some large storage tanks; and various public and commercial buildings such as the two stores.

Storms can hit Shishmaref extremely hard. Scarcely any of the island was left unscathed by the 1974 storm. Waves broke at 16 feet above sea level, and the island resembled a small mound in a boiling sea. Just 40 years ago the beach at Shismaref was significantly wider—nearly 120 feet according to village elders. Shishmaref first tried to hold back the Chukchi Sea as early as the 1950s by placing a series of sand-filled 55-gallon drums at the base of the eroded dune scarp. The drums provided little protection, although a few rusted drums can still be seen today.

After the 1974 storm, Shishmaref started serious efforts to hold back the hungry sea. The first option the community considered involved moving the entire town onto the mainland. This idea lost its appeal when the proposed site 5 miles away was found to be permafrost rich and the move was estimated to cost $2–4 million, an astronomical amount before the oil boom. Another option included partial relocation of shorefront property onto higher, more landward portions of the island. No action was taken on this suggestion, and the village remains tightly clustered on a small fraction of the island. Several engineering firms presented the people of Shishmaref with alternatives for hard stabilization of the shore bluffs. The engineering consultants clearly preferred rip-rap (see appendix C, *Shishmaref Erosion Control Engineering Studies*) but noted that sandbags were the cheapest method. Shishmaref residents chose the most expensive method, concrete mat.

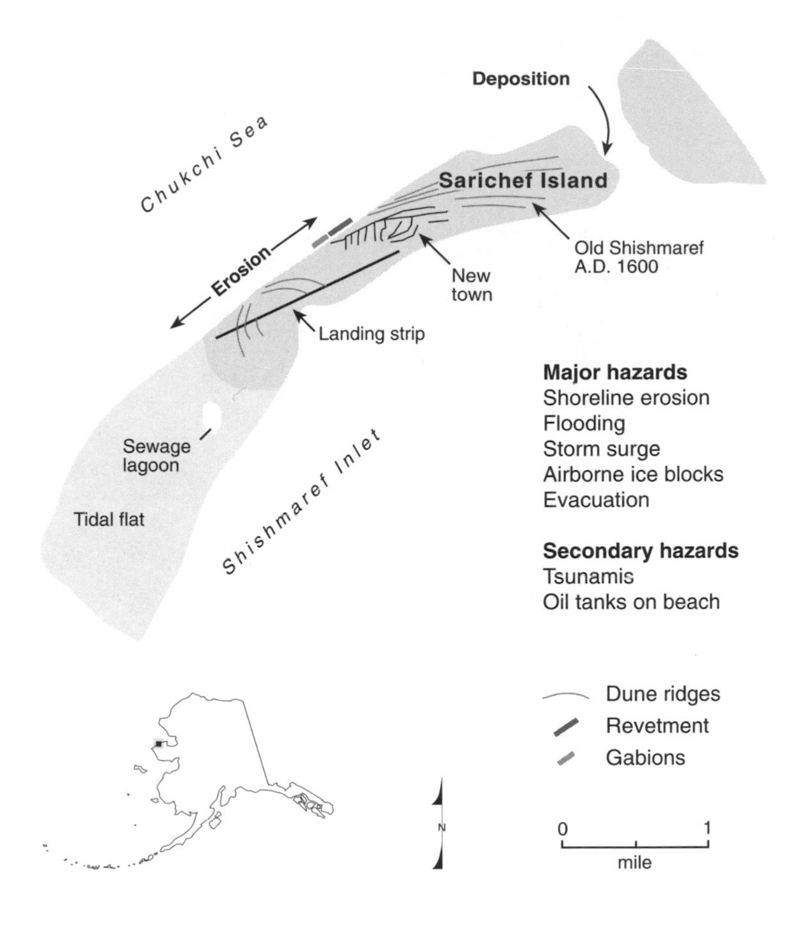

RM 9.5 Shishmaref

In the early 1980s Shishmaref employed two methods of hard stabilization (see chapter 6) that eventually extended over 1,700 feet of seaward dune face. The first, installed in 1982, was a series of gabions laid at the base of the bluffs. The gabion units were piled one on top of the other and attached by hand. The gabions remain largely intact along the south portion of town after more than 12 years of saltwater corrosion and wave action.

In 1984, more "serious" measures were employed in the hope of stopping coastal retreat. Called "the seawall" by residents, the Shishmaref revetment is an impressive testament to the futility of shoreline engineering (figs. 9.4 and 9.5). The revetment is a series of interlocked 25-pound cement bricks fastened together by wire, sitting on a plastic cover atop a mattress of sand. The concrete mat does not appear to extend below sea level, as the consultants proposed it should. Less than half of the structure is still standing, and most of that was rebuilt by residents in 1986.

Figure 9.4 The largely collapsed cement block-and-wire revetment in Shishmaref was built in 1984. The revetment was overtopped by storm waves shortly after construction when the underlying sand slipped out from underneath, and the entire structure collapsed. Easily mobilized fine sand was responsible for this spectacular engineering failure. In Kotzebue the same type of revetment has fared better because the beach sediment is coarser gravel that is less susceptible to flowing out from under the structure.

Figure 9.5 The shoreline behind the failed cement block revetment has retreated, threatening several buildings in Shishmaref. The shoreline in the background is protected by gabions that have been in place for more than 20 years.

The collapse of the Shishmaref revetment began during its construction. The edifice rests directly on sand bulldozed onto the beach from the tidal flats. The sudden catastrophic collapse of the revetment occurred when it was overtopped by storm waves in fall 1986. The water forced the supporting sand out to sea, leading to the collapse of the concrete blocks. Ice push at the base may also have been a factor in its deterioration.

Shishmaref residents continue to resurrect their revetment; local efforts in 1993 shored up the structure and added a 2–3-foot-high series of gabions on its upper portion. Owners of new houses elsewhere on the island are attempting to plant beach grass, trap sand, and build dunes as a defensive measure. Unfortunately, an October 1995 storm caused serious bluff erosion behind failed portions of the revetment. In the midst of the storm, residents were outside throwing discarded items—including old snowmobiles—into the surf zone in an unsuccessful attempt to halt the erosion. Careful maintenance will be required to keep the protective matting in place. Dune nourishment is an untested alternative for Shishmaref.

The Shishmaref erosion control efforts were expensive both economically and environmentally. Both the gabions and the revetment caused end-around erosion. In fact, the 1984 revetment overlapped this end-around zone, and this may have contributed to its undercutting by storm waters. The beach at Shishmaref is largely gone; it is now too narrow, rough, and full of obstacles to be useful for recreation and boat mooring.

At the present time the community is once again giving serious thought to moving to the mainland. The locations being considered are in an area on the south shore of Shishmaref Inlet that is underlain by ice-rich silt. Structures built here would require the emplacement of a thick and sizable gravel pad and the use of posts to provide a cold air trap beneath. The cost of moving Shishmaref would considerably exceed that estimated for Kivalina—more than $50 million given its 25 percent larger population. A minimum figure of at least $60 million seems a reasonable estimate. The move would also require building more than 20 miles of road from a rock quarry site. In the end, the cost of moving Shishmaref could well top $100 million!

The Final Word: Retreat from the Sea in Northwest Alaska

Storm activity over the past 20 years along the Chukchi Sea has occasioned near panic in residents of Shishmaref and Kivalina as they have witnessed their islands visibly constricting. However, the rates of change are not well documented. It is still unclear how rapidly the sea level is rising and how often large storms strike the coast. Quite possibly, existing engineering solutions could stave off the immediate threats. Building remediation using high posts, beach nourishment, and dune building is an untested possibility. In the long run, retreat is the best solution, but can the state of Alaska or the federal government afford to move 750 people several miles at the cost of well over $100 million? In an ideal world based on long-term planning, one could argue that the government couldn't afford not to, given the costs of repairing storm damage. However, the costs of the 1974 storm to Shishmaref were apparently minimal compared with the projected cost of moving. It should be remembered that Shishmaref proposes moving to a bluff location high above any storm surge threat; the Kivalina alternatives remain within the Wulik delta and floodplain. In asking for public support for relocation, residents of the two villages, or any community that proposes to relocate, should consider their responsibility to choose a safe location.

10 Southwest Alaska: The Bering Sea Coast

Introduction (RM 10.1)

Southwest Alaska is home to comparatively few people, but its waters are the nursery for the greatest remaining nonhatchery, wild salmon fishery in North America. Only four settlements along the southwestern shore have more than 1,000 people. Many more thousands live within 10–20 miles of the shore and are subject to such coastal hazards as tsunamis, ice override, storm surge flooding, and river delta flooding.

The Bering Sea is triangular; the Alaskan mainland forms its eastern side, the Siberian peninsula of Kamchatka the western side, and the Aleutian Islands on the south border the third side. Oceanographically, the Bering Sea has two components: a half-mile-deep basin in the west and a shallow, broad continental shelf less than 300 feet deep in its eastern third adjacent to Alaska. During the Pleistocene ice ages, terrestrial glaciers captured seawater, lowering the sea level and exposing the shallow Bering Sea platform as part of the land mass termed Beringia.

The south margin of the Bering Sea is formed by the arc of hundreds of Aleutian Islands, several of which are active volcanoes aligned with the subducting Pacific Plate and the narrow armature of the Alaska Peninsula. The treeless slopes of the Aleutians are home to abundant fur seal populations still recovering from overexploitation at the hands of eighteenth-century Russian fur traders.

Although not quite as stormy as the North Pacific, the Bering Sea still witnesses considerable fury from storms that originate in the Pacific. The storm season is primarily late summer through fall—August to early December—in contrast with the near year-round storminess of the Gulf of Alaska. Winter cooling allows the formation of nearshore ice that buffers the coast from storm waves for up to eight months of the year.

As in all of Alaska, winter here means dominance, to some degree, by the stable, cold air of high-pressure systems that descend over the land. This de-

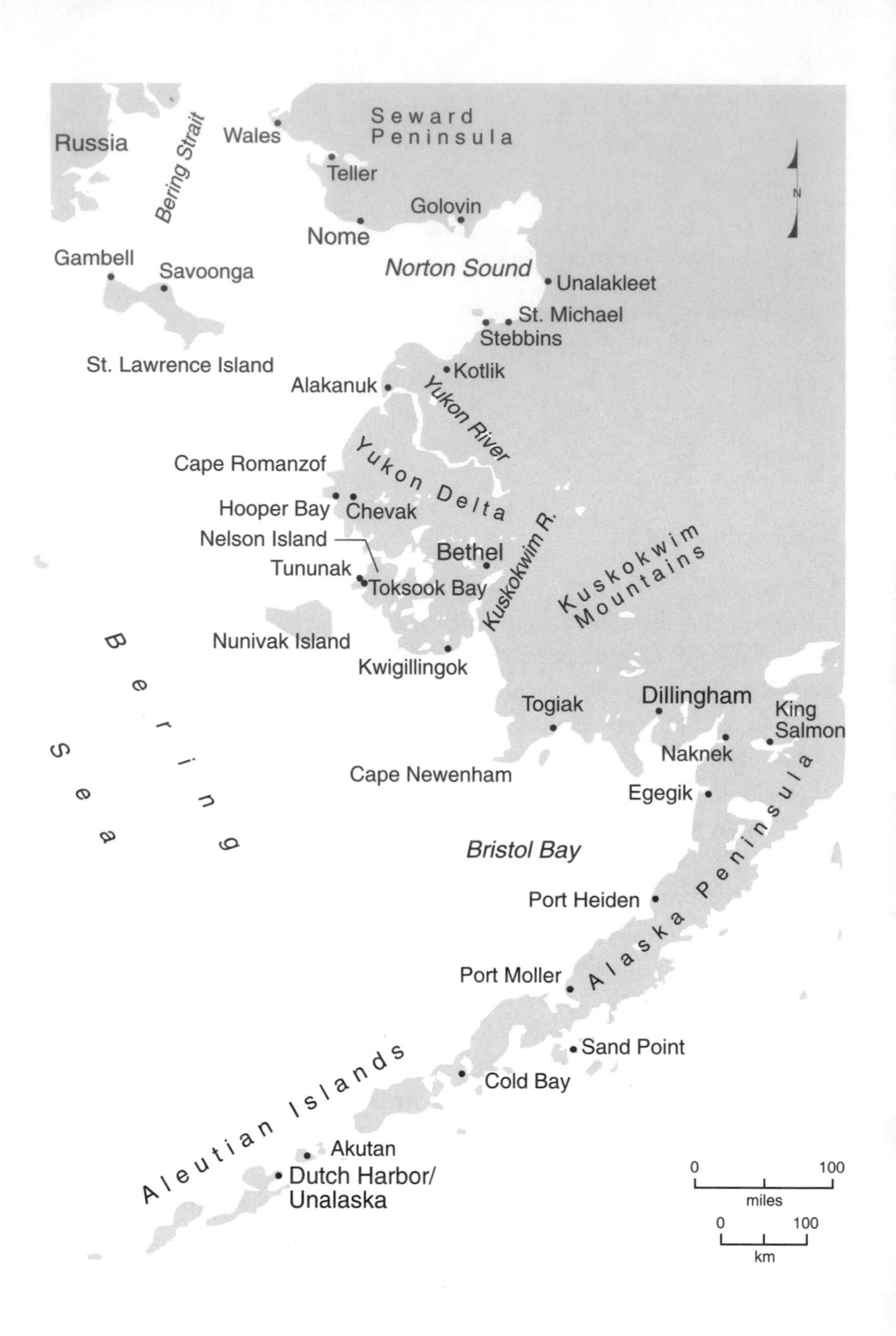

RM 10.1 Southwest Alaska

scending air is intensely cold, often -40°F or colder at the surface, and generates clockwise winds at its margins. On the Bering Sea, winds arise from the east and northeast quadrants to send sea ice southward, away from the coast. In this way, the Bering Sea coast acts as a conveyor belt for sea ice that can reach halfway down the Alaska Peninsula, and even farther in some years.

The flow of currents along the Bering Sea coast is to the north, following the coastline. The inshore areas are shallow and receive considerable fresh water from the numerous large rivers of the Yukon and Kuskokwim drainages originating in the Alaska Range to the east. The fresh water forms a lens atop the water column of the shallow basins and embayments of the coast that prevents upward movement of colder bottom waters and intermixing. In oceanographic terms the eastern Bering Sea is comparatively stable. The absence of upwelling means that nutrients are in short supply and rapidly depleted.

The flat shores of the Bering Sea are the geologic opposite of the rising mountains of south Alaska; material eroded from the mountains is carried to the sea margin and deposited in deltas. From Bristol Bay to Norton Sound, low bluffs of gravel, silt, sand, and clay predominate, although localized bedrock knobs protrude in relief, as at Cape Newenham, Cape Romanzof, and on the volcanic masses of Nunivak and Nelson Islands. On southern Bristol Bay, on the northern Alaska Peninsula, the shoreline sediments consist of gravels and sands pulverized and driven northward by the great Pleistocene glaciers.

Southern Seward Peninsula (RM 10.1)

Teller (RM 10.1)

Teller is a small native village (pop. 154) 70 road miles northwest of Nome that is jammed tightly on a gravel spit extending into Grantley Harbor. The closely spaced houses are located at low elevation on a spit that is eroding on both sides. Until 1990, erosion on the quiet (east) side of the spit was combated by transferring gravel from the tip of the spit, but this is no longer allowed. On the west side of the spit, a 100-yard-long seawall is attempting to halt shoreline erosion (fig. 10.1). The wall's astonishing array of elements includes junked vehicles (1960s telephone trucks), a rock revetment, sheet steel piling, 55-gallon drums, wooden bulkheads, and gabions (wire baskets filled with stones). Aside from the aesthetic problem created by this spectacular structure, sand and gravel transport on the narrowing beach seaward of it is being reduced. In the long term (decades?), reduction of sediment transport to the spit will increase both the rate of shoreline erosion and the need for seawalls.

Figure 10.1 Teller has attempted various shoreline protection schemes over the last 30 years, including a bulkhead, 55-gallon drums, and dumping abandoned telephone trucks and other junk on the shore. None of it has had much effect on the shoreline retreat problem.

The tide range here is low, and it makes little difference whether a storm strikes at high or low tide. Storms frequently bring water into the town. The 1974 storm brought waist-deep water into most of the town and floated some houses off their foundations. There are ample unused high-elevation building sites at the landward end of the spit. As the spit continues to narrow, storm damage may increase, along with construction and maintenance costs for the seawall. Moving to high ground may become a more appealing alternative with time.

Nome (RMs 10.2, 10.3)

Nome is a small (pop. 3,500), compact town huddled against the sea at the edge of a broad, treeless plain built of uplifted beach gravels. Established in 1900 as the nation's most famous gold rush town, Nome is now perhaps best known worldwide as the finish line of the Iditarod dogsled race across Alaska. In the early days, the beach was an important part of the social and commercial life of the community (fig. 10.2); the beach placers were the first source of gold. But big storms, usually arriving in the fall (1902, 1905, 1913, 1937, 1942, 1945, 1946), seriously disrupted the community. The 1945 storm hurled blocks of ice into the town, and the 1946 storm leveled six buildings. In the late 1940s the community considered moving buildings several miles inland, out of harm's way. The plan never materialized, however, due to opposition from powerful commercial interests—especially the Lomen family—that owned oceanfront property. Recent commercial development has shifted to the safer northern margin of town, a mile from the coast. Earlier adoption of this option might have saved Nome's beaches and spared U.S. taxpayers from spending millions of dollars on seawall construction.

In 1949–51 a seawall was constructed along 3,400 feet of the Nome shoreline (RM 10.3). The wall is 60 feet wide at the base, 16 feet wide at the top, and stands 18 feet above the mean low tide level (fig. 10.3). Granite boulders for the wall were obtained from nearby Cape Nome. The wall has done its job well. The big storm of 1974 did significant damage to Nome, but it would have done much more without the wall. However, the beach is gone.

There was an alternative solution to seawall construction besides relocation: beach replenishment. In the 1940s, when the wall was built, replenishment was rare in the United States. Nowadays we recognize the value of beaches, and attempts to preserve them are commonplace. Low-cost sand and gravel suitable for replenishment abound in the old mining areas around Nome.

The average lunar tidal range in Nome is 1.6 feet (maximum 2.4 feet), a rather strong contrast to the 30-foot tides of Seward and Anchorage. The low tidal amplitude means that it makes little difference whether a storm arrives at high or low tide. In addition, the Bering Sea surface here is frozen in winter, so the frequent winter storms do not create storm surge or storm wave problems.

RM 10.2 Nome

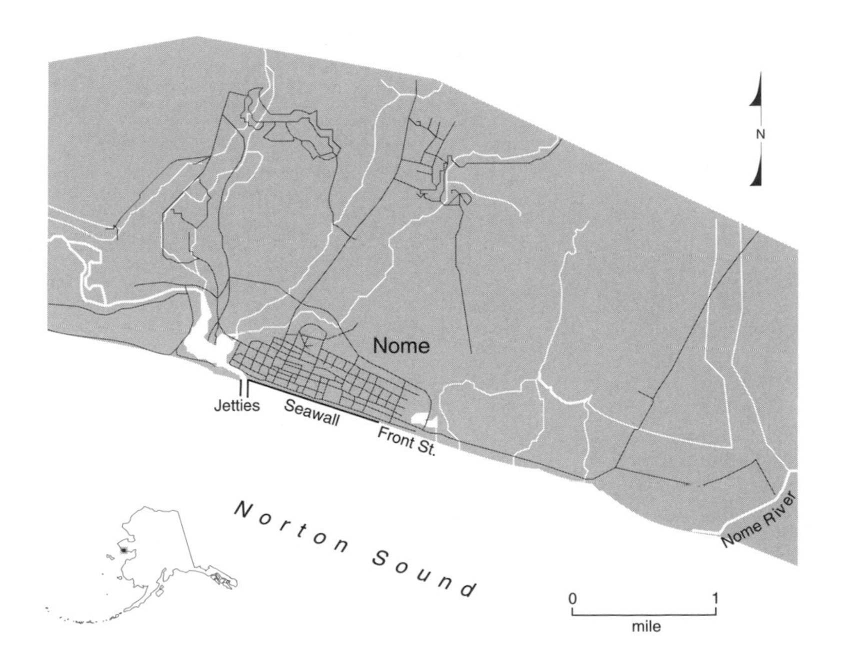

Figure 10.2 Coastal flood in Nome, October 25–26, 1946, on the beach as the storm was winding down, looking east. Note the 2-foot drop-off on the ocean side of Front Street created by erosion of the beach. The pilings in the foreground once supported various structures that were destroyed by the storm. The collapsed building on the beach in the background had been the Lincoln Hotel. Two people died in the building when the storm pounded it to wreckage. The extent of debris on Front Street suggests that the high-water line was about 2 feet above the street. Photo from Lomen Family Photograph Collection, Alaska and Polar Regions Archives, University of Alaska, Fairbanks.

There is a problem with the Nome seawall that will become evident in coming years. With time, the remaining beach in front of the wall (all of it now under water) will likely narrow and steepen. This means that reduction of wave energy by friction with the bottom will no longer occur, and the waves striking the wall during storms will become larger.

The city of Nome adheres to the Uniform Building Code, has a building inspector to enforce it, and participates in the National Flood Insurance Program. Nome residents can purchase federally subsidized flood insurance.

The city sometimes experiences high winds during the winter, and principles of wind-resistant construction should be followed in home construction or alteration. In addition, snow loading and permafrost foundation problems should be considered.

Residents of Nome tend to underestimate seismic and tsunami hazards. Because Nome lies on the north rim of the Bering Sea (RM 10.1), it could be exposed to seismic waves generated in the Aleutian trench 750 miles to the south, as well as in Kamchatka and Japan, even farther afield. Ethnohistoric records mention a tsunami in the 1830s; although rare, tsunamis are a real

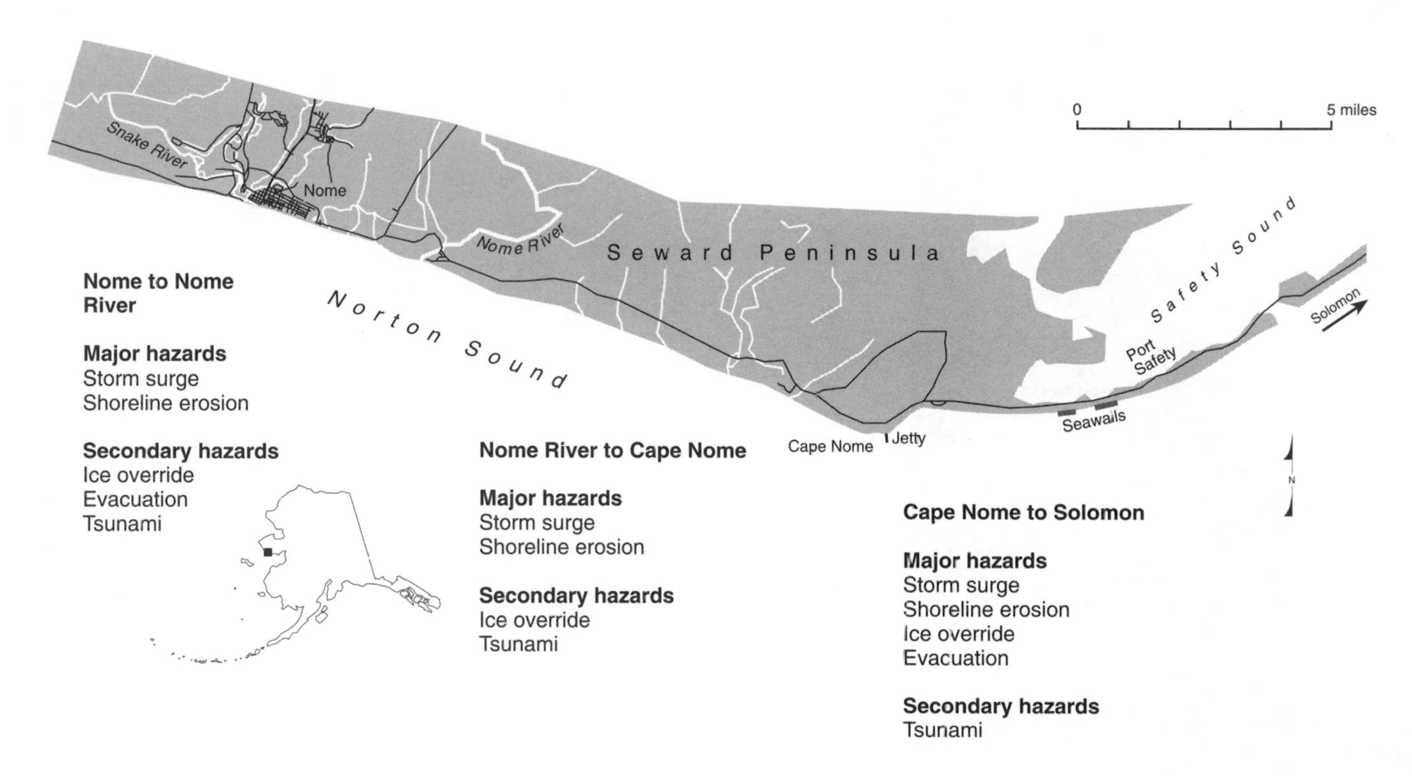

RM 10.3 Nome and Vicinity

Figure 10.3 The Federal Building in Nome overlooks the million-dollar granite boulder seawall built in 1951. The wall is occasionally overtopped by storm waves, usually in early fall storms.

possibility at Nome. Numerous faults have been mapped onshore near Nome; most trend north to northeast, and the closest are 2–4 miles offshore. Onshore, the Penny River fault is only 7 miles west of Nome. Seismic hazard planning studies place Nome in a comparatively low-risk category with a 10 percent probability of earthquakes measuring 3–4.5 on the Richter scale in a 50-year period. However, history indicates that sizable earthquakes are possible less than 100 miles from Nome; at least seven events of magnitude 5.0 or greater were recorded before 1975, and one earthquake was magnitude 6.0–6.4. When considering a time frame of thousands of years, residents should realize that the faulted Bendeleben Mountains north of Nome are subject to tectonic uplift.

River flooding is a comparatively marginal hazard for Nome residents, although the port and harbor facilities could be affected by peak discharge on the Snake River and Bourbon and Dry Creeks. Bridge washouts could be the most important hazard to Nome area residents.

Ice override may occur when storm wind conditions are coupled with sufficient open water. Norton Sound is usually an ice factory for the Bering Sea because the prevailing strong northeasterly winds generate offshore winds that carry newly formed ice out to sea. Ice is driven into Nome only when southerly winds hit the area, a comparatively rare event. The winds responsible for ice motion arrive from the southeast and are most likely to occur in November and December. Southwesterly winds are comparatively rare (less than 2.5 percent) but could cause substantial harm given the large fetch in that direction.

Sizable ice piles occur with considerable frequency in Nome. Ice override occurred a few years ago on the east side of town, but the ice sheet was turned back by a bulldozer. A pileup in the mild winter of 1980 reached a height of 20–25 feet, and one in 1974 was nearly double that, 30–40 feet. The seawall localizes the effects of ice override and prevents the transport of ice inland.

The Nome Harbor (RM 10.3)

The estuary of the Snake River debouches into the Bering Sea along the west boundary of the town of Nome. A sand spit blocks the estuary from the sea and was used as a barge lighterage point and tent settlement for several years until the storm surge of 1902 exposed its vulnerability. At present, the Snake River estuary is a port and storage facility for gas and fuel oil in several tank farms.

Dredging of the Snake River estuary and mouth started in 1919 and has continued over the last 75 years under the auspices of the U.S. Army Corps of Engineers. Most of the dredged material is dumped immediately offshore from town but has a minimal effect on beaches due to presence of the seawall. Engineers have proposed cutting channels through the sand spit to divert sand from the estuary. These proposals are ill advised. For several decades steel bulkheads have stabilized the inner shores of the estuary while two jetties of 200 and 400 feet maintain the position of the Snake River mouth. The jetties, built at intervals from 1919 to 1935, prevent sand transport and contributed to the catastrophic erosion of the downdrift beaches (and to the need for a seawall) by subsequent storms in the late 1930s and 1940s.

Nome Seawalls

Nome has expanded considerably since the seawall was constructed in 1951. For quite some time the easternmost five blocks of Nome were not behind the seawall but instead were protected by a revetment of lower elevation than the wall. While the seawall was freestanding with sand and gravel fill behind it, the eastern revetment was a pavement of smaller stones on the seaward-facing slope of Front Street. The eastern revetment was severely damaged by the October 1992 storm and was replaced by a large rock seawall (fig. 10.4). Thus, the hazardous nature of the revetment has been mitigated to some degree. As a further measure, dump-truck loads of sand and gravel could provide a protective beach, but it would probably have to be re-replenished every few years.

Figure 10.4 Gold miners on a Nome beach seaward of the revetment (foreground), with the seawall to the west (background). A major storm surge in 1992 partially destroyed this revetment and led to the extension of the seawall. Recreational use of the beach is probably ended forever here because the wall has effectively destroyed the beach.

A peculiar problem in the Nome area and northward concerns the effects of seawalls on the driftwood supply. Driftwood, traditionally and to some extent today, is an important resource for building and fuel. If seawalls reduce the supply of wood by refracting logs out to sea, then the coast may be deprived of a significant resource.

The narrower beaches created by seawalls also diminish the potential for avocational and small-scale mining of placer gold. Thus, a significant position of Nome's romantic luster and even tourist appeal has been lost irretrievably.

Nome to 1.5 Miles West of Nome River Mouth (RM 10.3)

This mile-long stretch of coast has numerous small subsistence or allotment buildings but few if any year-round residences. A repeat of the 1974 storm will certainly remove most of these buildings. The Cape Nome gravel road is very close to the beach along most of this shoreline reach, and small rock revetments protect the seaward side of the road at several locations. The road will be easily overtopped in big storms, but perhaps more important, all the small buildings seaward of the road will be destroyed.

Judging from experience elsewhere, the revetment along the road will grow longer and grander in coming years as shoreline erosion continues to threaten the road. But armoring the road may not be the best economic and environmental solution for the long term. At some point in the future the cost of protecting the road will rival and even exceed the cost of the road itself. Moving the road back now would spare future armoring costs and environmental damage to the beach. The minimum extent to which the road should be moved back (and the safest area for home and cabin construction) can be judged from the location of the lines of driftwood left stranded ashore after storms.

West of Nome and East of Nome River Mouth to Cape Nome (RM 10.3)

Along both reaches of shore the road is well back from the shoreline and is not likely to be harmed or blocked by major storms. Several small plywood hunting and fishing cabins next to the beach probably will be removed in the next big storm. Unfortunately, as time goes on, these cabins become more substantial and the amount of equipment stored there (generators, boats, snowmobiles, etc.) increases. It is best to site buildings well back from the shoreline, behind the most landward extent of driftwood lines.

Cape Nome to Solomon (RM 10.3)

Most of this reach is a barrier spit and barrier island. It is everywhere low in elevation and subject to overwash during storm surges. Most of the buildings here were affected by the 1974 storm. Most, however, are hunting or fishing cabins that are not occupied year-round.

No studies document the erosion problem in this area, but field indications are that the spit is slowly eroding on both the Safety Sound side and the open Norton Sound side. Overwash fans, low-elevation bodies of sand extending across most of the width of the spit, probably produced during the 1974 storm, are easily visible in most areas. You can get an idea of the extent of the storm flooding by looking at the location of driftwood lines (although, near buildings, the wood has often been removed and used for firewood or drying racks).

Some of the cabins in Solomon are partially protected from future storm overwash by the salt marsh, a mid-island low area in front of many buildings. Future storm overwashes may be halted at mid-island by the marsh.

This barrier spit may have a big problem in the long term. Mining at Cape Nome and dumping of the waste on the shoreline has obliterated any beach that may once have been present. More important, the recent (1986–87) construction of a large barge-loading facility extending across the shoreline in a seaward direction has probably completely cut off the sand supply to the spit from the west. The dominant direction of littoral sand transport here is from west to east. The cutoff of the spit's outside sand supply has already led to an erosional indentation near the terminus of the spit where it connects with Cape Nome. It is likely that the loss of sand will cause erosion all along the spit and may eventually increase the erosion rate of the entire spit complex. Whether or not this will happen and how soon cannot be predicted without detailed studies.

In a few locations, especially near the Cape Nome terminus of the spit, piles of gravel and mounds of bulldozed sand are lined up along the road (fig. 10.5), apparently to prevent overwash across the road during storms. In fact, the harm done by this bulldozed protection exceeds the results of storm overwash. Storm overwash merely transports a thin, wide wedge of sand over the road and the spine of the barrier island. High, bulldozed ridges along roads may actually concentrate storm wave attack at points of weakness, causing gullies to be eroded into the roadbed. Under the guise of restoring the road, in 1993 the Alaska Department of Transportation started placing large boulders adjacent to the Nome Council Road, in effect producing a seawall. This wall will irretrievably harm the spit in the long run.

Experience on barrier islands elsewhere indicates that it may be best simply to allow road overwash to occur. Overwash is very infrequent on this spit. Dikes or levees are costly, produce an unsightly wall in a beautiful natural environment, and rarely work anyway unless the entire road on the spit is protected by a well-maintained dike or levee.

Ice override is an additional hazard here, although most of the cabins are set far enough back to avoid any but the most extreme override. Finally, it should be noted that if a big coastal storm arrives before the Bering Sea ice cover is formed (e.g., a repeat of the 1974 storm), people probably will not be able to escape off the spit. Residents should evacuate immediately when a big storm is predicted.

A Summation

The consequences of seawall building continue to affect Nome residents. A major storm surge in October 1992 produced waves that overtopped the revetment protecting the newer eastern suburbs and led to its replacement by a freestanding seawall in 1993. The intensity of the 1992 storm produced a frenzy that led transportation planners to construct a rubble-fill seawall on the Safety Sound spit, despite the fact that the spit is subject to overwash only about every 10 years. This unwise solution is aimed at protecting the gravel road, not the cabins on the spit. Serious erosion will afflict unprotected portions of the spit in the future. Nome is headed down the primrose path of New Jerseyization of its coastline, fueled by too-ready access to seawall boulders from Cape Nome.

Other Villages and Camps (RM 10.1)

Villages along eastern Norton Sound primarily face storm surge and possibly tsunami hazards. Elevation of driftwood is a fairly good indicator of the magnitude of the threat, considering that the 1974 storm remobilized driftwood of less powerful and lower storm surges. Residents of *Golovin* (pop. 123) and *Unalakleet* (pop. 646) are well aware of the importance of wind direction in increasing wave heights during storms. Towns within narrow embayments, such as Golovin, may face higher storm surge levels than towns on the open coast. *Hooper Bay*, a large Yup'ik village (pop. 846), sits about half a mile inland from the bay atop a low mound but is still susceptible to storm surge flooding, as occurred in 1978. The airstrip runs to the ocean shore and is susceptible to flooding.

The Saint Lawrence Island communities of *Gambell* (pop. 548) and *Savoonga* (pop. 514) also face storm surge threats. Gambell lies atop a beach-ridge plain 21 feet above sea level but is safe from most storms. Savoonga is situated on the bluff and is at less risk from surges.

A number of subsistence activities are conducted at small, seasonally occupied camps in this area.

Figure 10.5 Gravel road on a barrier spit east of Cape Nome with Norton Sound on the left and Safety Sound on the right. Unwise decisions led to the protection of the gravel road with an aesthetically unpleasing dike. The long pile of rocks and sand protects the barrier spit road from inundation that occurs less than once a decade and interferes with barrier spit processes necessary for the barrier's long-term survival and migration.

The Yukon-Kuskokwim Delta (RM 10.1)

The Yukon-Kuskokwim (Y-K) Delta is about the size of West Virginia. Except for several bedrock knobs that form small hills or islands, the delta is extremely flat for more than 70 miles inland from the Bering Sea. By Alaskan standards, the Y-K Delta is well settled; villages of several hundred Yup'ik Eskimos are scattered along the sluggish distributaries of the delta. Many villages are within 10 miles of the sea.

The major geomorphic threat to people on the Y-K Delta is flooding, which can arise from the rivers, from the sea, or from both when the sea and tides flood the rivers. Flooding caused by snowmelt and ice breakup in May and June is fairly predictable and manageable. The course of breakup follows an upstream-to-downstream pattern and takes several weeks, and communities downstream can anticipate the scale and timing of flooding by listening to news reports. In most cases, communities are situated to avoid ice jams that may prolong flooding. As in more southerly locations, catastrophic breakup flooding occurs when a large snowpack melts suddenly during warm weather.

Storm surges can be considerably more destructive to Y-K Delta communities than snowmelt floods. Storm surges are reported to have traveled more than 40 miles inland during this century, and driftwood lines from recent storms extend several miles inland. Communities close to the coast also face elevated water levels associated with high wind and waves, forces destructive in their own right. Fortunately, most Y-K Delta villages are not on the open coast; exceptions are *Tununak* and *Toksook Bay* hugging the bedrock promontory of Nelson Island. Saltwater carried inland may adversely affect marsh vegetation and kill nesting geese, essential spring resources; flooding during June 1962 killed tens of thousands of goslings.

Just about every community along the margins of the Y-K Delta has moved once during the last 50 years in response to flooding. However, the recent construction of public buildings such as high schools and airports has increased the financial burden of moving entire communities. Informed building decisions by individuals will be increasingly important. Residents should build reasonably high above flood levels, both from river and coastal storms.

Bethel (RMs 10.1, 10.4)

The Yup'ik Eskimo center of Bethel (pop. 4,687) is roughly 100 miles upriver from the head of the Kuskokwim River estuary, on the low left bank of the tidally influenced Kuskokwim River. Although not strictly a coastal location, Bethel participates in the Coastal Management Planning Program.

Bethel was founded in the 1880s by Moravian missionaries who aimed to convert neighboring Y-K Delta natives. Today the town is a market and supply center as well as an air and government hub and the location of the regional hospital run by the Bureau of Indian Affairs.

Hazards in the Bethel area are confined largely to river flooding caused by ice jams or breakup, and bank erosion. Floods develop in two seasons: during breakup in spring (May and June) and in late summer and fall (August and September) when rainfall swells the river. Most of the developed part of Bethel lies within the 100-year floodplain. Floods have covered most of the town at one time or another; two eastern areas, lower Brown Slough and "Lousetown," are subject to flooding nearly every year. Floods in 1941, 1963, 1964, and 1974 almost reached the 50-year flood level, 16.5 feet above mean low low water (MLLW). Major floods increase river discharge by 10 times and velocity by 5 times. The tidal influence decreases when the river reaches what is normally a 20-year flood event. Thus, in normal years tidal influences predominate over river influences.

Bank erosion along the Bethel waterfront is the result of several factors, primarily high river levels, waves generated by boat traffic, and southerly winds. Because the silt bank is solidified by permafrost, collapse is furthered by warm weather. Erosion data collected since 1939 establish a comparatively rapid annual rate; erosion averages 8 feet per year along the waterfront on the east side, and up to 25 feet per year near the Chevron tank farm on the west side of town. According to the Corps of Engineers, the Kuskokwim River is shifting its main channel eastward, and erosion in Bethel will decrease. If this is true, then future erosion rates will be lower. Projected erosion rates are mapped in the Bethel Management Plan. The Bethel waterfront has a checkered history of bank stabilization methods, ranging from a timber bulkhead to submarine netting and junked cars. Presently, the bank is secured by a steel cell bulkhead. Erosion hazards may be complicated by the reemergence of barrels containing toxic chemicals used as erosion-control bulkheads in the 1950s, as occurred in summer 1993. Erosion can give rise to chemical pollution if toxic barrel contents leak into the river.

The Kuskokwim floodplain topography around Bethel is generally low and rolling, mostly under 100 feet in elevation, so slope-related hazards are not a consideration for residents.

The Bethel region is the one of the least seismically active areas in Alaska, as documented by the absence of measurable quakes in the last 100 years. Anecdotal accounts indicate that the 1964 quake shook the area, however, and there may be a moderate potential for earthquake-related damage. Tsunamis generated by Aleutian seismic events could, in theory, be hazardous to Bethel if coupled with a high tide or flood, but tsunamis are considered a low probability for Bethel.

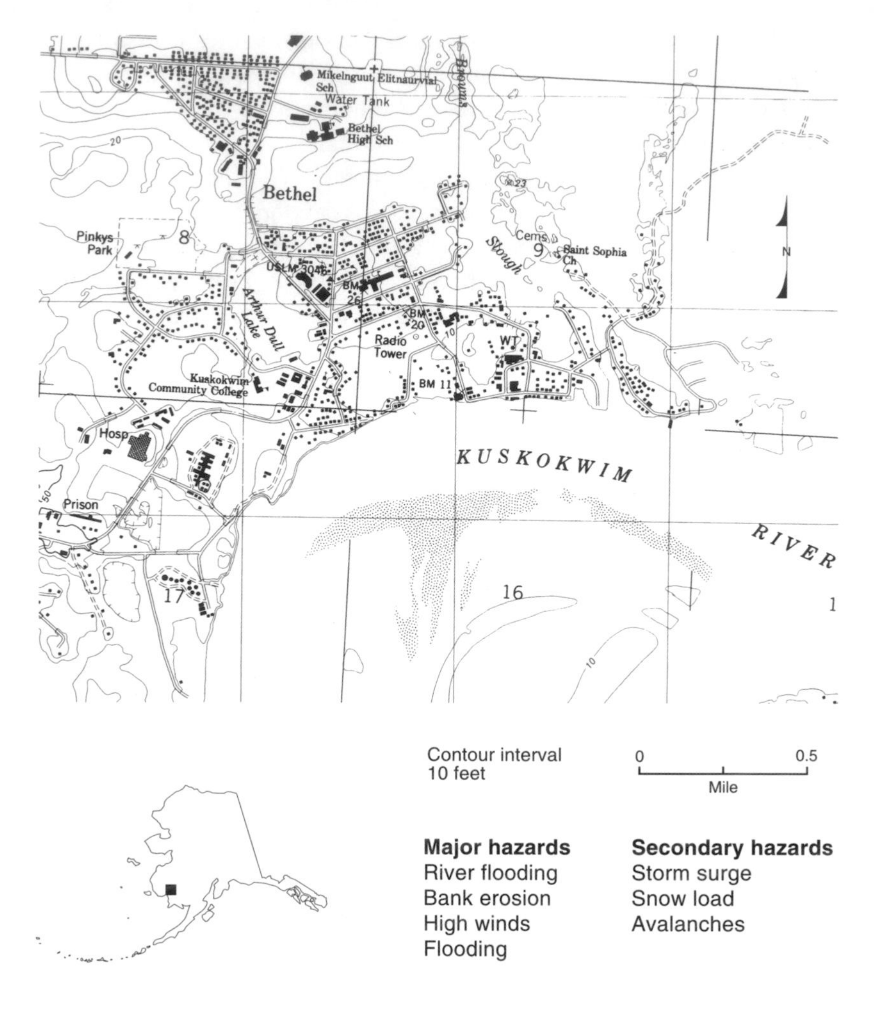

RM 10.4 Bethel

Bristol Bay (RM 10.1)

Dillingham (RMs 10.1, 10.5)

The rapidly growing town of Dillingham (pop. 2,017) lies at the confluence of the Wood and Nushagak Rivers, at the head of Nushagak Bay. Dillingham is both a fish-processing center and a trade and local air travel hub with a sizable Native American community (more than 51 percent of the total population). The population of the town swells by 25 percent seasonally when canneries hire hundreds of temporary workers and fishermen operate out of the harbor. As is the case in many Alaska villages, the government is the second largest employer in Dillingham. The town rests on 70-foot-high eroding bluffs comprised of sands, silts, and clays deposited by the last widespread glacial expansion over the Nushagak lowland over 40,000 years ago. Although Dillingham is close to the southern limit of perennial sea ice, ice pans usually form over Nushagak Bay from early November to early May, with some ice persisting until June. Tidal influences are keenly felt at Dillingham, which sees a maximum tidal range of nearly 28 feet and tidal currents of up to 5 knots.

Hazards here coincide with people's desire to locate buildings close to eroding bluff margins. Recent bluff erosion threatened Snag Point sewage lines in the northern part of town and could send untreated sewage into the Wood River (*Anchorage Daily News,* November 30, 1993). Bluff erosion is the result of fall storm activity and, in the absence of permafrost, does not involve large-scale block collapse as it does in Barrow. Bluff erosion may be accelerated by the revetments on the harbor entrance, updrift from the bluffs. Seawalls tend to beget more seawalls, and in the future, Dillingham may require extensive seawall protection.

Fall storms present a serious threat both to the erodible bluffs and to any marine craft not sheltered in the man-made boat harbor west of town. Dillingham is considerably more sheltered than many western Alaska communities, but persistent south or southeasterly winds can generate fierce waves rolling toward the town. Wind measurements at King Salmon, on the Alaska Peninsula, are a good indicator of likely storm waves at Dillingham. Calculations show that waves of a 100-year storm reach only 4.5 feet at Dillingham, while storm waves likely every 2 years reach 3.5 feet. Water elevations of up to 12 feet may be associated with surges accompanied by 6-foot waves. For most of the town, storm surges will not produce flooding, although individual structures, including the boat harbor and property, may be affected. In 1980 a severe storm brought sustained winds of 70 mph and seas of up to 6 feet and damaged several vessels.

The boat harbor is a unique "half-tide" facility accessible only at mid-tide levels due to the intensity of high-tide currents and severe winter ice problems. The harbor basin, constructed in the early 1960s, is subject to shoaling due to the high sediment concentrations in the Nushagak River. Engineering structures at the mouth of the harbor include a rock revetment.

Although most planners ignore it, the tsunami hazard may be significant for Dillingham. Located at the northern margin of Bristol Bay, the town is in a direct line with several Aleutian volcanic centers. Geological evidence indicates that a tsunami may have overtopped bluffs south of Dillingham as recently as 3,500 years ago. The frequency of tsunamis is unknown, but an informed builder expects the worst.

Naknek (RM 10.1)

Storm surges and bluff erosion threaten several small villages on the southeast shore of the Bering Sea, from Naknek to Point Moller. Bristol Bay, the most productive wild salmon fishery in the world, owes its economic survival to fishing and fish processing. The first canneries were built in 1884; by 1900 a dozen were operating along the shores of Bristol Bay. Two towns, Naknek and King Salmon, are sufficiently developed to require a borough government. The two are linked in the small Bristol Bay Borough, which encompasses only 500 square miles. The cannery and fishing towns of *Naknek* (pop. 593) and *South Naknek* (pop. 133), and the upriver village of *King Salmon* are on bluffs that offer protection from storms but have localized erosion problems. The bluffs are highest at Naknek (75–100 feet) and

RM 10.5 Dillingham

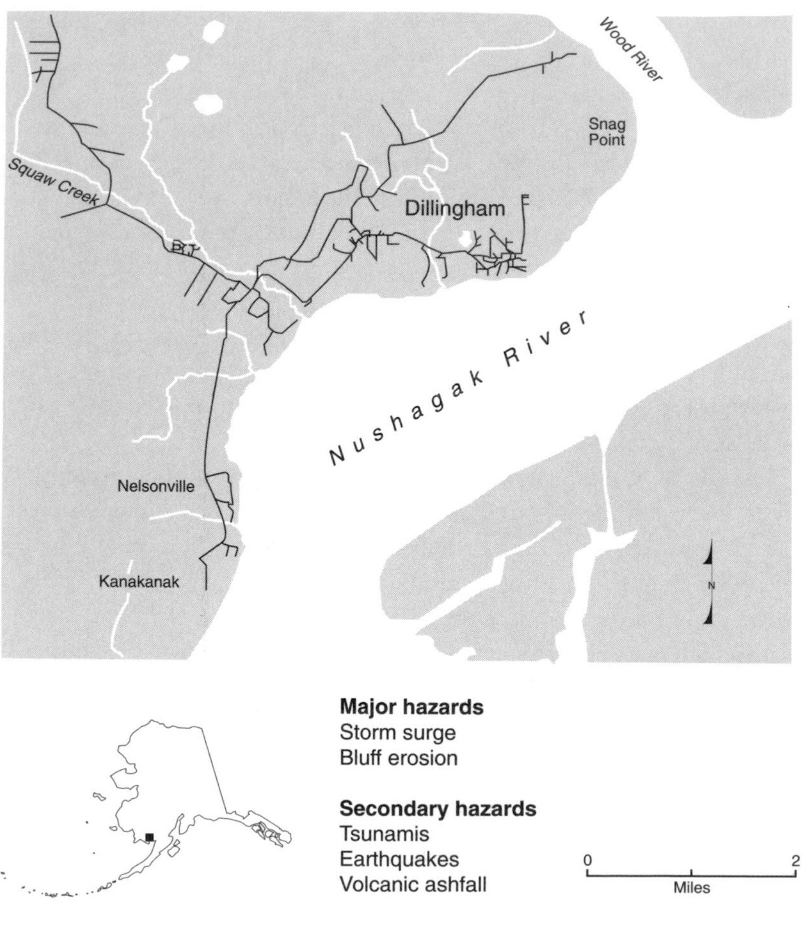

decrease upriver. The bluffs of the Bristol Bay Borough are sand and gravel and are of glacial origin. They are moraine or outwash features deposited at or beyond the farthest margins of glacial advance from the Alaska Peninsula to the east. Naknek and King Salmon are within the zone of discontinuous permafrost, which means that fine-grained sediments are likely to contain ice. Bluff erosion is most likely in association with southwest and west winds, high water, and waves during fall storms, especially when the bluffs have thawed, but winter storms may cause bluff erosion as well. The landslide hazard is considered significant by borough planners, who advise precautions in future developments.

Naknek is located along the outer reaches of shallow Kvichak Bay, which is choked with sand and silt from the Kvichak River. Tides are particularly strong in this area, which adjoins the upper reaches of Bristol Bay and the outer margins of the Kvichak estuary. Tides reach up to 18.5 feet in the estuary. Storm waves vary considerably according to the tidal stage during the storm and are rendered unpredictable by the myriad tidal channels and sandbars within Kvichak Bay. Storms arrive when persistent winds are generated from the west or southwest, generally during the fall. The ice cover along Kvichak Bay is limited to floating pan ice; shorefast ice cannot form due to intense tidal currents. Ice override is virtually impossible in Naknek.

Volcanic eruptions pose a threat to residents; 40 active Alaska Peninsula volcanic centers are within several hundred miles of the Bristol Bay Borough. Major ashfalls have covered the Naknek–King Salmon area repeatedly over the last several thousand years. The 1912 Katmai eruption covered the area with about an inch of ash. As elsewhere, the extent of a particular ashfall is determined by the winds prevailing during the eruption. The most likely scenario for ashfalls at Naknek involves easterly winds, which occur mainly during the winter. Eruptions during the summer months will likely occur along with low-pressure systems that produce southerly and westerly winds and send ash to the east and north.

Naknek and King Salmon are north of the major centers of seismic disturbance on the Aleutian subduction zone. Local earthquake activity is apparently restricted to considerable depths, but may affect residents by initiating landslides within the unconsolidated gravel and sand bluffs of the borough.

The Aleutian Islands (RM 10.1)

The spectacular arc of Aleutian volcanoes extends Alaska's territory southwest into the Eastern Hemisphere a distance of 1,000 miles. The Aleutians are the southernmost part of Alaska, reaching south to latitude 50 degrees north, nearly the same as the Canada–United States border. All the islands are beyond the tree line due to the fierce and persistent winds that blow there. Located within the influence of the powerful Aleutian atmospheric low, the islands are subject to nearly perpetual horizontal rain during all seasons. Each year there are several days, sometimes in a row, that are totally clear and present unsurpassable vistas. Once home to tens of thousands of Aleuts, most present-day residents are either crabbers and bottom fishermen of diverse nationalities or U.S. Air Force personnel stationed at the remote post of Shemya. Adak Air Station, recently home to several thousand servicemen, once boasted a McDonald's restaurant but faced personnel cutbacks in the 1990s. The discussion here is limited to nonmilitary towns.

Dutch Harbor and Unalaska (RMs 10.1, 10.6)

While many cities can claim to have a Hard Rock Café, only Dutch Harbor has the Falling Rock Café, so named because shortly after its completion a large rock crashed through the wall and into the dining room! Perhaps symptomatic of the hazards of Aleutian living, the café was built immediately adjacent to a recently excavated rock cliff, which was still unstable from blasting. With 3,100 people, the twin cities of Dutch Harbor and Unalaska are the largest settlements in the rocky, volcanic arc of the Aleutian Islands.

Unalaska and Dutch Harbor are side by side on Unalaska Island and Amaknak Island, respectively, connected by a short bridge known as "The Bridge to the Other Side." The benefits of the harbor were first realized by Aleut people more than 4,000 years ago, as testified by the 50-foot-deep midden that lies near the bridge on the Amaknak side. The Aleuts developed an extremely specialized maritime hunting technology with elaborate hunting hats and watercraft. They prospered despite episodic burials by volcanic ash and periodic shortages when earthquakes altered the productivity of sea urchin beds or fishing grounds. By 1700, when Russian fur traders and missionaries arrived, the Aleuts had established eight villages among the adjacent islands. Aleuts served the Russians as fur seal hunters and translators. In return, the Russians offered Christianity and inspired the architecture of the dual-spired Orthodox church in Unalaska built in 1894. During World War II, Dutch Harbor suddenly became strategically important when the Japanese occupied two of the western islands (Attu and Kiska, more than 1,000 miles away), probably as a diversion to draw U.S. military forces away from the South Pacific. Whatever the reasons for the occupation, the Japanese threat was responsible for an overnight population boom in this area from a few hundred to 60,000 American servicemen.

Unfortunately, this enlarged population did not include the Aleuts, who were forced off to internment in former fish canneries, some as far distant as southeast Alaska, for the duration of the war. Now Unalaska–Dutch Harbor is a booming fishing center; hundreds of vessels call the harbor home during the season.

A great deal of dredging, filling, excavating, and flattening went on here during World War II, and has continued into recent years with the construction of modern port facilities (fig. 10.6). Liquefaction may occur in fill areas during an earthquake, increasing the extent of the damage. In addition, the military left behind toxic wastes in sufficient quantity to justify declaring the old facilities a superfund cleanup site. Evidence of the pollution problem is a small but perpetual oil slick in Iliuliuk Bay.

Residents recall feeling frequent small earthquakes, but these have caused little damage. The area has not had a major earthquake in recent times, although Atka and Adak suffered a damaging quake in the past. Large earthquakes are quite possible in Dutch Harbor–Unalaska, and plans should be made accordingly.

In an earthquake-prone area such as the Aleutian chain, tsunamis are also a real possibility. The Dutch Harbor–Unalaska area has an advantage in being located on the Bering Sea side of the volcanic chain, and thus is somewhat protected from the strong tsunamis generated throughout the Pacific rim. The fire department informs residents of tsunami alerts by sirens and loudspeakers urging all residents to move to elevations of at least 100 feet. The 1964 earthquake may have caused a seiche rather than a tsunami. Another seiche is possible if cliff collapse along the shoreline produces a large wave that sloshes back and forth in the bay. At least one such event occurred in the distant past, as evidenced by the rockpile at the base of the cliff along the east side of Iliuliuk Bay.

Every fall and winter, high winds—often in excess of 100 mph—cause damage in the community ranging from broken windows to lost roofs. The strongest winds are from the southwest. The damage could be much greater, but longtime construction practices here emphasize the principle of continuity (i.e., strong connections between all elements of the house from the foundation to the roof).

The thin tundra cover on the volcanic rock is normally quite stable, but human activities often destabilize it. For example, incorrect placement of culverts changes drainage patterns away from natural drainage channels and may initiate mudflows and severe erosion during very heavy rainfalls. A good example of this problem can be seen on the slope to the north of Unalaska Lake. A misplaced culvert on the road going up the side of the hill created a large and rapidly widening gully, which generated a mudslide that caused minor damage to the town garage. Another phenomenon that may

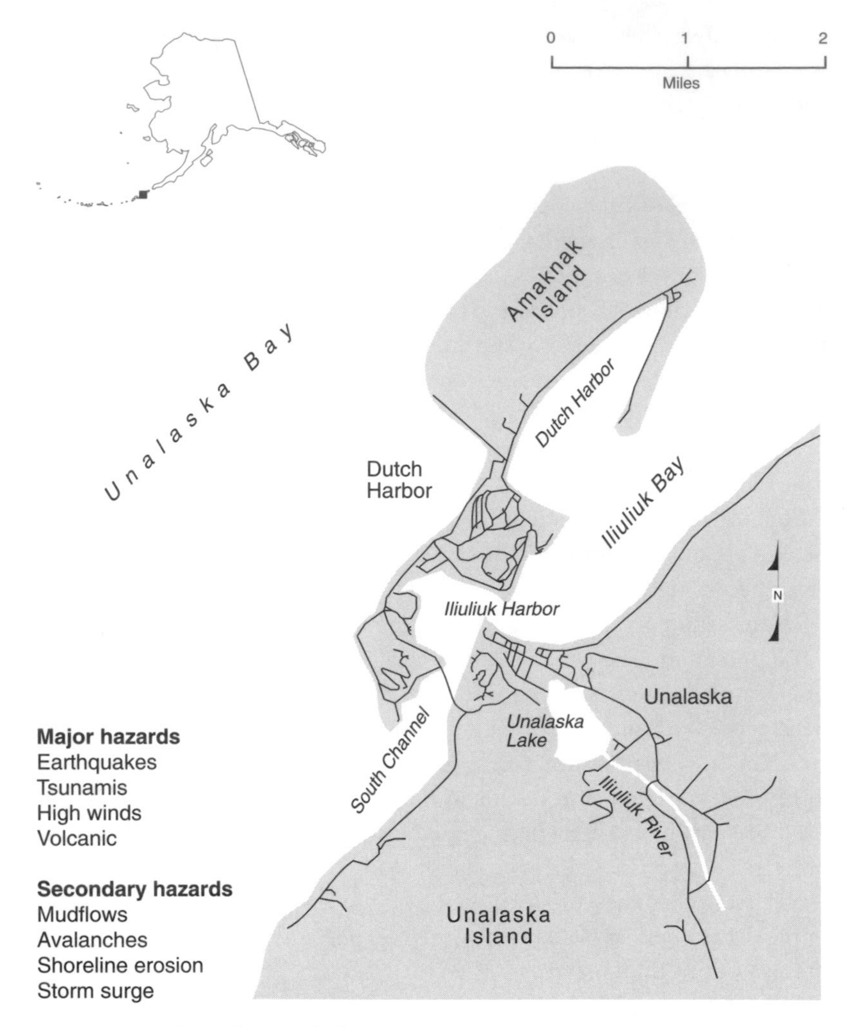

RM 10.6 Dutch Harbor/Unalaska

add to tundra destabilization is the burrowing activity of introduced ground squirrels and rats. Elsewhere, local slope failure was generated by notching the hillside for a homesite. Rather than notching, it might be better to build split-level houses while maintaining the original slope. Tundra erosion can also be minimized by building a rock revetment on the eroding cut face of a notch.

The Iliuliuk River, a small stream flowing through Unalaska, apparently offers only a minor threat, and then only under the most extreme rainfall conditions. Sheet flooding during heavy rains has caused some problems for flat-roofed buildings.

Many of the area's roads have spots at low elevations immediately adjacent to a bay shoreline (fig. 10.7). The roads are protected by dumped rocks

Figure 10.6 Fuel storage tanks in Dutch Harbor are susceptible to inundation during extreme storm surges and tsunamis.

Figure 10.7 The tundra-covered hillsides in Dutch Harbor are subject to minor slumping, endangering homes built on hillsides. The rocks dumped along the shore are probably inadequate to protect the road from erosion during storms.

(as opposed to placed rock) along the edge of the road, and moderate erosion damage is fairly common in storms. A quick look at the map of Unalaska–Dutch Harbor reveals that there are shorelines oriented in virtually every direction; any wind will form waves that impinge on some shore. The potential for shoreline erosion (and for storm surge, discussed below) on a particular shoreline segment is a function of fetch (the distance of

Figure 10.8 Unalaska has not witnessed major destruction due to storm surges for at least a century; the Russian Orthodox church (upper left) built in 1894 is still standing.

open water in the direction the wind is blowing from) and water depth in the bay (the deeper the water, the bigger the waves and the bigger the erosion and storm surge threats).

Storm surges accompanied by waves occasionally spill over into Unalaska but are rarely damaging to structures or property. The continued presence of the Orthodox church built in 1894 adjacent to the shoreline is proof that storm surges (and tsunamis) have not been major events on this shoreline in the recent past (fig. 10.8). However, the transport of drift logs (with attached barnacles) up hillsides and into valleys indicates that storm surge is a threat that should be taken more seriously by the population. In addition, storm surge waves have to return to the sea, and sometimes they do so very rapidly, even catastrophically, causing very rapid and dangerous currents, perhaps in a direction different from the upsurge direction.

In general, the slopes are too short and the snowfall is insufficient to produce large, damaging avalanches. Small avalanches, however, may be fairly common. One avalanche in 1990 knocked a parked car off a road, and several homeowners in the Iliuliuk Valley have reported avalanches striking their homes, but with little structural damage. Since development is increasing on the sides of this valley, this hazard may become more important in the future. The road to Summer Valley is sometimes closed by avalanches, and has been closed by a combination of shoreline erosion, mudflows, and avalanches.

Dutch Harbor and Unalaska sit atop an island arc constructed entirely by volcanic activity. The arc is still a very active volcanic area. Two small, fresh-appearing volcanic cones can be seen from town, and two active volcanoes, Akatan and Makushin, are in the immediate vicinity. The most immediate volcanic hazard is ashfall, and possibly acid rain. Major ashfalls have not occurred here in recent years, but layers of old volcanic ash are preserved in the tundra, indicating that they have occurred repeatedly in the past.

11 The Gulf of Alaska and South-Central Coast

Introduction (RM 11.1)

The southern coast of Alaska is formed by the huge circular curve of the Gulf of Alaska, defined by the narrow Alaska Peninsula on the west and the islands of southeastern Alaska on the east. The arc serves as a backstop for a major current system of the North Pacific Ocean. Currents in the gulf move counterclockwise, with the Alaska current heading north off southeast Alaska and then being diverted by the mainland to the southwest to form the Alaskan Stream. Local reversals that change the current direction to clockwise occur off Baranof Island and south of the Copper River delta.

The Gulf of Alaska is one of the world's major low-pressure centers and a cauldron for storms. Energy generated in the gulf affects the entire Pacific coast of North America. Storms are a constant feature of the gulf; a storm crosses the gulf nearly every five days during the six months of winter. These weekly storms, which can pack winds up to 80 mph, maintain a continuous cloud cover and generate a constant supply of relatively warm and water-laden air. The high mountains fringing the gulf act as a backstop, causing storms to stagnate, and precipitation in the form of snow or rain can cascade to earth relentlessly for days.

The Gulf of Alaska coast can be subdivided into five regions: (1) the Kodiak Archipelago and the very lightly populated south shore of the Alaska Peninsula; (2) Cook Inlet and the Kenai Peninsula, including Anchorage; (3) Prince William Sound; (4) the Yakutat–Icy Cape coast; and (5) the Alexander Archipelago. We will cover only the first four regions in this chapter; the fifth is discussed in chapter 12.

The western Gulf of Alaska coast takes two different forms: low bluffs of unconsolidated sandy silts and steep, rocky cliffs. The sheltered areas include both the sandy shores of Cook Inlet and the rocky and glaciated northern shores of Prince William Sound. The exposed shorelines include

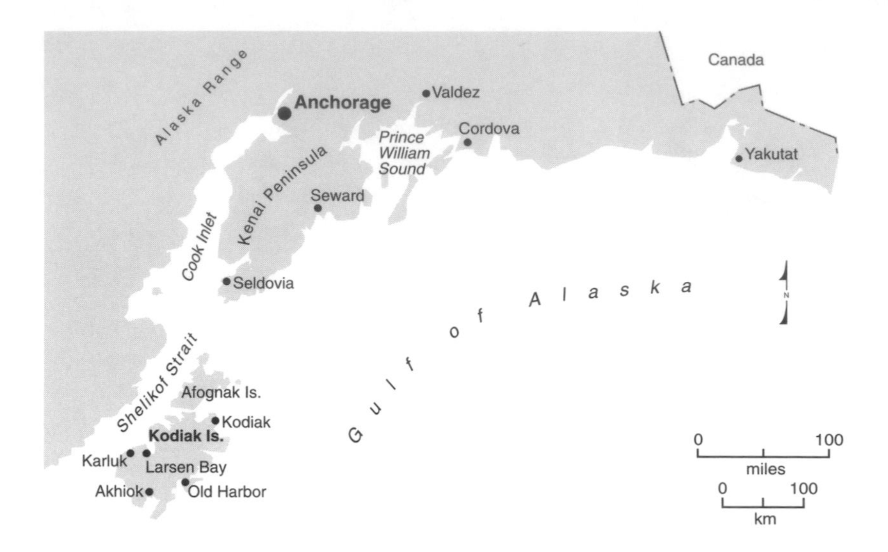

RM 11.1 Gulf of Alaska

the outer Kodiak Archipelago, the southern Alaska Peninsula, the outer Kenai Peninsula coast, and outer Prince William Sound. This region, which receives the full fury of the North Pacific, may be further subdivided by the western tree line of the temperate rain forest on northeast Kodiak Island. This division also marks the former limit of glacial expansion along the outer coast. The mountain arc of the Gulf of Alaska coast was heavily scoured by glaciers during the Pleistocene ice ages. When the continental glaciers melted and the sea level rose, the valleys became fjords.

Much of the nearly treeless Kodiak Island Archipelago was an ice-free refugium during the last full glacial expansion. However, valley glaciers did push southward from the Alaska Peninsula onto the western margins of Kodiak to discharge gravelly sediments that now form low, eroding bluffs. Kodiak Island is cut by numerous narrow arms of the sea, and its numerous estuaries are nurseries for salmon. Before overfishing damaged populations in the 1920s, the Karluk River on southwest Kodiak supported million-plus salmon runs. Swift and treacherous currents, coupled with fog, render Shelikof Strait between Kodiak and the Alaska Peninsula extremely hazardous, and it is deservedly legendary in navigation circles. The west shore of Shelikof Strait is the Alaska Peninsula, which is formed by the volcanic peaks of the Aleutian Range. This rugged and forbidding coast is largely uninhabited.

Cook Inlet is a constricted fjord estuary that extends more than 125 miles into the interior of south-central Alaska. Sheltered from open-water storms and tsunamis, the inlet is subject to high tides and seasonal ice cover. Large glacier-derived streams feed the upper inlet a steady diet of fresh water, sand, and silt. Much of this sediment is trapped to form tidal flats in the narrow arms of upper Cook Inlet.

Prince William Sound, east of the Kenai Peninsula, is sheltered from the open sea by a series of large islands to the south that impart a lakelike quality to the water body. The sound differs markedly from north to south and east to west. On the north, the mainland side, several large tidewater glaciers discharge icebergs into the sound; on the south, the islands are ice free. Glaciers once overtopped parts of the islands of Prince William Sound, as its uplifted shorelines indicate.

The comparatively straight Yakutat coast is backed by 15,000–18,000-foot mountains and immense piedmont glaciers, especially the Bering and the Malaspina, which is the size of Rhode Island. This coast is virtually uninhabited save for the small community of Yakutat. At the end of the Little Ice Age, 200 years ago, Captain George Vancouver described the Yakutat coast as a fearsome Greenlandic facade with numerous calving ice fronts in its embayments (e.g., Icy and Lituya Bays). Although the glacier front is several tens of miles farther inland today, the dangers from calving ice remain.

Kodiak and Adjacent Island Communities (RMs 11.1, 11.2)

The Kodiak Island Group consists of seven major islands and hundreds of smaller ones (RM 11.1). The largest is Kodiak Island (13,588 square miles), followed by Afognak (700 square miles) and Sitkalidak (117 square miles). Humans have occupied the Kodiak Archipelago for 8,000 years, and its archeology records an increasingly more complex succession of sea mammal hunters and fishers in larger and larger villages. By 1750, several thousand Koniag Eskimo occupied the most productive stream mouths. The Koniags were organized into small "nations" that featured elaborate status differences ranging from slaves captured in warfare to chiefs who organized trade and diplomacy with the Aleuts and Tlingit. The first European mission to encounter Kodiak was Vitus Bering's in 1741. Russian efforts at colonization did not solidify until Grigory Shelekhov (or Shelikov) established the first permanent settlement on southeast Kodiak Island in 1781, defeating and enslaving the Koniags.

Today, 15,000 people inhabit the islands, more than half of them in Kodiak City and its environs (RM 11.2) and the remainder either in rural areas or in the small towns of *Akhiok* (pop. 77), *Karluk* (pop. 71), *Larsen Bay* (pop. 147), *Old Harbor* (pop. 284), *Ouzinkie* (pop. 209), and *Port Lions* (pop. 222). The U.S. Coast Guard base at Monashka Bay has a population of 2,129. Many of these towns are on eroding bluffs susceptible to storm activity and lie within the 100-foot contour that indicates great susceptibility to tsunamis. Each community is examined in detail by the Kodiak Borough Planning Study.

The Kodiak region is one of the most active earthquake areas in the world (see chapter 3). The granddaddy of them all was the 1964 Good Friday earthquake. A discussion of the effects of this earthquake is the best way to review the nature of the earthquake hazard for Kodiak and vicinity. The quake was caused by movement along an unusually long stretch of the subduction zone, which explains why the damage was so widespread. For a variety of reasons, especially the fortunate occurrence of the earthquake at low tide, loss of life in the Kodiak Islands was slight. Much of the Kenai Peninsula and the Kodiak Island Group subsided from 1 to 6 feet. The degree of ground shaking depended on the nature of the underlying material. Buildings atop thick, water-saturated, unconsolidated deposits received the most severe shaking; buildings on top of solid bedrock suffered the least. Structural damage to buildings due to ground shaking was slight. Most of the damage at Kodiak from the 1964 quake was due to tsunamis (see fig. 2.4) and subsidence. Some liquefaction of unconsolidated sediments occurred, and this is believed to be the main cause of damage to the Kodiak Fisheries Cannery. Strong currents in the port channel off Kodiak removed the thick sediment layer in which pilings were buried and made future construction of port facilities more costly. In all, the quake caused $45 million in damage in the Kodiak Archipelago.

Tsunamis have been and will continue to be a major hazard on Kodiak. They can alter and transform the shoreline catastrophically within minutes. In 1964, spits, barrier islands, and bay-mouth bars were considerably modified. At least nine tsunamis, mostly small ones, have been recorded in the Kodiak Island Group since 1788. The tsunami that struck Kodiak in 1964 was the best documented and also the largest in historic times. Before that, the largest recorded tsunami wave height in Kodiak was 2.3 feet, the result of the world's largest recorded earthquake, the 1960 quake in Chile. The waves generated by the 1964 tsunami were 2 to 15 times as high, ranging from 5 to 31.5 feet along the southeast-facing coasts. A great deal of beach sand was lost, but much of this sand later returned naturally during calm seas.

All seven Kodiak Borough communities experienced some shoreline inundation during the 1964 quake, and Kodiak and Ouzinkie suffered major property damage. In Kodiak, 215 structures were destroyed, among them 30 homes from Shahafka Cove that were washed into Potatopatch Lake by a 30-foot wave. The principal damage downtown was caused by boats that slammed into town on the tsunami wave. Old Harbor was almost completely destroyed; only the church and school remained. The very small village of Kaguyak was destroyed and abandoned. The highest run-ups from the tsunami in Kodiak ranged from 13 to 30 feet above mean low water. In Old Harbor, the maximum run-up was 22–30 feet, and in Ouzinkie, 16–22

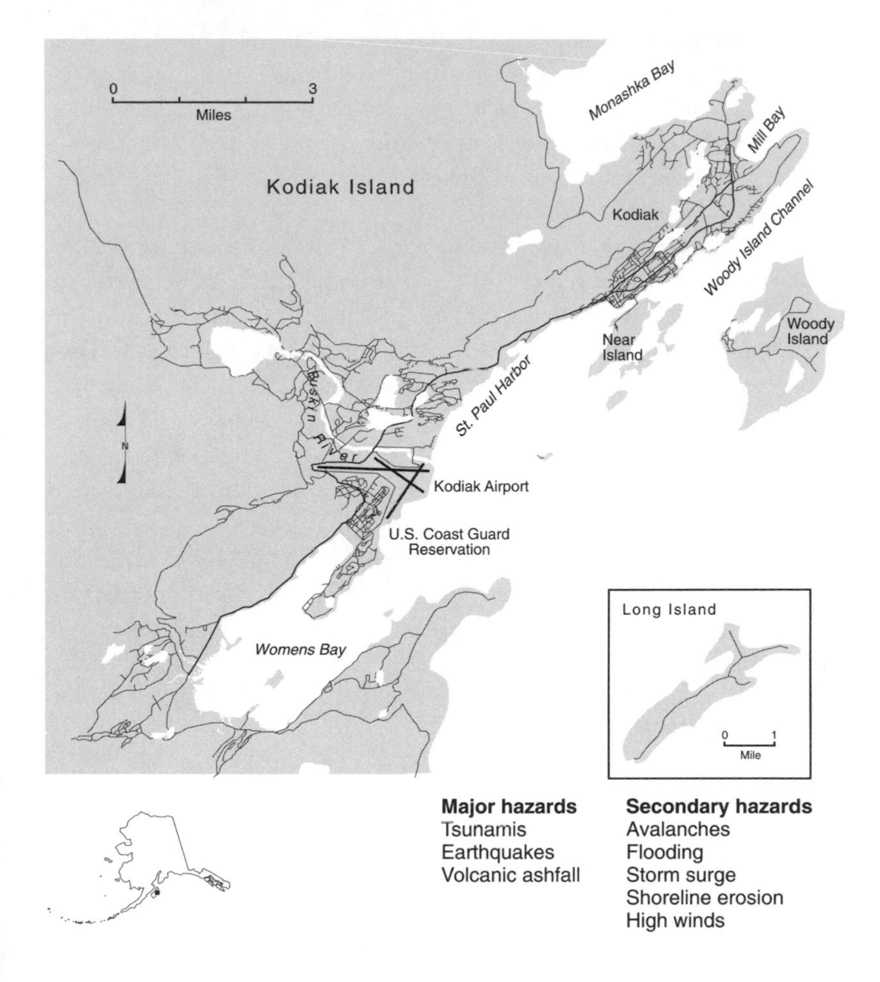

RM 11.2 Kodiak

feet. The other communities of the Kodiak Borough are not on the southeast-facing coast and received much less inundation.

Locally generated tsunamis are also a possibility in Kodiak Borough. The key to avoiding tsunami damage is elevation; the higher the better is the rule for homesites. Remember that if a future tsunami hits at high tide—or even worse, at a high spring tide—the damage could be much more severe than in 1964. Residents should also remember that Kodiak is poised to receive tsunamis from the Ring of Fire across the eastern Pacific, as happened in 1960.

The volcanic hazard is another aspect of Kodiak life familiar to all who know the area's history. In 1912 the most violent volcanic eruption in the recorded history of Alaska occurred: the misnamed Katmai event (it was the adjacent Novarupta volcano that erupted). Six cubic miles of ash and rock were produced and blown or washed away to the north and southeast. The northern tip of Kodiak Island, more than 100 miles away, received a foot of ash. Ash thickness within the town of Kodiak ranged from 1 to 3 feet—enough to collapse some roofs. Corrosive acid rain fell after the eruption, some of it as far away as Cordova.

There was doubtless a great deal of respiratory distress among local inhabitants at that time, similar to the problems experienced by Kenai citizens in the more recent eruption of Mount Redoubt. The ashfall from the Novarupta eruption killed most of the plants except those, such as willows, that were able to extend through the ash, although a great deal of the vegetation recovered within three years. The eruption affected world climates for at least two years afterward. Residents of many south-central Alaska towns find disruption of air travel to be one of the most threatening aspects of volcanic eruptions.

Landslides pose another threat to residents of the Kodiak Archipelago. Landslides are of many different types, determined by all sorts of factors, including the slope angle, the moisture content of the soil, and the type of surface rock and soil cover (see chapter 4). In general, landslides can be triggered by rainfall or caused by the shaking action of earthquakes, as is often the case in Kodiak.

The 1964 earthquake triggered thousands of landslides and avalanches, mostly along the southeast side of the island and predominantly in areas underlain by rocks of Tertiary age. The landslides were not only abundant, they were often larger than any previously witnessed. In addition, landslides continued at a much higher frequency in the months immediately following the earthquake than before it. The Pillar Mountain landslide, well known to the residents of Kodiak, destroyed a portion of Rezanof Drive in the early 1970s. The path of the slide is still easily visible along the highway today. If the slide had been larger, it might have created a local tsunami in the harbor. Fear of such an event caused local engineers to notch the base of the slide and construct a wall made of gabions, providing space to accommodate a future slide rather than allowing it to flow into the harbor.

On Halloween night 1991, the soil cover on the hillside above Hillcrest Avenue in Kodiak failed after 7–9 inches of rain fell within 24 hours. Saturated soil and rocks slid down the mountain as a muddy, jumbled mass, destroying three houses and injecting mud into the living rooms of a number of others. Scars at the headwall of the 1991 slide are still distinguishable on the steep hillside. Whether such events are important elsewhere in the Kodiak Archipelago can best be judged by the presence or absence of scars on the hillsides (see chapter 4). Landslides, despite sporadic occurrences, do not appear to be a major constraint on development.

In general, snow and ice avalanches are not a hazard to the communities of the Kodiak Islands because most of them occur in the interior where no one lives. Occasionally roads or tracks in the interior have been closed by avalanches.

Streams in the borough are characterized by small drainage basins, short lengths, and steep gradients. Floodplains tend to be narrow. Floods are usually flash floods that result from heavy rainfall or, in winter and spring, from melting or breakup of ice dams impounding temporary lakes. The Womens Bay (a name inherited from the early Russian settlers) neighborhood, or Bells Flat, is the only part of Kodiak in danger from river floods. Some houses are in the floodplains of two streams: Russian Creek and Sargent Creek. The worst-case scenario for flooding in Kodiak is the failure of the Lake Bettinger Dam (the lower reservoir), which would inundate a number of buildings along a 2-mile path to the bay. Currently, the reservoir is used only for flood control and is empty most of the time. The lower basin is there to act as a catchment reservoir if the dam of the smaller, upper reservoir should fail. A major earthquake would have to occur during a time of unusual rainfall, causing both dams to fail, for the worst-case scenario to come to pass. Other than Kodiak City, Port Lions is believed to be the only outlying town with an important river flooding problem. Here, channel icing can clog the river and cause flooding.

Storm surge flooding is a problem only on the shoreline and in bays facing the open ocean. In general, the adjacent continental shelf is narrow and steep, and storm surges are expected to be small—perhaps on the order of 2 feet. Although the prevailing winds issue from the west-northwest, high winds from the east occasionally pound Kodiak's shores (see fig. 5.1). High east winds can make it necessary to evacuate small boats from the harbor.

Shoreline erosion accelerated along many Kodiak shorelines following the 1964 earthquake due to the relative sea level rise as the land subsided. In Karluk this led to erosion of the spit, paving the way for its eventual breaching during a storm in 1970. The breach accelerated erosion in front of the town. As a result, the whole community was moved upriver to less erosive ground. As of 1994, only the old church, built on high ground, was still in use in the old townsite. Virtually all of the communities in the Kodiak Island Group have some shoreline erosion problems. In most cases, the erosion threatens only a few buildings and the problem is most efficiently and cheaply solved by moving the buildings. Elsewhere, in undeveloped locations, erosion proceeds at a varying pace, but no problem exists because no buildings or roads are threatened.

During the fall and winter, Kodiak may be beset by very strong sustained winds ranging in force from 75 to 115 mph. During the worst winds, roofing may be blown off; sometimes entire roofs have been lifted off, along with other pieces of houses. The resulting airborne debris is hazardous to adjacent buildings and unwary pedestrians. Perhaps equally hazardous are the shallow-rooted spruce trees, which can be blown over, although much of Kodiak is beyond the tree line. Spruces are especially prone to blowdown when forests have been thinned by development. Falling trees have caused serious property damage and have also been responsible for power outages. Good construction practices that follow high-wind building codes are the best way to mitigate the wind hazard, but preparing for a mature spruce to fall on your house is more difficult!

At Port Lions, a large breakwater protects the harbor entrance and marina. The breakwater was constructed in the early 1980s by the U.S. Army Corps of Engineers at the cost of several million dollars. The breakwater failed, residents and borough officials report, within two months of completion. Local residents had cautioned about the likelihood of collapse in the strongest terms, but the Corps persisted in the project nonetheless. The breakwater was subsequently rebuilt as a shorter structure because of money problems.

The oil spill hazard to Kodiak arises because of its position downwind of the shipping lanes out of Valdez and Cook Inlet. The direct effects from the *Exxon Valdez* spill were comparatively minor in terms of oily beaches and dead animals, but residents felt the catastrophe very emotionally, and a sense of violation still persists in the native community. Subsistence resources are used with trepidation by some residents.

Cook Inlet: Anchorage and Kenai Peninsula (RM 11.3)

Geologic Setting

Cook Inlet defines a large estuary connected to the Gulf of Alaska; it is 20–50 miles wide and 175 miles long southwest to northeast. The inlet, the size of Maryland, is a funnel-shaped wedge of saline water that reaches far into central Alaska and mixes with fresh water from the north. The waters in the upper inlet can be extremely fresh; the Tanaina Indians considered the upper inlet part of the rivers to the north. These rivers impart a deluge of fresh-ground glacial silts and sand into the two arms of the inlet: Turnagain and Knik. The sediment overwhelms the tidal currents in the inlet and settles in vast and treacherous tidal flats.

Cook Inlet occupies a structural basin formed about 50 million years ago during the Tertiary period. The coasts along the margins of the inlet differ drastically. The mountainous Alaska Peninsula on the west side exhibits an irregular outline because the shoreline is pockmarked with small, drowned glacial valleys; on the east side, the Kenai Peninsula is composed of relatively steep, straight bluffs formed from sands and silts deposited by glacial

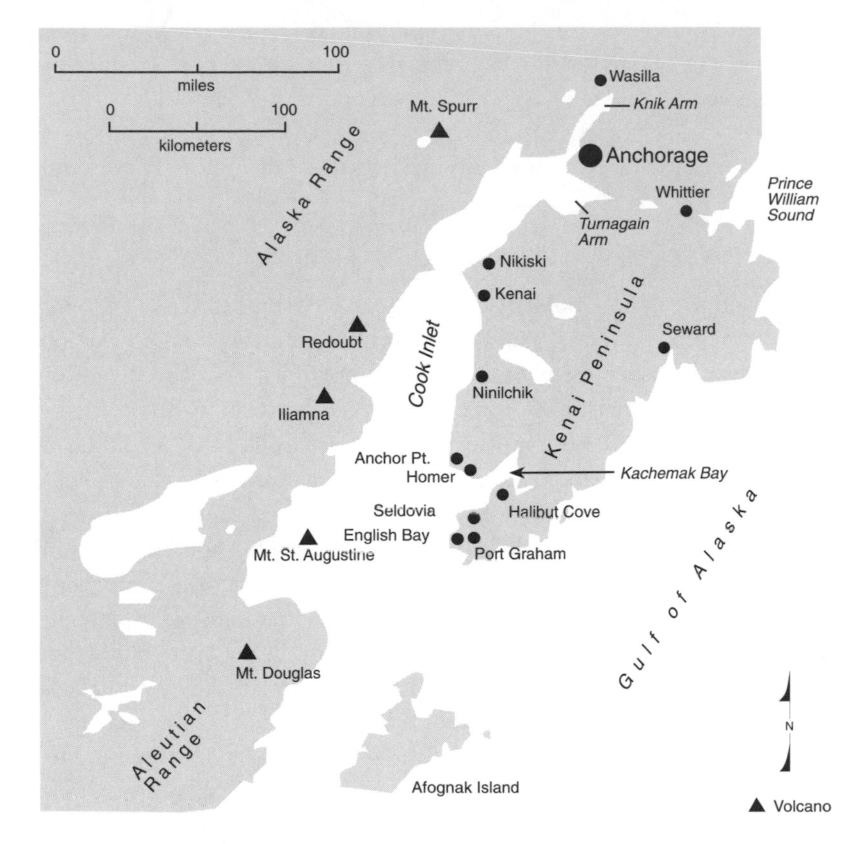

RM 11.3 Cook Inlet and West Kenai Peninsula

rivers and moraine-dammed lakes. A series of active faults stitch together the mountainous massifs surrounding the inlet—the Aleutian chain to the west, the Chugach and Kenai Mountains to the east.

Five active volcanoes break through the Aleutian Range just west of Anchorage; a cluster of 10 more is located farther south. Our own human scale is easily overwhelmed by the cataclysmic power of the steam clouds and ash that issue from these mountains. In 1986, a Dutch jetliner careened down 15,000 feet in a matter of seconds after ash from Mount Redoubt entered its engines (which were restarted just in time). Several times in the last decade, Anchorage and Kenai Peninsula streets and skies were transformed into a Pompeiian interlude of ashen twilight and shifting sands. The geologic record indicates much larger events than these in recent millennia. In 1912, the Alaska Peninsula's Novarupta volcano produced one of the largest eruptions this century, releasing 30 times the volume of ash produced by Mount Saint Helens! Geologists continue to accumulate evidence on the internal dynamics and history of the Aleutian Range volcanoes, some of which are quite old. Mount Redoubt started to build 900,000 years ago, while the oldest Mount Spurr flows are about 250,000 years old. Although Mount Saint Augustine had erupted with some regularity every 40 years or so since the first observed eruptions in 1812, the 1986 eruption still came as a surprise. The Cook Inlet volcanic eruptions may not be as catastrophic as those at Pompeii, but their more frequent occurrence makes them just as hazardous.

Anchorage (RM 11.4)

In terms of population, politics, and finance, Anchorage is Alaska's preeminent metropolis. It contains nearly 45 percent of the state's total population. Mountains are visible from the city on all horizons. The jagged but low peaks of the Chugach Range form the eastern horizon. To the west, the Aleutian and Alaska Ranges appear to float above Cook Inlet. Anchorage is the late bloomer of Alaska's towns; as recently as 1940 only 4,000 people lived in the area. The city owes its start to two circumstances: it is the last ocean-navigable location on the Alaska Railroad, built in 1914 from Seward to the Fairbanks gold fields; and it was among America's northernmost outposts during World War II and the cold war years that followed. The population explosion began during the war years, and by 1960 the population had grown to 44,000. The 1964 earthquake devastated the city, but little trace of that calamity is evident today, except for regraded slopes and an occasional abandoned building. Most of the housing tracts destroyed in 1964 were rebuilt in nearly the same locations as if nothing had happened. The influx of oil money in the 1970s generated another profound expansion in Anchorage. The population had swelled to nearly five times its previous size by 1980, leading to widespread development on the sizable peninsula formed between the two arms of Cook Inlet.

In the 1990s, the Anchorage skyline is dominated by multistory hotels and buildings bearing the logos of oil companies and banks (fig. 11.1). Government of all stripes provides a considerable amount of the economic energy in town. The U.S. Army maintains a strong presence at Fort Richardson just north of downtown, and the U.S. Air Force has a base at Elmendorf. Anchorage International Airport is the nexus for most Alaska travel, despite its loss of international flights following the opening of Soviet air space. Malls and shopping centers are scattered across the landscape. Anchorage's pride is its extensive park system, which parallels small creeks that cut through the bluffs that outline the city. Bike paths are ubiquitous throughout town, including a coastal bike path that hugs the bluffs.

The hazards that threaten Anchorage (fig. 11.2) are among the most ferocious on the planet: earthquakes, avalanches, high winds, tides, and even storm surge, to some extent. While development in the last 30 years has brought new structures untested by seismic forces, the 1964 quake also brought the town into the modern age. Anchorage now has a disaster plan, a

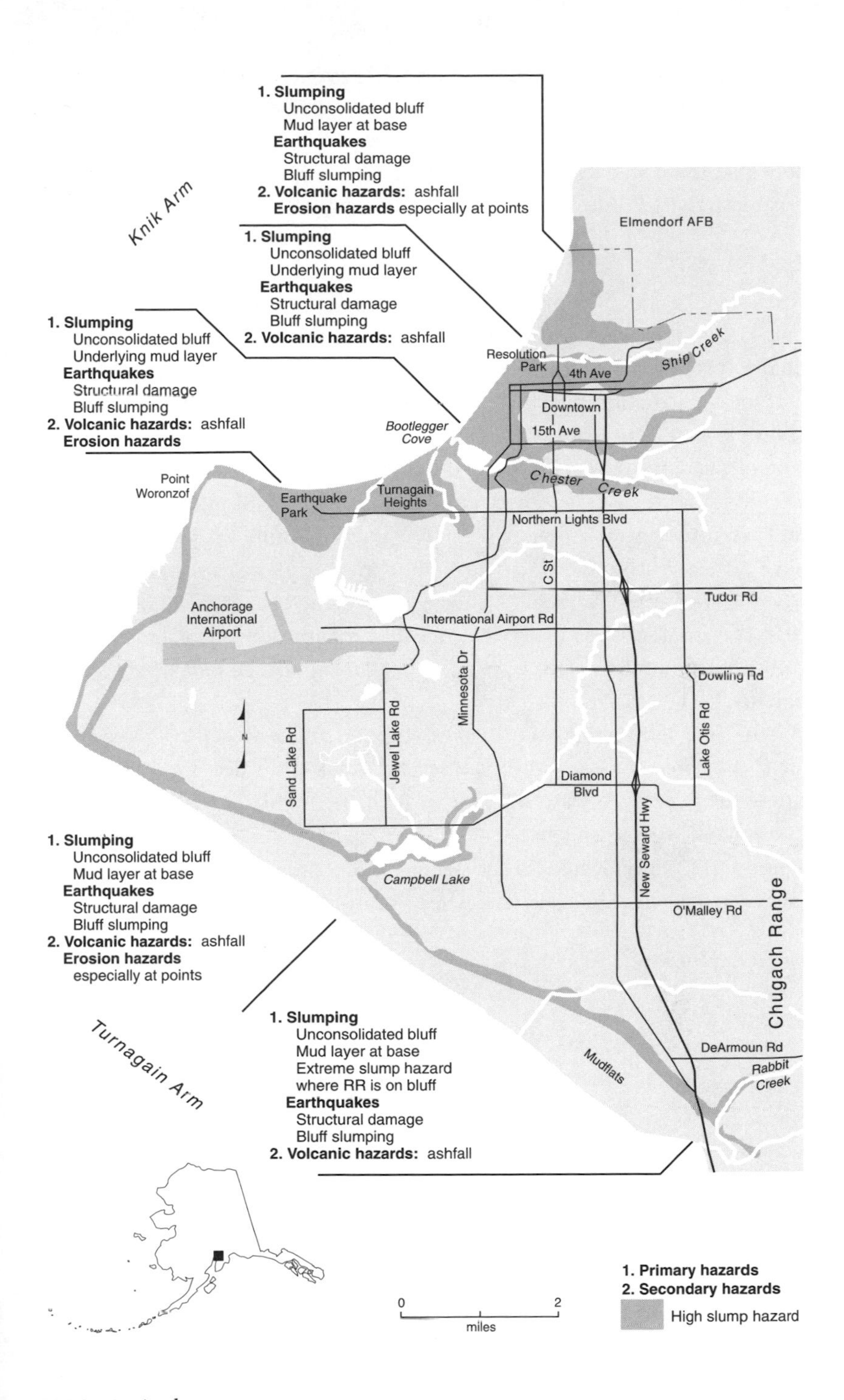

RM 11.4 Anchorage

civil defense coordinator, and comparatively strict building codes. However, Anchorage residents have constructed a culture of denial and dissociation. "At least we don't live in California," they say. In fact, Californians are probably glad to be able to say, "At least we don't live in Alaska." The amount of seismic activity in Alaska puts California to shame.

Geologic Setting

Anchorage residents sit on a geologic hot seat just west of the site of the second-largest recorded earthquake in North America: the 1964 Good Friday quake. It is not widely realized that Anchorage has the same earthquake rating, for construction purposes, as Los Angeles. Anchorage owes its vulnerability to its position at the eastern margin of the Cook Inlet forearc basin, part of two fault systems associated with the Pacific Plate as it is being subducted under the North American Plate (see fig. 3.1). More than 20 powerful (magnitude greater than 6.0) earthquakes and more than 100 small ones have rattled Anchorage since 1900. Four magnitude 7.0 or greater quakes have hit within 100 miles of Anchorage at a recurrence interval of only 17.5 years. Seismologists estimate that another magnitude 7.0 or larger quake could strike anytime (see appendix C, *Earthquake Alaska: Are We Prepared?*).

While the subducting-zone quakes of the Aleutian Megathrust (see chapter 3) may pose the greatest threat, several other faults deserve serious consideration as well. At least three major faults lie within 30 miles of Anchor-

Figure 11.1 Downtown Anchorage, set on unconsolidated bluffs and fronted by wide tidal mudflats, silhouetted against the Chugach Mountains.

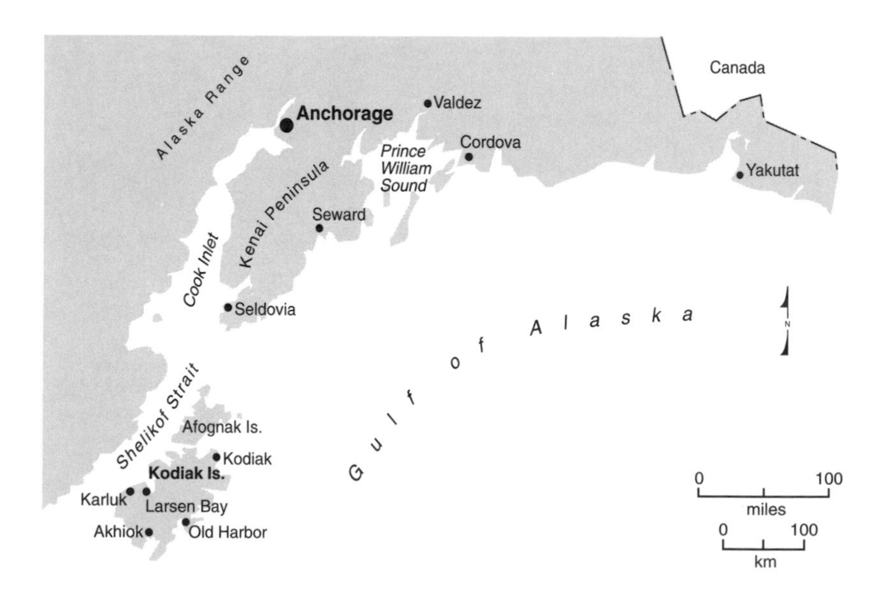

Figure 11.2 Development adjacent to the Elderberry Park area, downtown Anchorage. The L Street slump was the result of extensive bluff collapse in the 1964 earthquake, and the area is still subject to earthquake-related slope failure. Note the rock revetment in the foreground.

age: the Bruin Bay fault to the southwest, the Castle Mountain fault to the north-northwest, and the Border Ranges fault at the eastern margin of Anchorage. Geologic evidence indicates that major displacement of the Castle Mountain fault occurred between 225 and 2,100 years ago; a 5.7 magnitude quake was recorded in 1984. Too little is known to predict when and if future activity will occur. Seismologists do know that most earthquakes are associated with the plate boundary, and while some quakes occur closer to the surface than others, the relationship with known faults is not exact. Many of the recent small quakes have occurred in areas between the faults, both west of Anchorage and north of the Castle Mountain fault near Palmer. Geologists lack instrument data for two 7.0+ quakes that occurred in the 1930s and 1940s; these could have been associated with the Castle Mountain fault. Anchorage residents should bear in mind that the Northridge, California, earthquake of 1994 was caused by a previously unknown fault.

How will Anchorage residents weather another major quake? That will depend on the precise location, the type of building construction, and the type of underlying sediments. Let us first consider the geologic background of the Anchorage area.

Solid bedrock can only be found far beneath the surface of modern Anchorage, at depths of several hundred to 1,000 feet. For the most part, Anchorage rests atop glacial and near-glacial deposits that are comparatively young and unconsolidated. Poor consolidation means that the sediments are soft and are subject to considerable deformation by earthquake-induced vibrations. The degree of deformation depends on the particular mixture of grain sizes and the distribution of the rock layers. Some layers act like a slippery, greasy surface on the bottom of a stack of dishes while others grip like absorbent cotton.

One sedimentary layer is the major source of problems for Anchorage builders: the Bootlegger Cove formation (see fig. 3.2). It is an extremely fine-grained clay that originated during the last glaciation and is named for the adjacent Bootlegger Cove. Because of its lack of strength, the clay is a plane of weakness for slope failure and movements. The sand overlying the clay is saturated by water and can undergo a rapid buildup of pressure within its interparticle spaces when subjected to earthquake vibrations.

Understanding the complexities of the geologic past can help residents understand and predict how individual homesites may respond in an earthquake or other catastrophe. With this in mind, consider the following brief history of landforms in the Anchorage area. Chapter 1 discusses Alaskan geology in more detail.

During the Pleistocene Ice Age, from 28,000 to 12,000 years ago, large glaciers entered the Anchorage bowl from the north and east. The largest glaciers originated in the north but only occasionally deposited moraines at their margins. More frequently, silty sand and gravel came from the ice front that calved into shallow sea water. The Anchorage of 15,000 years ago resembled the fjords of Prince William Sound. However, the present surface of Anchorage was considerably lower then—by 240 feet—due to the weight of the ice masses on the surrounding hills. As the glaciers receded, the newly uncovered land started to rise. This process of recovery, or isostatic rebound, was most rapid in the years just after the glaciers receded, but it still continues today at a small but perceptible rate (inches per decade). The sea level rose worldwide as ice melted from the glaciers. Although the glaciers were no longer bringing material to the rising Anchorage surface, up-valley glaciers continued to bring sediments to the adjacent valleys, Turnagain Arm and Knik Arm. In fact, more than 1,000 feet of sediment has filled Turnagain Arm over the last 14,000 years. Part of this filling occurs because the ground subsides several feet after each earthquake, allowing the tides to redistribute the material; in just 20 years, the sediments brought in by the tides have matched the subsidence of the 1964 earthquake.

Anchorage sits atop a complex array of sediment types because of its former position at the mouth of a tidewater glacier (fig. 1.5). As you might imagine, in this setting, sediment-laden waters discharged from glaciers interact with tidal currents, wind-generated waves, and a host of other factors. Much of the sedimentary material formed as a subaqueous delta when sand

and silt from beneath the glacier encountered the ocean. Icebergs carrying larger materials such as pebbles and rocks may have occasionally rafted in sediments. During winter, the water was probably stilled by its ice cover, and fine particles had time to settle out of the water column. The result is a complex melange of sands, silts, and clays.

The prospective Anchorage home buyer or resident confronts a bewildering array of geologic maps and pamphlets from the U.S. Geological Survey, the state of Alaska, and the municipality of Anchorage (appendix C includes an annotated list). The most useful maps show slope stability and foundation conditions. An important consideration for a property owner in Anchorage should be the distance of the property from the slope break.

Most property "with a view" in Anchorage lies within 100 yards of the spectacular sand and gravel bluffs above Cook Inlet. Bluff margins are the weakest sites in the area. As mentioned in chapter 3, the effects of the 1964 quake were greatest along the northern margin of downtown, where several very large slumps occurred. The potential for future slumps in this area still exists. Although insurance agents, bankers, and property owners may pretend otherwise, the land under the downtown hotels is the most hazardous in Anchorage. While most structures are designed to meet earthquake codes, major vertical shifts could occur under the buildings. The goal of the code is to ensure that buildings are able to withstand major quakes without collapsing—but that is not to say without sustaining any damage. Further, vibrations set up in one building by seismic waves may affect buildings nearby. Other earthquake complications Anchorage residents may face include widespread power outages, disrupted roads, collapsed bridges, and a damaged and closed airport.

Other Hazards in the Anchorage Area

Prospective residents should realize that a number of hazards other than earthquakes and volcanic ashfall may also affect property "with a view." These include slope failure caused by storm erosion. In an area west of Anchorage International Airport, several hundred yards of bluff have eroded in this century, much of it since 1949. The bluffs erode when high-energy storms enter Cook Inlet and generate large waves at their bases. Storms arriving in the fall are the most dangerous because the bluffs are not yet frozen and their sediment can be easily eroded.

Tidal forces in Cook Inlet are legendary—among the greatest in the world. Tides in Turnagain Arm extend over a range of 27 feet from low tide to high, the height of a two-story building. The tides occur twice a day (two high, two low). The Cook Inlet tidal flats are treacherous in the extreme. Although the silty clays appear firm, any pressure can cause them to liquify and turn into a quicksand-like mush capable of swallowing pedestrians. In the late 1980s, a woman trapped on the flats could not be rescued, even by a helicopter and trained emergency personnel; she eventually drowned at high tide. Her death should be a warning to the casual explorer.

Since the oil boom of the 1970s, people are occupying the lower slopes of the Chugach Range, east of Anchorage, in increasing numbers, both for housing and for short-term uses such as skiing and hiking. Slopes contribute to hazards by channelizing airflow and providing a source of potential energy for destabilized ground or snow. Debris slides are not as great a problem in Anchorage as they are in southeast Alaska, although they do occur. Intense winds and avalanches pose greater hazards for people. High winds (more than 100 mph) can extend far into the Anchorage bowl area and should influence construction in areas that are mapped as high-wind hazards by the municipality of Anchorage.

Avalanches are a particular danger in the mountains east of Anchorage. Major avalanche activity in Anchorage occurs when heavy snows follow a wet snow that fills irregularities on the surface and is succeeded by a cool, clear period that changes the structure of the snow. Fortunately for mountain residents and users, the municipality contracted for analysis and mapping of high-risk areas. Eagle River has witnessed repeated avalanches; a number of them occurred during 1979–80. The Mile High subdivision and milepost 9 on the Eagle River road appear to be particularly susceptible. Avalanche tracks, as marked on U.S. Geological Survey topographic maps, often follow topographic features that channel flow. Home buyers on the Eagle River road are advised to examine these topographic maps carefully. A second area of avalanche risk is on the south margin of Anchorage along the north shore of Turnagain Arm. This area includes several drainages, notably Rainbow, Indian, Bird, and Glacier Creeks. Glacier Creek is west of the Anchorage ski resort of Girdwood, itself the site of several large avalanches. The first avalanche to reach the base of the ski resort was an earthquake-induced event in 1964; subsequent events in 1973 and 1975 damaged or derailed chairs and cables on the lifts. To the west, near Indian Creek, an avalanche in 1975 blocked the Seward Highway near milepost 105.

For information on individual homesites or travel routes, home buyers, owners, and recreationists should consult the series of large-scale (1 inch to 500 feet) avalanche hazard maps prepared for the municipality of Anchorage that are available at the Department of Community Planning and area libraries.

North Shore of Knik Arm (across from Anchorage) (RM 11.4)

This long reach of shoreline is lightly developed at present, but increased development is expected in the future. The area is immediately south of the Castle Mountain fault and has a strong likelihood of future quakes; some may be sizable. Most of the shoreline is lined by 30–50-foot bluffs of unconsolidated glacial material. Where they are eroding at a significant rate (ca. 2

feet per year), the bluffs are unvegetated and steep. Where bluff retreat is minor to negligible, the seaward-facing slope is gentler and extensively vegetated. The nature of the Knik Arm shoreline adjacent to the base of the bluff determines whether the bluff is eroding. Wide salt marshes and mudflats adjoining the upland prevent wave and tidal current attacks on the bluff and promote gentle, well-vegetated slopes. An example of this is the Settlers Bay Village development upstream from Knik. At the town of Knik, a tidal channel abuts the shoreline and waves and currents remove bluff material at high tide. No marsh or mudflat is present in front of the steep slope, and the bluff is retreating.

Independent of the erosion problem is the problem of slope stability on the ubiquitous bluff along the north side of Knik Arm. The bluff is made of unconsolidated glacial sediments of highly variable grain size. The bluff margin and face are susceptible to slumping, especially in earthquakes. The best construction sites are well back (100 yards) from the lip of the bluff. This will not allow a view of the sea, but it will provide some security for your home. So far, new development in this area, as elsewhere in America, has crowded as close to the sea as possible. A number of homes in Settlers Bay Village are built on the seaward slope of the bluff. This is not a safe location. If you still wish to buy a site with a view of the sea, foundation design should be a major part of your overall architectural thinking. This entire area is subject to earthquakes, and earthquake-resistant construction is highly recommended everywhere along the north shore of Knik Arm.

The possibility of glacial outburst floods is a real one for residents of the Mat-Su valley, the Knik River region and adjacent shores of the Knik Arm. Glacially dammed Lake George in the Chugach Mountains has repeatedly filled and emptied catastrophically during the last 30–35 years.

North Anchorage: Birchwood Subdivision (RM 11.4)

Much of this coast is low in elevation and densely wooded, and bordered by broad marshes and mudflats. There is little or no slumping hazard, and shoreline retreat is minimal. Current development is well back from the shoreline, and access to the coast is infrequent. The low elevation makes shoreline development here susceptible to storm surge from very major storms or a large tsunami at high tide. Subsidence on the tidal flats may accelerate erosion of the low slopes in the area. Earthquakes are a hazard here as they are in this entire region, and all construction should be earthquake resistant.

Anchorage: Government Hill and Port Area (RM 11.4)

The crest of the bluff here is largely covered by buildings. The "shoreline" consists of harbor facilities and is not subject to erosion. The bluff was the site of a major slump (the Government Hill slump) during the 1964 earthquake. This 11-acre, 1,180-foot-wide slump involved 900,000 cubic yards of material. A significant possibility exists that a slump will occur during the next earthquake as well. Future construction should avoid the 300-yard-wide zone extending landward from the base of the bluff. Earthquakes are a real possibility here. Planners and residents should prepare for them by choosing "safe" sites and by insisting on earthquake-resistant construction.

Facilities in the port area face major problems during and following earthquakes. The docks and cranes would probably collapse and be rendered useless for days or longer. The numerous storage tanks and hazardous chemicals are susceptible to rupture and, of course, fire. Disaster planners and casual visitors should keep both possibilities in mind.

Downtown Anchorage: Ship Creek to Chester Creek (RM *11.4*)

Most of this reach was affected by the massive L Street slump in 1964 (fig. 11.2), which displaced material seaward from the net of fractures at the crest of the bluff. A great deal of damage was done to downtown buildings and roads, some of it extending 4 blocks inland. The slump affected all or part of 30 city blocks (72 acres) extending 4,800 feet along the bluff top. The total volume of material involved in the slide was about 6 million cubic yards. The smaller Fourth Avenue (fig. 11.3) landslide, which moved material northward into Ship Creek, involved only 14 city blocks (36 acres) and 2 million cubic yards of material (see appendix C, *The Alaska Earthquake Series,* 542-A). The Alaska Native Health facility, which emerged relatively intact from the 1964 earthquake, is at risk from another quake.

Foundation excavations near Elderberry Park reveal geologic evidence of multiple slides, including a chaotic mass of sediments and buried trees, indicating that this area has been unstable for a long time (see appendix C, *A Field Guide to the Geologic Hazards of Anchorage and Turnagain Arm*). Many geologists believe that renewed movement of this slide is highly likely in the next large earthquake. The *Slope-Stability Map of Anchorage and Vicinity* (see appendix C) classifies the stability of a 300-yard-wide area on the bluff slope and crest of this shoreline reach as only moderate but with risk of large landslides. The band of moderate slope stability extends inland along the slopes facing Ship and Chester Creeks, but includes pockets of high risk slopes. Foundation conditions are poor in the two 1964 slump zones, and special design precautions must be taken before construction there. The 300-yard strip of moderately stable land, especially the part landward of the railroad tracks, should be considered high risk for construction sites. The entire reach is buffered from erosion and flooding by the Alaska Railroad right-of-way, which follows the base of the bluff.

Residents and property owners should remember that Anchorage had few high-rise buildings in 1964, and the effect of a large quake on the present towers is uncertain. It is noteworthy that the oil companies have

Figure 11.3 The bluff along 4th Avenue slumped severely during the 1964 earthquake and has since been regraded. This area remains a high-hazard slump zone in future earthquakes.

built their buildings well inland from the slump margins—a standard worthy of emulation.

Before purchasing or further developing any property in this portion of Anchorage, study the hazard maps in the *Anchorage Coastal Resources Atlas* (available at the city planning department and libraries; see appendix C) or the maps available from the U.S. Geological Survey office in Anchorage.

West Anchorage: Turnagain Heights and Earthquake Park (RM 11.4)
This shoreline reach comprises the entire 8,500-foot-long, 130-acre Turnagain Heights landslide. The massive slide occurred during the 1964 earthquake and destroyed 75 houses in one of Anchorage's most upscale neighborhoods (see appendix C, *The Alaska Earthquake Series,* 542-A). Some of the houses were carried more than 500 feet and ended up well seaward of the original shoreline. The actual cause of the landslide was slippage within the Bootlegger Cove clay formation, which acted as a slippage plane for the massive slide.

From the hazards standpoint, the situation is best summed up by the National Research Council's report on the earthquake, *The Great Alaska Earthquake of 1964* (see appendix C), which recommends no new building or rebuilding in the Turnagain Heights area. It is immediately obvious that this advice has not been followed. Future earthquake activity is a real possibility here. Prepare for it by proper site choice and earthquake-resistant construction—or by buying elsewhere.

Figure 11.4 Shell hunting along a sandy beach in the Turnagain Heights area of Anchorage. The beach is supplied by rapid erosion of the vegetated earthquake-disturbed sediment to the right.

The landslide advanced the shoreline more than 1,000 feet seaward and made it highly irregular in outline. Today, the shoreline is once again a fairly smooth line (fig. 11.4), essentially back to its preearthquake position. Currently, the shoreline is retreating landward, probably at a rate of 1–2 feet per year, and a sandy beach has been formed by the erosion of the unconsolidated landslide material. Unfortunately, piles of construction or earthquake rubble plus a rock revetment built by the state to protect the bike path are increasing local rates of shoreline erosion by blocking lateral beach sand transport.

Before purchasing property here, study the hazards map in the *Anchorage Coastal Resources Atlas.* Do not build or buy in the zone of high slump risk. In this area, if you can see the sea from your house, you may be in trouble.

Southwest Anchorage: Earthquake Park to Campbell Lake (RM 11.4)

This bluffed shoreline is composed of unconsolidated, easily eroded materials with varying rates of shoreline retreat. At Point Woronzof, the bluff is 30 feet high and has only a narrow beach protecting its base. The lack of vegetation on this bluff, its unconsolidated nature, the lack of a pile of talus at the base, and the lack of a protective mudflat are sure signs of relatively rapid shoreline retreat; in this case about 2 feet per year. The higher (100+ feet) bluff at Point Campbell is eroding at a slightly lower rate. Except for Points Campbell and Woronzof, the bluff face here is heavily vegetated. Along the entire bluff top and face is a narrow zone of very unstable material with a strong risk of landslide. This zone is widest (200–300 yards) for a distance of 2 miles along the bluff top southeast of Point Campbell. A zone at high risk from landslides, especially during earthquakes, surrounds Campbell Lake (and is covered by expensive homes). A view of the sea or the lake may not be worth the potential cost to the homeowner. Better to play it safe and build or buy back from the slump danger zones. Earthquakes are a real possibility here. Prepare for them by proper site choice and with earthquake-resistant construction. Refer to the *Anchorage Coastal Resources Atlas* in the local library or in the city planning office *before purchasing property here.*

South Anchorage: Campbell Lake to Rabbit Creek (RM 11.4)
On its south margin, a low (25–35 feet), vegetated bluff backs the shoreline, which is fronted by a wide mudflat, exposed at low tide, and a broad freshwater marsh. Shoreline retreat here is very gradual, perhaps less than 1 foot per year. The bluff top and face are susceptible to landslides (fig. 11.5), especially during earthquakes. Expensive houses crowd to the edge of this bluff, and new developments advertising a view of Turnagain Arm are at risk. At the southeast end of this shoreline reach, the railroad is located on the bluff face. The cut for the railroad bed steepens the slope and significantly increases the likelihood of slumping for buildings located on the bluff top (fig. 11.6).

The bluff along this shoreline reach (which was undeveloped in 1964) did not undergo significant slumping in the 1964 earthquake. It is important to note, however, that development activities in a slump-prone area tend to increase the probability of slumping. Grading, removal of vegetation, changes in drainage patterns, and lawn watering all are slump-promoting activities. Earthquakes are a real possibility here. Prepare for them by proper site choice and with earthquake-resistant construction.

Before purchasing property or constructing a house, visit the library or the city planning department and study the hazard maps in the *Anchorage Coastal Resources Atlas.* Do not buy in the high-risk slumping zone.

West Coast, Kenai Peninsula (RM 11.3)

The entire west coast of the Kenai Peninsula is accessible by road, and most of it is privately owned and available for development. State Highway 1, which hugs the shore, is proclaimed the westernmost highway in the United States. The entire coastal area from East Foreland at Nikiski to Anchor Point, near Homer, is a plateau of sediments deposited by glaciers and by rivers flowing from glaciers to the north and east. A bluff of unconsolidated material is present where the plateau meets the sea, ranging in height from 20 feet to 200 or more. Where no recent slumping or erosion has occurred,

Figure 11.5 Muddy tidal flats along Turnagain Arm south of Anchorage. The bluff, which is steepened by a railroad cut, is highly susceptible to landslides, especially in the event of earthquakes.

the bluff is vegetated. Where slumping or erosion has occurred within the past few years, the bluff is gray to brown and lacks vegetation. Most of the bluff is retreating landward. Clearly, the best strategy is to buy or build well back from the bluff edge.

Residents of the Kenai Peninsula should be equally concerned about tsunamis, earthquakes, volcanoes, coastal erosion, coastal bluff slumping, and flooding—a rather daunting list of natural hazards. Recognizing the hazards for any area can lower the risks for you and your family, however—a most worthwhile endeavor.

The erosive nature of the bluff of unconsolidated material along the Kenai Peninsula is readily apparent to even a casual visitor to this shoreline. In some areas, trees are leaning seaward, and at a number of sites (e.g., Kalifonsky Beach) the uppermost soil layer overhangs the bluff. Indentations in the generally straight bluff are also indications of recent slumping, as at Clam Gulch. The bluffs are probably retreating at a rate of between 1 and 3 feet per year. But a given location may see no retreat for several years followed by a single event (slump) that moves the shoreline back 10 feet virtually in an instant.

Bluff erosion along the west Kenai Peninsula coast was greatly increased by the 1964 earthquake. The land subsided a few feet along the entire coast, and beaches at the base of bluffs narrowed or disappeared as a result. Waves begin to attack the bluffs directly, causing slumping and bluff retreat. In a few more decades, if local people do not interfere with coastal processes, the material derived from bluff erosion will likely build the beaches out again.

Tsunamis are always a possibility on the west Kenai Peninsula coast and could derive from either submarine landslides or from volcanic activity on the other side of Cook Inlet or from Mount Saint Augustine (see chapter 3). Tsunamis are mainly a threat to developments facing Cook Inlet in areas of low elevation, mainly at river mouths. These low areas are often jammed with recreational vehicles (RVs), campers, people fishing, and birdwatchers during the summer months (fig. 11.7). There are camping areas at the mouths of the Kenai, Kasilof, Ninilchik, and Anchor Rivers. The bulk of year-round housing and commercial development on the Kenai Peninsula is on bluff tops, out of immediate danger from all but the largest tsunami. Storm surges can cause flooding at river mouths, which are also susceptible to flooding from upriver during high rainfall and spring ice breakup.

In addition to initiating tsunamis and causing structural damage to buildings, earthquakes may also produce slumping on the Kenai Peninsula bluffs—as occurred with the 1964 earthquake. The good news is that, unlike the Anchorage area, there usually is not a mud layer at the base of the bluffs to lubricate and enhance slides.

Hazards from the volcanoes to the west, besides the aforementioned tsunamis, include ashfall and acid rain. In December 1989, total darkness reigned at noon in Kenai and Soldotna due to ashfall from the Mount Redoubt eruption. The ash halted air traffic, blocked drains, and short-cir-

Figure 11.6 This railroad cut, also shown in fig. 11.5, has severely reduced the bluff's stability. The houses built on top of the bluff are set back from the edge, but not far enough. In terms of earthquake risk, this may be the most dangerous building site in the Anchorage area.

cuited electrical transformers. Perhaps the most dangerous volcano in Cook Inlet, from the standpoint of risk to the west coast of the Kenai Peninsula, is Mount Saint Augustine, which rises from the seafloor to form an isolated island. Earthquakes, landslides, and eruptions on Augustine immediately and directly affect the waters and shores of Cook Inlet.

Kenai (RMS 11.3, 11.5)

The mouth of the Kenai River is the entrance to the spawning grounds for tens of thousands of salmon. Unfortunately, it seems that about the same number of fisherfolk are strung along the riverbanks on a typical summer day. Residential development and heavy pedestrian traffic are threatening the very resource that feeds Kenai, the eternally returning salmon. Developers are virtually mining wetlands, increasing bluff erosion, lowering the water table—and eventually river levels—and altering the river's chemistry by adding chlorinated runoff (*Anchorage Daily News*, September 4, 1994).

The present settlement is a direct descendant of Russian colonial occupation, as evidenced by the century-old Holy Assumption Orthodox Church in Old Town. The town was primarily a military installation, a role that continued briefly in 1870 after the United States took possession of Alaska. In the 1880s, the first canneries were placed at the nearby mouth of the Kenai River. Rapid growth did not arrive until 1957, after the discovery of oil just north of Kenai at the Nikiski field. Subsequently, Kenai has expanded inland, miles distant from the original townsite.

Old Town rests on a bluff about 100 feet high; the bluff is bare and is undergoing constant erosion from waves, tides, wind, and rain. During a 3-day period in early October 1994, a 3-foot-wide strip of bluff collapsed into

Figure 11.7 A congested campground at a river mouth on the western Kenai Peninsula is in an area at high risk from tsunamis. RV parks are typically located at low elevations to allow access to beaches.

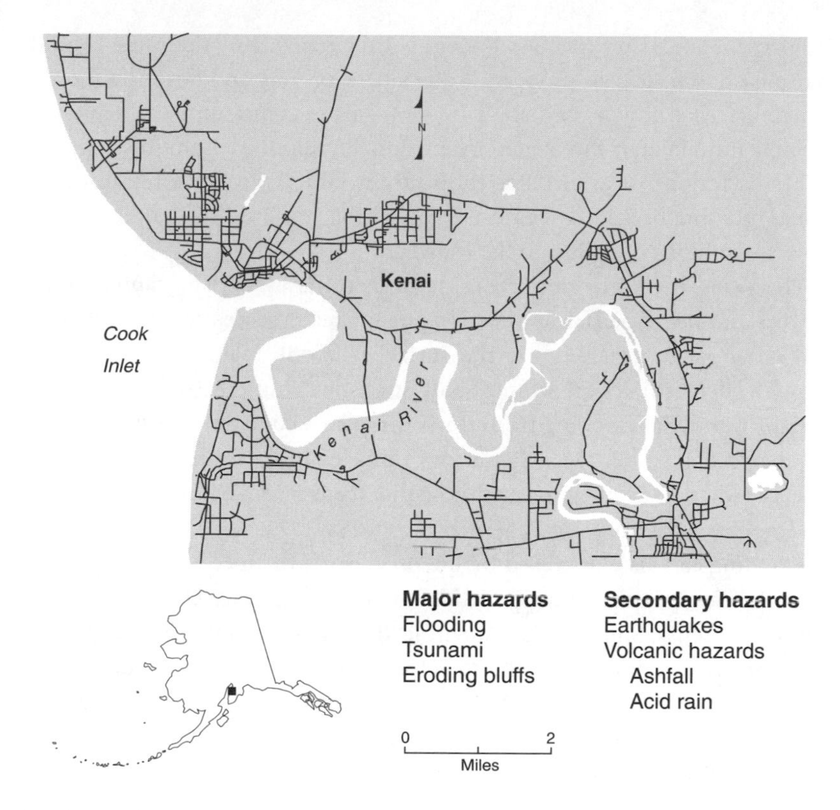

RM 11.5 Kenai

Cook Inlet. Protecting the bluff is a constant concern for Kenai Borough planners. Residents feel that a "vast" amount of real estate has been lost to the inlet. John Williams, Kenai's mayor, told the *Peninsula Clarion* in 1994 that "an entire road and two or three platted lots" have eroded away since 1953. Since the early 1980s, the city of Kenai has thought seriously about erosion control and even contacted the U.S. Army Corps of Engineers to seek solutions (*Peninsula Clarion,* October 10, 1994).

The solution favored by some Kenai residents is a seawall, responding to the gut-level appeal of hard stabilization. Residents should bear in mind the finality of this "solution" and realize that it might merely transfer the effects of wave action from one part of the bluffed coastline to another. A seawall will cut off the sand supply to the beach, causing the loss of beach in front of the wall and increasing rates of shoreline erosion on adjacent unwalled beaches. Further, erosion from the top of the bluff may not even be affected by a seawall. Much bluff erosion is the result of rainfall and runoff, which is increased by asphalting, ground cracking, and other processes unrelated to the sea. The residents might be better off calling a truce with Mother Nature and retreating back the 25–50 feet that is eroded every 20–30 years.

Homeowners and prospective buyers within 100 feet of the bluff should take a hard look at their property. Voters in Kenai City and Borough should remember that only a very small fraction of the community's inhabitants are now immediately threatened by erosion. Should the community pay for the indiscretions of a few? Retreating from Old Town (moving an occasional building back) may spare the rest of the community—and U.S. taxpayers—millions of dollars in the long run.

The Kenai River presents several hazards to its users (e.g., cannery employees and fishermen). The low-lying delta area is susceptible to tsunamis and other waves generated on the surface of Cook Inlet. A tsunami produced by debris slides on Mount Saint Augustine could wreak havoc on the Kenai River delta. Flooding from upriver is also a concern, as residents discovered in September 1993 when heavy rains led to the draining of a glacial lake, the infamous Snow River pothole that feeds into the head of the Kenai River (*Anchorage Daily News*, September 17, 1993).

Flooding can also be caused by intense storms that occasionally spin off from Pacific typhoons, as happened in fall 1995 (*Anchorage Daily News*, September 22, 1995). Since the last storm in the 1980s, several hundred people had constructed homes on the floodplain, and they suffered the consequences as the waters rose, with damage estimated at more than $11 million (*Anchorage Daily News*, October 12, 1995). Lack of floodplain planning, zoning, and enforcement played a big role in this "natural disaster." Residents along the Kenai River are advised to take responsibility and not rebuild in harm's way, even if the Federal Emergency Management Agency (FEMA) offers flood insurance.

Earthquakes are a threat to Kenai residents, as they are to all south-central Alaskans. The good news is that the bluffs beneath Kenai are sandy and perhaps less susceptible to catastrophic slumping. If the land subsides, however, as it did in 1964, then bluff erosion will accelerate. The best advice for Kenai residents is to stay away from the margins of the sea and river. Volcanic ash from nearby volcanoes across the inlet poses a significant health risk to residents.

Ninilchik (RMS 11.3, 11.6)

Ninilchik was founded in 1846 by Russian colonists, pensioners who had married Native American women. The small community of Ninilchik holds the distinction of being the only Alaskan town to have had a series of groins emplaced along its shoreline by the Army Corps of Engineers. Although nearly all of the town lies atop the bluffs of the Kenai Peninsula, the harbor is situated on a sandy spit at the outlet of the macrotidal (32-foot range) Ninilchik River. As soon as the boat harbor was dredged in the early 1960s, a continuing series of engineering solutions was necessary to maintain the spit in its unnaturally fixed position. The engineers first tried jetties to sta-

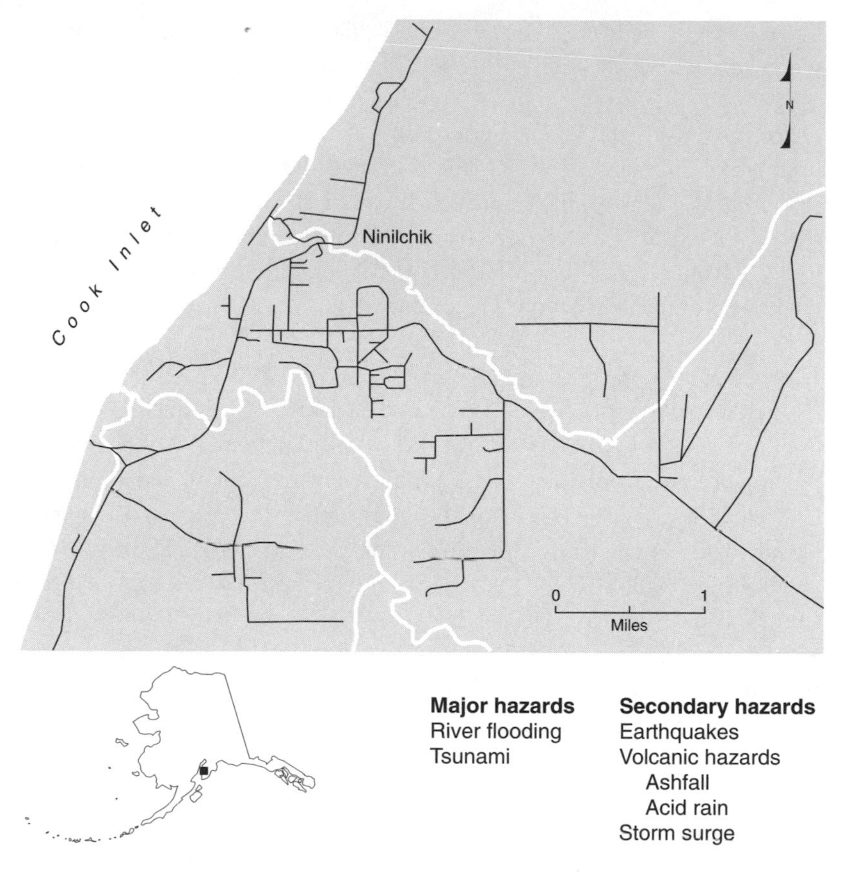

RM 11.6 Ninilchik

bilize the river outlet and finally added a series of groins to trap sand heading north toward the inlet. The groin system, spaced at intervals of 150 feet, is backed by a bulkhead of spruce beams.

Ninilchik faces hazards similar to Kenai's, although few people live close to the bluff margin here. Fall storms can hit hard at the beach campgrounds, the spit, and the boat harbor, as happened in fall 1994. Earthquake, ashfall, river flooding, and tsunamis are also considerations for Ninilchik residents. The tsunami threat is a particular concern during the summer camping season because of the many RVs camped at the river mouth area.

South Kenai Peninsula (RMs 11.3, 11.7)

Anchor Point to Homer (RMs 11.3, 11.7)

From Anchor Point south, the coast is significantly higher than most portions of the west coast of the Kenai Peninsula. This southern part of the

Kenai coast is bedrock—consolidated Tertiary sandy siltstone with thin seams of coal, which often ends up as black chips on local beaches. There are two sets of steep slopes in the region. The lower cliffs are hard rock and are comparatively stable. The upper slopes are silt and sand gravel of glacial origin, like the bluffs of the northern part of the peninsula, and are very susceptible to catastrophic landslides. Traces of one such slide that occurred several thousand years ago can be seen from the highway near Diamond Ridge. Earthquakes are also a hazard on this stretch of the coast, and earthquake-resistant construction is recommended.

Homer (RMS 11.3, 11.7)

Homer, a major sport-fishing locale and once a coal-mining center, faces daunting coastal hazards. The town lies within Kachemak Bay, a sheltered portion of lower Cook Inlet, and is composed of a mainland and a spit area. Glaciated, jagged mountains on the opposite shore present an inspiring backdrop to Homer, one of the most scenic areas in Alaska. Homer and its vicinity (i.e., Halibut Cove, RM 11.3) are the farthest outliers of Anchorage suburban development. Homer has a sort of laissez-faire attitude about planning, as witnessed in its freewheeling approach to seawall construction.

Perhaps foremost among the hazards here are tsunamis. A tsunami map, available with some persistence at city hall, uses the 100-foot contour as the likely run-up from a very big tsunami. Of course, most tsunamis are not nearly that large, but the Big One, according to the map, would cover all of the spit as well as the town's large lower-elevation fringe. The biggest single tsunami hazard for Homer may be Mount Saint Augustine in Cook Inlet, which can be seen from Homer on a clear day. This volcano is so close that there will be very little warning time after a volcanic explosion or a large landslide sends a wave on its way. A major factor determining the amount of damage from a tsunami will be the tide level at the time the wave arrives. The tide range between low and high tides is on the order of 30 feet. Thus, a large tsunami arriving at low tide would do much less damage than the same wave arriving at high tide. The 20-foot tsunami produced by the 1964 earthquake arrived at close to low tide, which spared the town much damage, especially on the spit.

Bluff erosion is a major problem for some homeowners in Homer, and it will become increasingly important in the future as bluff-front sea-view development proceeds. Bluff erosion is most severe in the Bishops Beach part of the old town; bluff tops there are retreating at a rate of 2–3 feet per year. A number of buildings and cabins have been moved or allowed to fall in. On any given day, fresh slumps of bluff material can be seen on the beach. The Elks Club recognized the erosion problem and attempted to halt it, but managed only to prove that nature bats last at the shoreline (fig. 11.8). Initially, the club put out gabions. These were smashed about in storms, so in

RM 11.7 Homer

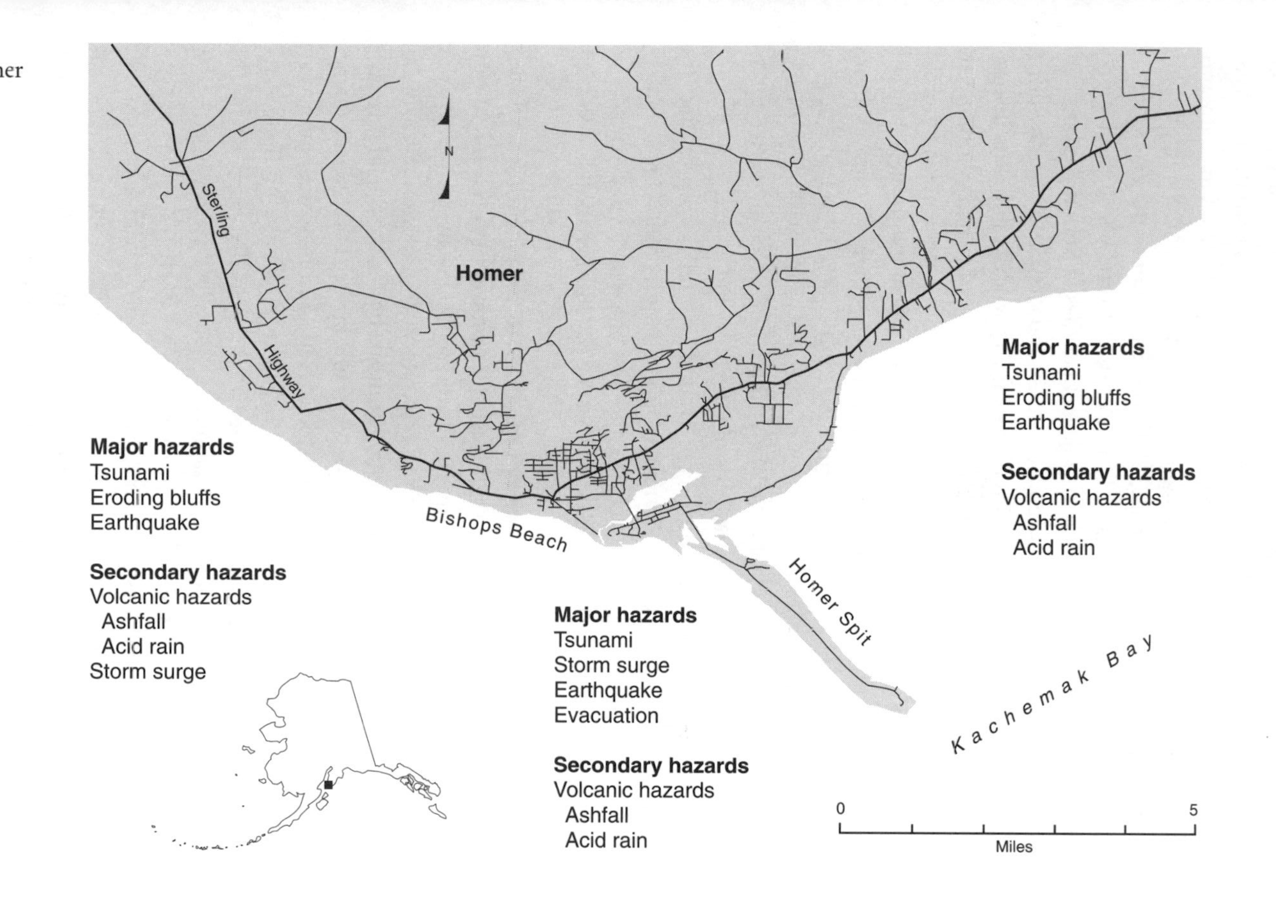

N
Homer
Sterling
Highway
Bishops Beach
Homer Spit
Kachemak Bay
Major hazards
Tsunami
Eroding bluffs
Earthquake
Secondary hazards
Volcanic hazards
Ashfall
Acid rain
Storm surge
Major hazards
Tsunami
Storm surge
Earthquake
Evacuation
Secondary hazards
Volcanic hazards
Ashfall
Acid rain
Major hazards
Tsunami
Eroding bluffs
Earthquake
Secondary hazards
Volcanic hazards
Ashfall
Acid rain
0
5
Miles

1991 large concrete blocks were placed at the base of the bluff. Storms have already begun to move these blocks about, but they will nonetheless help reduce erosion for a few years. There are three problems with the Elks Club shoreline armor: (1) only the base of the bluff is protected, but many processes (e.g., rainfall, frost wedging, and slumping) work on the upper bluff and will continue to cause retreat; (2) the rocks in front of the Elks Club are slowing down the contribution of bluff erosion to the beach sand supply and causing accelerated erosion on adjacent beaches in both directions; and (3) the rocks act like a groin and create a sand deficit in the downdrift (southeast) direction. The rapid retreat of the downdrift shoreline belonging to the club's neighbors is directly attributable to the club's shoreline armor. The Elks Club seawall is a classic example of the problematic nature of shoreline armoring. The club has temporarily solved its problem but will have to spend much more money to make it a "permanent" solution. Eventually the Elks will spend more than their club is worth and will destroy their neighbors' property; and sooner or later the armor will affect Homer Spit. It would have been better to armor the entire beach or, better yet, to replenish it by hauling in truck loads of local sand and gravel.

All decisions regarding armoring of Homer's shorelines must reflect the need to avoid reducing the sediment supply to Homer Spit. Unfortunately, halting bluff erosion on the beaches to the west of the spit will eventually do just that. The sediment contributed to beaches by bluff erosion is necessary for healthy, wide beaches.

Bluff erosion in Homer is largely caused by storms. The most damaging storms usually come from the southwest and occur at high tide. Local residents believe that the most important erosion occurs in the fall. During winter, the bluff is frozen and erosion is retarded somewhat, even in big storms. As is the case with other Kenai Peninsula bluffs, the rate of retreat of the Homer bluff shoreline increased as a result of the shoreline subsidence caused by the 1964 earthquake. Subsidence left the bluffs without a wide protective beach, and high-tide waves began to attack the bluffs directly. Some of the bluff faces are heavily vegetated, especially along Kachemak Road, indicating temporary bluff stability; *temporary* is the key word.

Other factors controlling bluff erosion rates are fetch and sediment type. Fetch is large to the west of Homer Spit, and open-ocean waves coupled with strong tidal currents erode the shoreline. To the east, up Kachemak Bay, the fetch is less and the waves are smaller, although katabatic winds off the Grewink and Harding glaciers produce some significant waves from time to time. At Millers Landing, spray from winter waves smashing into the frozen bluff coats nearby houses with ice.

Sediment type, especially easily eroded mud layers, can play a major role in increasing bluff erosion rates. Mud layers are present at some locations at the base of the bluff. A quick walk on the beach can be very informative in this regard. Large mud layers at the base of bluffs may increase the likelihood of bluff fracture during earthquakes.

Homer residents face the standard south-central Alaska volcanic hazards such as ashfall and acid rain. Chances are that such events will not be life threatening if residents seek shelter. Finally, the entire region is susceptible to earthquakes. Suitable construction is strongly recommended, even though earthquake-resistant construction is not required by the city.

East Homer (RMS 11.3, 11.7)

Numerous individual property owners have taken the bluff erosion problem on the eastern shore of the Homer area into their own hands. One landowner with a metal-beam pier created a large boulder revetment at the base of the bluffs. Unfortunately, this structure seems to have caused substantial erosion on adjoining portions of the bluff, and those property owners also have emplaced small boulder breakwaters for protection. Erosion will doubtless accelerate in this area, and political pressure will build to New Jerseyize the East Homer shoreline with continuous seawall protection. It is still not too late to avoid this situation by retreating and removing the existing rock walls. In addition, East Homer residents should consider the usual volcanic, earthquake, and tsunami hazards.

Figure 11.8 Erosion has increased in this section of Bishop's Beach in Homer due to updrift armoring of the bluff. Armoring of the bluff base at the Elks Club reduced the sand supply to this downdrift beach, narrowing it and causing the rate of bluff retreat to increase.

Homer Spit (RMS 11.3, 11.7)

Homer Spit is a unique geologic feature. The spit hangs off the north entrance of Kachemak Bay like an appendix (fig. 11.9). Its position represents a combination of modern littoral transport and subsurface factors such as the presence of a glacial moraine. Except for Sandy Hook, New Jersey, across from New York City, this is perhaps the most heavily used spit in the United States. On any summer day, the spit bristles with hundreds of RVs, trailers, campers, tents, cars, and strolling tourists pursuing souvenirs or patronizing the numerous halibut charters, restaurants, bars, and hotels. Boardwalks and shops are built on pylons on the seaward side of the spit. A large boat harbor is carved into the spit gravels on the bay side, adjacent to a small cannery. Although the 1964 earthquake caused 4–7 feet of subsidence on Homer Spit and produced a 20-foot tsunami, the quake's occurrence at low tide mitigated a potential catastrophe. Today, the spit is much more heavily used, and a tsunami that arrived during a high spring tide or a fall storm could be deadly to thousands.

At 19 feet elevation, Homer Spit is significantly lower than the maximal potential storm surge height calculated by the Coastal Engineering Research Center of the Army Corps of Engineers. The Corps reports that a likely 100-year storm would top out at 30 feet above sea level. Although such a storm has not yet hit Homer Spit, it is clearly a possibility to be reckoned with by planners and residents.

Tsunami waves—generated either locally or far out in the Pacific—pose a drastic threat to Homer Spit. Oceanic tsunamis may take hours to reach Homer, and the Tsunami Warning Center likely will provide adequate

Figure 11.9 Heavily populated (during the summer) Homer Spit is less than 20 feet above sea level and is at extreme risk from tsunamis and storm surge, especially if such events occur at high tide.

warning time for preparations and evacuation. Far more threatening are tsunamis from local sources. Should Mount Saint Augustine generate a tsunami, there would be only minutes of warning. No action could be taken to protect people on the spit. Hence, a tremendous evacuation hazard exists at Homer Spit.

The spit is also threatened by erosion control schemes such as those at Bishops Beach, updrift from Homer Spit, and on the spit road itself. Erosion control structures starve the spit of its construction materials: gravel and sand from the eroding bluff. Homer Spit has a long history of attention from engineers, starting with a railroad built onto the spit in the 1890s for hauling coal. This and succeeding roadbeds have required a series of erosion control devices over the last 50 years. Most of the erosion problems on the spit center on its neck near the mainland. The erosion has been fostered rather than retarded by a series of crude structures built from materials ranging from junked cars and wooden pylons to cement blocks. These devices starved the downdrift portion of the spit, and the problem requires further attention by the Corps.

South Shore of Kachemak Bay (RM 11.3)

Several communities lie on the deep fjords indented along the south shore of Kachemak Bay (across the bay from Homer), which is accessible only by air and sea. The small communities of *Halibut Cove, Seldovia, English Bay,* and *Point Graham* are increasingly subject to residential and commercial development. Halibut Cove is an exclusive settlement of high-priced estate homes and second homes. Seldovia, English Bay, and Point Graham are older settlements and native villages with densely packed small frame houses and boardwalks. Hazards on the south shore of Kachemak Bay are somewhat less severe than those in the Homer area. Bluff erosion and slumping are insignificant threats to the bedrock shoreline.

The tsunami hazard is potentially serious at Halibut Cove, which lies close to the head of Kachemak Bay. In 1964, a 24-foot tsunami hit Halibut Cove and a series of complex refracted waves up to 15 feet high hit Seldovia and its embayment.

The tsunamis generated by the 1964 earthquake hardly affected Halibut Cove and Seldovia because the quake occurred at low tide, although the subsidence of several feet affected many shoreline structures. Any structures in Seldovia or English Bay located near the high-tide line face tsunami threats similar to those on low-lying Homer Spit, and human safety depends on the efficiency of local authorities in instituting warning and evacuation procedures. Know your evacuation route to high ground. Earthquake-resistant construction is recommended.

Seward (RMS 11.3, 11.8)

Seward (pop. 2,699) is nestled at the head of the fjord of Resurrection Bay amid steep slopes and glacier-scoured uplands. The geomorphic setting of the harbor and working town of Seward poses a number of hazards, some serious, for the resident and casual visitor alike. The city rests on a combination alluvial fan and delta at the mouth of the glacially scoured basin of Lowell Creek (fig. 11.10). Such deltas are subject to episodic shifts in the position of active channels as catastrophic rains overtop channel banks and as landslides contribute sediment to rivers. Extremely high rains in 1939 and 1986 produced floods in the Seward area, and some homes were inundated by the Resurrection River.

After the 1939 flood, the Corps of Engineers dammed Lowell Creek and directed its flow through a bedrock tunnel into the bay south of the city. This ingenious drainpipe solution staved off major flooding in 1986, which is said to have been a 150-year flood. Nevertheless, a major threat to human life and property is still present from flooding through Lowell Canyon. This is especially true because the hospital and a retirement home are directly below the dam, along with a portion of the city (fig. 11.11). Significant flood erosion also occurs on river floodplains in the northern part of Seward, which is not on the Lowell Creek alluvial fan and delta.

The flood hazard in Seward has several peculiarities. Consolidated clays or silts on steep bedrock valley walls are subject to saturation during rains and may collapse or slide downslope. Some slides may block small rivers or creeks and create temporary lakes that fill and are then subject to rapid draining when the natural "dam" is eroded. Flooding is also a problem in the northern outlying areas of Seward that lie at the margins of the Resurrection River and on alluvial fans along valley walls. Avalanches are another possible hazard on the steep slopes in the Seward area. Although rare, avalanches have occurred just above Seward—in 1917, for example.

The greatest natural calamity to befall Seward was the 1964 earthquake, which induced subsidence and generated a small tsunami in Resurrection Bay that caused substantial damage to the rail yard and docks. Although the areal extent of damage did not approach that in Valdez or Anchorage, the economic losses were substantial. The Alaska Railroad never replaced the 12 lines of track and buildings that were destroyed. Some, which literally slumped away into the sea, are still evident as ruins in the tidal zone. The seaward margins of Seward are not developed; instead, they are maintained as a park and campgrounds. The setback of houses may help to avoid future losses; however, in view of the unconsolidated nature of the Seward fan and delta, future earthquake slumping is a possibility.

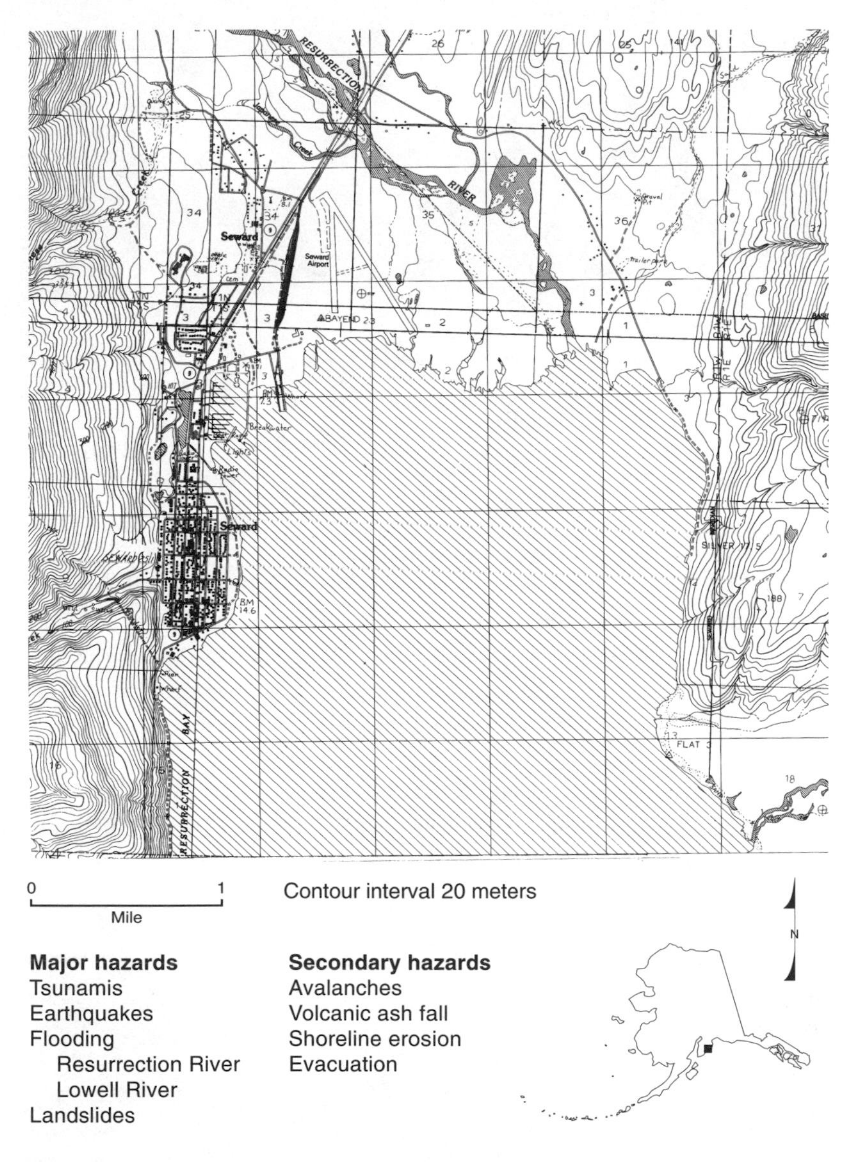

RM 11.8 Seward

The tsunami hazard is widely recognized by Seward officials, who regularly issue warnings for the occasional low-level tsunamis that strike the shore. Ten tsunami warnings were issued in a recent year, and officials fear that their "crying wolf" is causing some residents to ignore the tsunami hazard. The state Emergency Office maps the 100-foot contour, which includes nearly the entire town, as the zone of maximum tsunami hazard.

Figure 11.10 Seward is located on an alluvial fan at the mouth of Lowell Canyon. Lowell Creek has been diverted around the town by a tunnel. The elevation afforded by the fan reduces the tsunami risk for part of the town, but a significant flood risk exists should the stream diversion tunnel become clogged during a storm. The community hospital (fig. 11.11) and a retirement home are built in the former stream valley. Both could be flooded if the diversion tunnel became clogged.

Figure 11.11 The retirement home (shown here) and general hospital in Seward are located in the potential flood path below the Lowell Canyon Dam. In the worst-case scenario, the stream diversion tunnel becomes clogged, causing the stream to flow into town, right through the hospital and the retirement home.

This elevation may be excessive in estimating the zone of potential damage, but for evacuation planning it is best to be conservative.

Significant lower-order tsunamis will certainly affect the casual camper or RV owner who parks on the park shoreline on the east margin of Seward (see fig. 6.7). A tsunami several feet tall could affect hundreds of RVs and campers on a typical summer day; the cumulative losses in RV property alone could total in the millions of dollars. The number of lives lost could be considerable. More visible warning signs are clearly necessary to warn casual tourists to walk to high ground when the alarm sounds.

Earthquake hazards are relatively well addressed by the city of Seward, which enforces the Uniform Building Code with inspectors. The city of Seward is also a participant in the National Flood Insurance Program.

Ashfalls from Alaska Peninsula volcanoes pose a significant, although relatively minor (compared with Cook Inlet), threat to Seward. Beach erosion is a limited threat to the city's waterfront park, which is already showing an occasionally active scarp downdrift from a major revetment near the city dock. Beach nourishment (with gravel) and no further seawall extension are logical solutions to maintain the recreational value of this beach. Gravel brought in by dump truck might be the best alternative to armor.

Prince William Sound (RM 11.1)

Valdez (RMs 11.1, 11.9)

Valdez (pop. 4,074) is a city whose fortune is tied to nature—its bountiful resources and its deadly hazards. The visitors' bureau and museum capitalize on the city's earthquake history and climatic extremes, touting a Man versus Nature frontierism. Established as a jumping-off point for Klondike gold rushers in 1897–98, the town became the supply depot for a route that was established through Keystone Canyon, and today Valdez is the beginning of the Richardson Highway. In the early 1900s, gold at Fairbanks and copper mines in the vicinity were the basis for the town's economy, along with Fort Liscum across the bay (now the terminus of the Trans-Alaska Pipeline). As the mines played out, Valdez declined, and the great earthquake of 1964 sealed the old town's fate. The rebuilding of Valdez involved selecting a new townsite several miles to the west. The choice of Valdez as the deepwater port for the Trans-Alaska Pipeline brought about its resurrection in the 1970s. Another miniboom followed the *Exxon Valdez* spill and the cleanup of 1989–91.

Although in the past settlements were readily moved after disasters, the relocation alternative for hazard reduction is generally dismissed by modern society. Valdez is an exception, and a good example of what can be done. The town was the first modern U.S. community to completely relo-

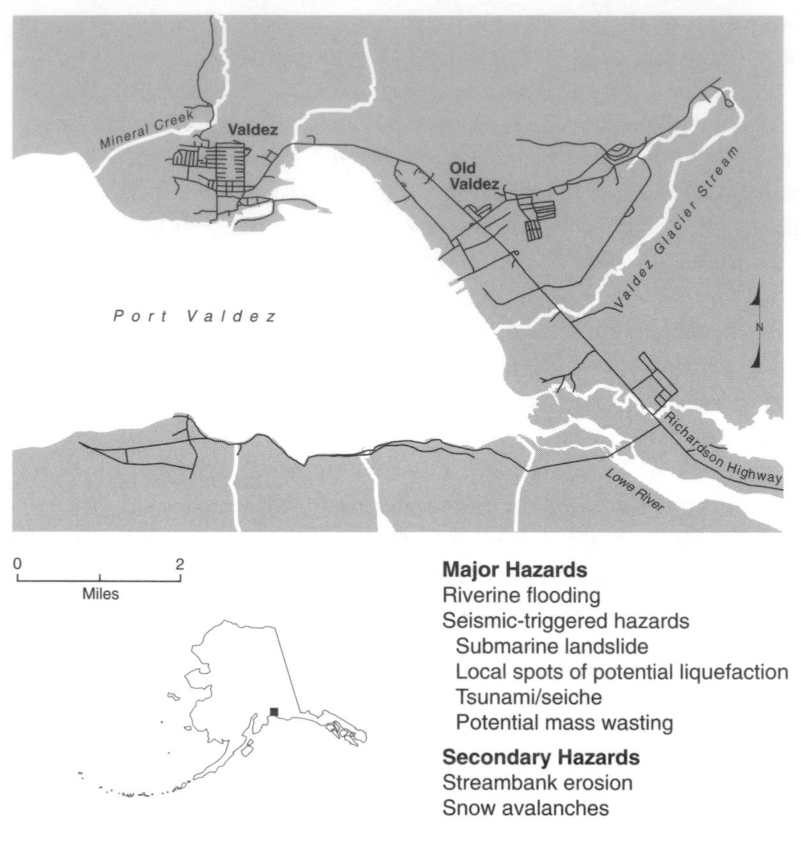

Note: for site specific information, see Valdez Coastal Management Program hazard maps.

RM 11.9 Valdez

cate after a disaster. The devastating March 1964 earthquake, submarine slide, and ensuing tsunami were not the first in the Valdez area, but in the wake of the destruction people recognized that future events might have even worse impacts—so the community was moved. As tempting as it might have been to say, "That was the big one, and now it's out of the way," those farsighted people knew that nature had not changed. The geology of the area was the same, the oceanography of the fjord was the same, and the weather was the same. So Valdez was moved off the unstable glacial-outwash delta to a site 4 miles to the west, on higher ground and somewhat firmer footing, with natural topographic protection. Map 11.9 summarizes the hazards in the vicinity of Valdez.

Valdez lies just 45 miles southeast of the epicenter of the great earthquake of 1964. A long episode of ground shaking (4–6 minutes) triggered a massive submarine slide off the front of the delta on which the town was located (fig. 3.5). The slide displaced a 98-million-cubic-yard mass of earth that carried a slice of land 4,000 feet long and 600 feet wide, including the docks, into the depths of the harbor. Most of the 32 casualties were people on the docks who were sucked down by the slide or washed away by the wave of water it sent up. *The Great Alaska Earthquake of 1964* (see appendix C) provides an unforgettable description of the event:

> The harbor became a maelstrom and the big dock began to break up; mounds of water hit the *Chena*. When Captain Stewart reached the bridge, the ship was lying over to port 50° to 70°. [The harbor was draining.] The noise was tremendous, and witnesses saw incoming waves raise the freighter 30 feet higher than the dock's warehouse. Captain Stewart looked down to see people running on the dock, but as they ran, the dock disappeared. The warehouses, the packing plant, the cannery, the bar, the people plunged into the boiling water.

Beyond the town, 8,000 feet in from the shore and south of Valdez Glacier Stream, the ground cracked and liquefaction occurred, even though the coarse soil there would not be expected to liquefy. The fact that the ground was frozen and impermeable may have contributed to the failure. Liquefaction is one of the greatest potential hazards to Valdez, particularly in the area of the old townsite—land that reverts back to the city in the year 2007. Dredge and fill sites in the port area also may be prone to liquefaction.

At least 11 major fault systems within 150 miles of Valdez can generate earthquakes of magnitudes between 6.6 and 8.6, causing ground shaking and liquefaction, and generating submarine slides and seiches. Such quakes have occurred before in Valdez's short history, in 1899 (magnitude 8.3), 1908 (magnitude unknown), 1911 (6.5), 1912 (7.2), and 1925 (magnitude unknown). Two moderate earthquakes in 1983 (July and September, magnitudes 6.4 and 6.3), centered near Columbia Bay, caused minor damage in Valdez. The history of Valdez should dispel any misconception that the occurrence of one large earthquake reduces the likelihood of another. Three great earthquakes associated with the Chugach–Saint Elias fault system occurred within one year. Builders in Valdez must continue to select safe sites and construct earthquake-resistant buildings.

The story of the 1964 earthquake overshadows the tsunami hazard in Valdez. Earthquakes and submarine landslides into the constricted fjord basin may set up seiches that affect the local coast. The 30-foot wave of 1964 came at low tide. The same wave at high tide would have been 44 feet above mean sea level. Submarine slides at the west end of the basin generated a 170-foot wave run-up (high-water mark) in Shoup Bay. Sea waves, probably seiches, also were reported in association with the earthquakes of 1899, 1908, and 1925. Tsunamis are more likely to occur on shores facing the open waters of Prince William Sound. The village of *Chenega* was completely destroyed by a tsunami that killed 23 of the 72 inhabitants and left only the school building standing.

Steeply dipping rocks, glacial sediments, and thin soils on steep slopes may be destabilized by ground shaking or water saturation (precipitation or meltwater). Debris slides and rockfalls are not uncommon but are local in extent. Some roads are likely to be blocked or damaged from time to time, but most existing development is not threatened. New development expanding to the base of steep or unstable slopes may be in extreme danger, however.

Snow avalanches are more common in Valdez than debris slides, and the north-northwest growth of the town up the alluvial slope has resulted in development in the potential avalanche zone. During the winter of 1993 a snow avalanche piled up against the back of the high school and several houses. Fortunately, the buildings were not damaged and school was not in session, otherwise the avalanche might have destroyed cars and, perhaps, people.

The combination of high rainfall, high snowfall, and steep slopes increases the likelihood of flooding. Flood potential is serious for several locations in the Valdez area, particularly in developments beyond the main townsite. The bridge across Mineral Creek opened the area west of the stream, including the floodplain, to development (the Cottonwood subdivision). Engineering reports had suggested that development on the west side of the creek should either be removed or flood-proofed, and even that the bridge be abandoned because it is in an unstable area and subject to outflanking by flood erosion. Nevertheless, development went forward, necessitating dikes on the streambanks to reduce the flood threat and revetments on the banks next to the bridge to reduce the erosion threat. Ironically, the protective dike resulted in the development's reclassification on the Flood Insurance Rate Map published by the National Flood Insurance Program from an A-zone to less hazardous B- or C-zones, but now the protective dike and revetted banks must be protected, more property is at risk, and more people depend on the bridge and are likely to be cut off during a flood. Gravel mining within the stream channel provides an opportunity for some artificial management of the channel; however, an upstream landslide during a heavy rain could generate surge flooding that will overwhelm the channel.

Along the Lowe River, east of Valdez and just off the Richardson Highway, is the development of Alpine Woods Estates and the Nordic subdivision. The former experienced flooding in the 1980s. At the time, engineering reports recommended relocating the subdivision; however, the wisdom of this option was apparently lost in the 20-year interval that followed the relocation of Valdez. Instead, the community opted for an engineering "solution" by diking the borders of the development, thereby encouraging continued occupation of a hazardous flood zone. A 1990 report ("City of Valdez Lowe River Stabilization/Relocation Study"; see appendix C) identifies 95 of 148 properties as being below the 100-year flood elevation. These properties rest with false security behind a dike system that may or may not provide the expected protection during a major flood. Farther east along the Richardson Highway, potential glacial outburst flooding is a threat.

Combined storm surge and high tide flooding may affect the Port Valdez shoreline, particularly the low-elevation areas of the river deltas. The estimated combined high tide and storm surge 100-year maximum is 10.6 feet above sea level (see appendix C, *Valdez Coastal Management Program*). The rocky shorelines resist coastal erosion, but storm surge waves may attack the delta shoreline.

The native name for the Valdez area is Tetaluk, which means "windy place." From October through April, winds tend to be from the northeast and may be gusty or reach gale force. Extreme winds of 80 knots have been recorded. Williwaws, rapid downhill flows of cold, dense air from surrounding mountaintops, develop locally. But it is snow that is the bane of Valdez. The average snowfall is about 300 inches per year. The winter of 1989 produced 560.7 inches, breaking the old record of 517 inches from 1928–29. The record snowfall for a single day is 46.7 inches. New Valdez included lots of open space in its layout to accommodate wintertime snow storage. The heavy snow cover also creates problems for fighting structural fires (although the cover protects adjacent buildings).

High winds and heavy snow loads pose serious problems for the integrity of structures. The city of Valdez utilizes the Uniform Building Code and has minimum requirements for 115 mph wind loads, including tie-downs, and 90-pound snow loads. Valdez sometimes receives windborne ash during volcanic events, but this is a minor hazard that has less loading potential than the high snow accumulations.

Accidents related to the harbor and oil terminal sea traffic are a significant hazard. Although the *Exxon Valdez* oil spill was outside this area, the potential for a large oil spill remains. The harbor boasts the world's largest floating concrete dock facility for oil tanker repair.

Cordova (RMs 11.1, 11.10)

Cordova (pop. 2,114) straddles a wedge of glacial clay attached to bedrock between Eyak Lake to the east and Orca Inlet to the west. To the south the town has an expansive vista of Orca Inlet, an arm of Prince William Sound. Fish, particularly salmon, are the mainstay of Cordovans. As the local bumper stickers say: "Quit Beefing Eat Fish." The boat harbor is the focal point of Cordova, and most of the area's hazards may be defined in reference to it. Earthquake-induced changes in water level constitute the principal geophysical hazard. The 1964 quake uplifted the port by several feet. While this necessitated dredging, it also rendered storm waves less threatening. The tidal amplitude here is more than 20 feet.

Old Cordova is a compact area restricted to less than 10 blocks within a square mile. The main commercial-government district rises abruptly from the harbor; the largest hotel (aptly named The Reluctant Fisherman), the post office, and the city hall are at the base of an excavated cliff beneath First Street. First Street, still the town's mercantile hub, contains the liquor stores and bars, banks, and stores. All these structures lie at elevations beneath 100 feet—within the tsunami run-up zone. The likelihood that a large tsunami run-up will reach 100 feet depends on local quakes and the tide stage. In 1964, although the first tsunami hit just after the quake, another 20-foot-high tsunami hit at high tide (adding 13 feet) about 7 hours later, fortunately after the land had uplifted 6 feet. Minor damage was reported. In the absence of records of other tsunamis or mathematic models, it is wise to assume the worst. One resident has his cabin tethered to large logs in front; this is probably not a wise precaution for others to emulate. Tsunamis and seiches of local origin are a serious threat to much of low-lying Cordova, the southern half of the town. A special threat from the waves caused by seiches exists on Lake Eyak to the east.

Most of north Cordova sits solidly on bedrock (fig. 11.12) covered by several feet of glacial till; foundations of concrete or pilings here are comparatively stable. In fact, Cordova sustained little structural damage in 1964. Despite this, the effect of the earthquake on Cordova was profound. The 6-foot uplift had substantial economic costs because the all-important boat harbor, cannery docks, and other facilities were rendered high and dry except at the highest tides. In addition, much of Orca Inlet was impassable due to the uplift of shoals; this required a long dredging program. Extensive reconstruction of the harbor was required at considerable cost—all told, about $3 million.

South Cordova rests on unstable clays or silts deposited in proglacial ponds (fig. 11.13). These sediments are likely to shake and settle during earthquake shocks and are not suitable building sites. A major threat to the entire town derives from the placement of the new hospital on artificial gravel fill atop these silts. Ominously, hospital workers can feel even the slightest tremor. Storm waves present little danger to most of Cordova, which is protected by rock revetments and seawalls. However, numerous houses in the Odiak Lagoon area, in South Cordova, are only 5–10 feet above high tide and might be affected by high waves if not for the tidal flat to seaward, which damps most waves. Watch out for a storm at extreme high tide!

Steep slopes can produce flooding if unusually heavy rains fall in a short period. To rain-drenched Cordova, heavy rain means 12–15 inches in several

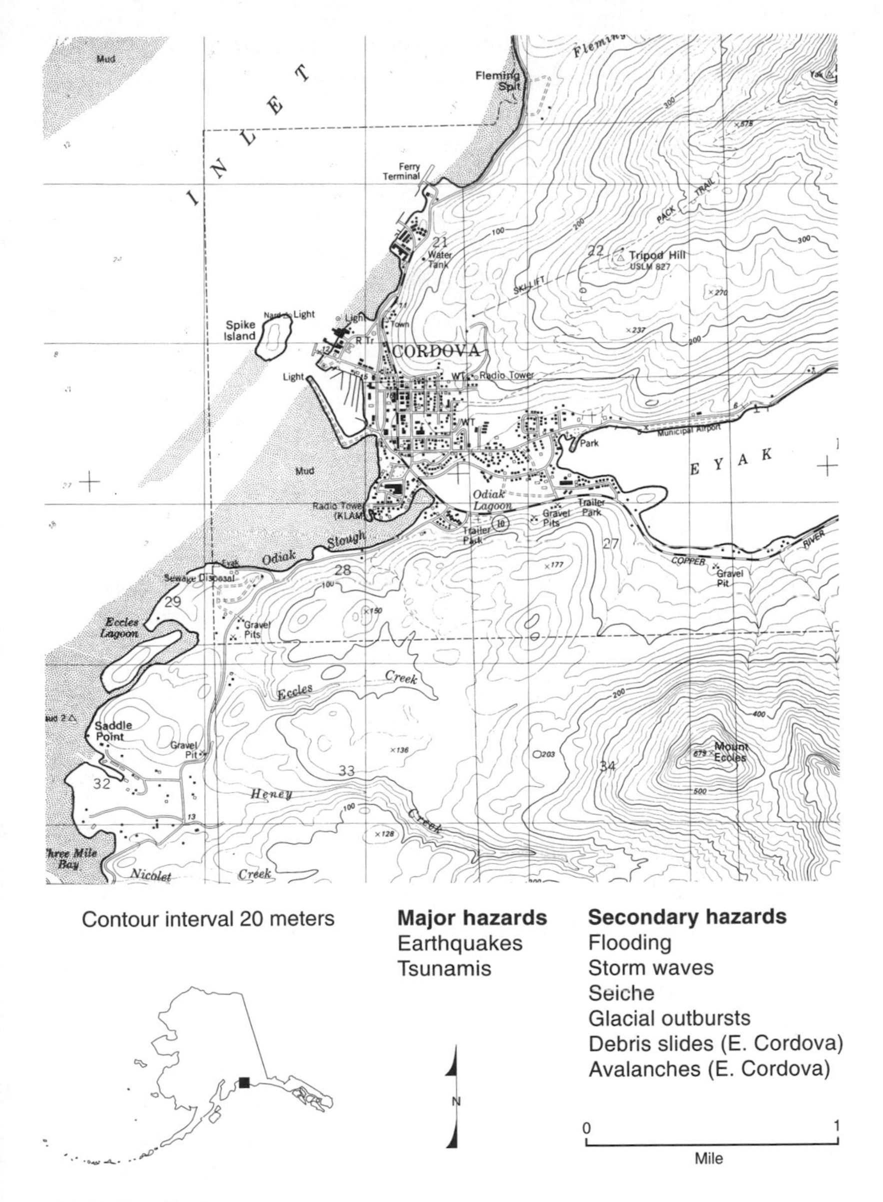

RM 11.10 Cordova

days, as with the flood of August 1981. Still, no major debris slides occurred in the town proper, although the potential hazard is clearly there.

East Cordova: Eyak Lake

Several hazardous conditions accompany Cordova's 1970s expansion eastward to the shores of Eyak Lake. Although slope problems are limited to rain-induced flooding in town, debris slides and avalanches are common along two points on the Copper River Highway out of town. Avalanches at milepost 5 have resulted in life-threatening situations.

Eyak Lake is fed upstream by a channel from the Scott Glacier. A shift in distribution in 1983 occasioned a greater input of glacial water, with its high load of suspended sediment. Although the water level in the lake was affected, no profound effects have yet occurred. A higher water level could flood shoreline properties. A glacial outburst flood could present a serious threat, especially if water levels rose too suddenly for evacuation. The lake is 2 miles wide and has considerable potential to generate sizable seiches or tsunamis under the effects of earthquakes.

Eastern Gulf of Alaska (RM 11.1)

Yakutat (RMs 11.1, 11.11)

One of the most physically isolated towns in the state, Yakutat (pop. 544) lies within the inner curve of the Gulf of Alaska on a coastal plain of glacial outwash discharged from the Saint Elias Range, which towers 15,000 feet above sea level. Several large glaciers discharge into Yakutat Bay from Russell and Hubbard Fjords. The town is situated on a peninsula of glacial moraine and outwash that encloses Monti Bay, a small embayment of Yakutat Bay. Yakutat Bay is the only protected harbor in the 400-mile stretch of coast between Cordova, to the northwest, and southeast Alaska. Despite its harbor, development in Yakutat is generally limited to seafood processing and fishing. Since the oil boom, local government has flourished, due in part to airport improvements and some tourism. Yakutat is primarily a Tlingit community founded comparatively recently in the wake of the withdrawal of glacial ice during the late eighteenth century.

Despite its cartographic attractiveness, the entrance to Yakutat Bay is blocked by a submarine ridge that produces extensive breakers and swell, rendering it dangerous to navigation. Expansion of the town's port facilities is a major goal of local entrepreneurs, but requests for coastal improvements should be assessed with caution. Yakutat is close to several major faults, including the epicenter of the large quake of 1899, which often bears its name. Despite this proximity, the underlying coarse glacial sediments may mitigate some of the earthquake threat to residents. Larger clasts, such as cobbles and boulders, are less sensitive to seismic disruption than smaller sediments. Little direct earthquake damage has occurred in the town, except during the 1958 quake, which damaged a steel-frame airport hangar. However, submarine landslides can produce waves in Yakutat Bay; a 60-foot wave followed the 1899 quake. Considering that the wave run-up record—1,740 feet—is held by nearby Lituya Bay, Yakutat residents should

Figure 11.12 Most of north Cordova is set on solid bedrock capped by glacial till, which generally provides good building foundations. The boat harbor was dredged after the 1964 earthquake to compensate for 6 feet of uplift.

Figure 11.13 Many houses in south Cordova are only a few feet above sea level and at risk of flooding from storm waves coupled with high tide. The houses here rest on clays and silts, which are potentially very unstable in earthquakes. The new hospital is built on such a foundation.

be wary of the enclosed fjords! Tsunamis from the open waters of the Pacific may be somewhat dissipated by the submarine ridge off the Yakutat inlet. Further, the position of the town within Monti Bay may provide some shelter from tsunamis and waves generated by the collapse of ice onto its surface. As elsewhere, however, even medium-sized tsunamis or seismic

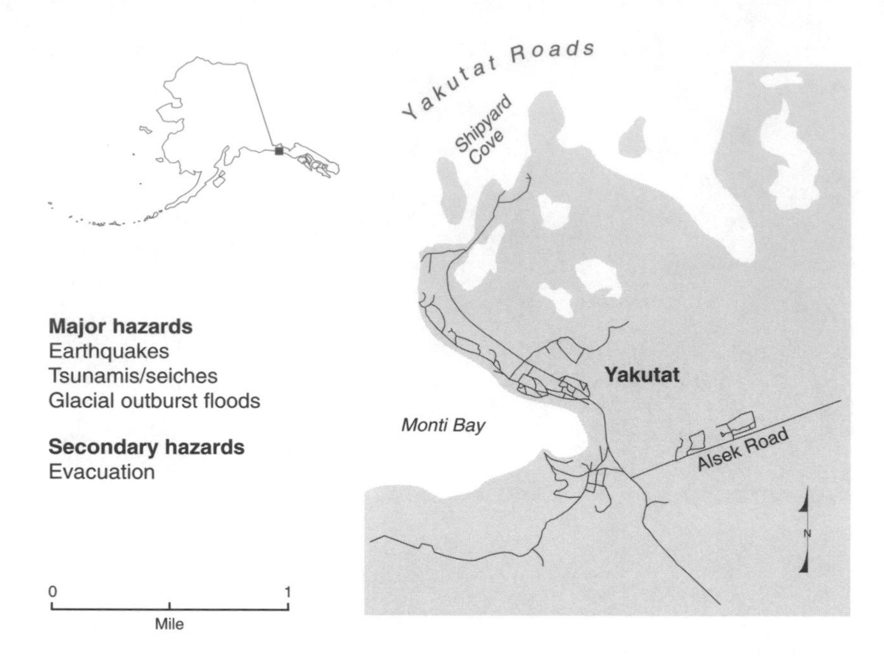

RM 11.11 Yakutat

waves may be dangerous if coupled with high tides. Studies of the tsunami hazard at Sitka are thought to be applicable to Yakutat. If so, then there is a 65 percent chance that a 20-foot tsunami will hit the town within any 100-year period. A significant number of structures in the northern part of town lie below 60 feet in elevation; some structures and waterfront facilities sit below 30 feet. Catastrophic release of water from glacier-dammed lakes in the upper Yakutat Bay is a persistent danger. At times, the Russell Glacier impounds water from the bay fronting the Hubbard Glacier, as in 1987. This temporary dam may break and force a sudden rise in the water level in Yakutat Bay. In 1845 an icefall from the glacier into the bay produced a maximum wave run-up of 115 feet and killed 100 people, and a lesser-scale event in 1905 produced a 12-foot wave in a fjord in upper Yakutat Bay.

In view of the morainal topography, which is comparatively flat, little immediate danger exists from slope-related processes. The collapse of unperceived bodies of glacial ice might present a danger to Yakutat.

12 Southeast Alaska

Introduction (RM 12.1)

The geologic lay of the land has produced an assemblage of hazards in Alaska's panhandle that vary significantly over short distances. A maze of waterways carrying names such as strait, sound, canal, passage, channel, arm, and fjord separate rugged island land masses serrated by ancient and modern glaciers. These elongate bodies of the sea trend northwest-southeast to north-south, with smaller arms more or less at right angles (southwest to northeast). The major passages reflect the geologic fabric of the Coastal Range and the Alexander Archipelago. This pattern, created by the great expenditure of energy along a former boundary between crustal plates, influences the region's commerce and politics as well as its natural processes. Much of southeast Alaska lies in a high-risk seismic zone, and tsunamis threaten the open-ocean and waterway coasts of the western areas.

The fundamental feature repeated throughout the southeast coast is the U-shaped glacial valley, carved by the incredible erosive power of alpine glaciers and flooded by the post–Ice Age rise in sea level. These flooded coastal valleys are characterized by steep rock walls with high relief and floors of old glacial-marine sediments and younger debris. The unconsolidated clays, sands, and heterogeneous soil mixtures (till and colluvium) are often water saturated, weak, and unstable. Along the waterways, only a narrow strip of flat to rolling land is accessible, as are wetland meadows and muskeg on the upper valley floors.

Town and village development here is dictated by this elongate geography. Buildings, roads, utilities, and row houses along the valleys are all located within a narrow corridor at the lower elevations, often on the shore and paralleling the waterway. The valley's form, or geometry, also dictates a specific set of geologic hazards. Variability in orientation, elevation, slope

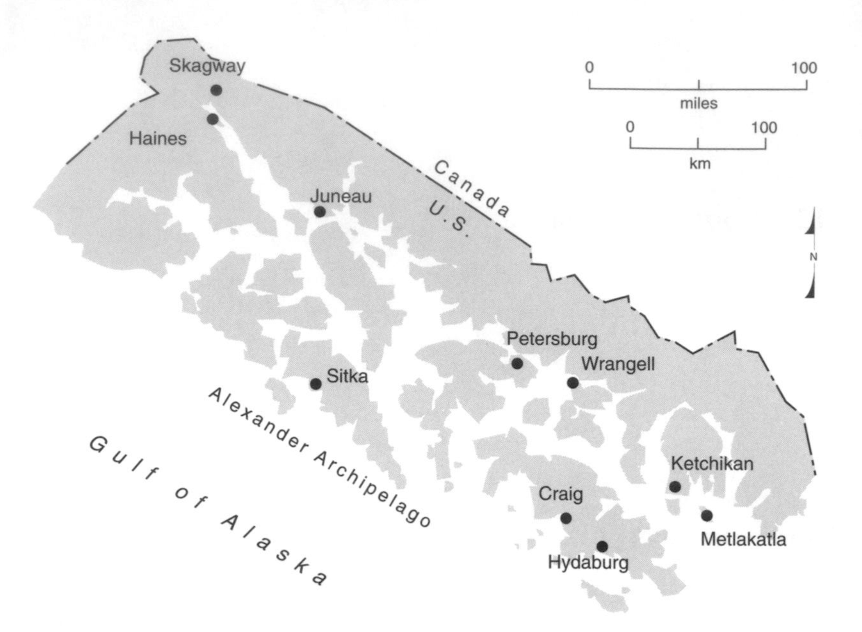

RM 12.1 Southeast Alaska

angle, vegetation, glacier or snow cover, and other factors results in rapid lateral changes in the degree of risk and specific hazards likely to affect any given site.

Steep rock walls and thin soils favor debris flows and slides down the valley walls, perpendicular to rows of houses, main streets, and service lines. The same is true of avalanches. The narrow, restricted arms of the sea favor high tidal ranges and strong, complex currents. Storm winds from the southeast that push water into the passages atop high tides generate storm surges and occasional seiches, resulting in flooding and shoreline erosion. Winds blowing seaward, out of the waterways, can create confused seas and treacherous currents (e.g., Ford's Terror).

The highlands between the U-shaped valleys are home to snowfields and glaciers. These, when combined with the rugged topography and restricted waterways, create unique climatic hazards. Takus, localized winds with hurricane-strength velocity, create strong wind shear forces as cold air masses rip down the mountainsides. Climatic factors are highly variable along the valley axes. Rain and snowfall may vary by tens of inches over short distances, and wind velocity, temperature, fog, and microclimate change rapidly along the lower valley as well as from valley floor to mountain ridge. Wind patterns are often complex and unpredictable, but where currents meet, strong turbulence can result—as evidenced by forest blowdowns. The skies of the region are as fraught with danger as the seas.

Side valleys are both a resource and a hazard. Snow alternating with rain in the higher valleys or on glaciers can release torrents of water that flash-flood into the main valley, again striking perpendicular to the trend of the town or village occupying the main valley floor. Cirque lakes and artificially dammed streams are utilized for water supplies, but aging dams and spillover caused by high rainfall, runoff, or mass wasting into these lakes also create the potential for flash flooding.

As population centers grow in southeast Alaska, a Lower 48 mentality of development is usually applied. Grid-plat residential development may work just fine on the flatlands of eastern Montana or northern Illinois, but such an approach is both inappropriate and dangerous when carried out on a mountain creek meadow on the debris slope of a U-shaped valley. Low-cost development here is the antithesis of safe development. House designs that work in other geographic settings, even within Alaska, may not be appropriate here. A basement in rugged glacial terrain is not compatible with the weak, poorly drained soils that must support basement walls.

When locating and building in southeast Alaska, stricter is better. As much care should be given to site evaluation as to site preparation and actual construction. A general summary of the hazards for the region's communities follows. Site selection in each should be based on careful analysis of the local hazards. Keep in mind that high-risk sites may exist in low-risk areas, and vice versa. Site-specific engineering must deal with drainage, soil creep, foundation integrity, high snow load, wind-bearing load, and resistance to seismic events, flooding, or shoreline processes where applicable.

Haines (RMs 12.1, 12.2)

The fishing, timber, and mining village of Haines (pop. 2,117) lies on a triangular wedge of glacial and river gravels on the south margin of the Takshanuk Mountains adjoining tidal flats and the Chilkat River delta at the head of Chilkat Inlet. Haines was founded as a Presbyterian mission in 1879 to convert the inhabitants of several Chilkat Tlingit villages. The growth of the town followed the discovery of gold both nearby and in the Klondike, the construction of Fort Seward in 1902–4, and the establishment of several canneries in the region. Most houses in Haines are on comparatively level ground. Slope failures and landslides occur frequently, but few houses are built on or near slopes. Although Haines, like Skagway, is adjacent to a delta, it is not on deltaic sediments and is less subject to subsidence than Skagway. Soils in the Haines area vary from well-drained gravels atop glacial sediments, which provide good foundation material, to sandy or silty clay deposits that are susceptible to earthquake-induced failure. The placement of homes, septic systems, and other facilities should be governed by maps of soil properties.

Earthquakes are thought by planners to pose the greatest potential hazard in the Haines region. The seismicity of the small faults concealed by vegetation or sediments is an unknown quantity; effects from the larger systems like the Queen Charlotte fault may be lessened with distance. Although seismic records indicate more than 100 significant quakes regionwide in the last 100 years, the largest local event was a 5.3 magnitude quake in November 1987. An earthquake of magnitude 8 on the Richter scale cannot be ruled out; in fact, a magnitude 6.9 event is very likely to occur here within a 100-year period. The quake would have its greatest effects on the poorly consolidated materials just up from downtown and Portage Cove. Sudden tectonic uplift or subsidence of up to 10 feet, theoretical possibilities not documented in the Haines area, could produce significant alteration of the waterfront and its facilities.

As in any community in earthquake country, a particular parcel's response is determined by its degree of consolidation. Areas of man-made fill are most susceptible to seismic disruption; in Haines, these include the air

RM 12.2 Haines

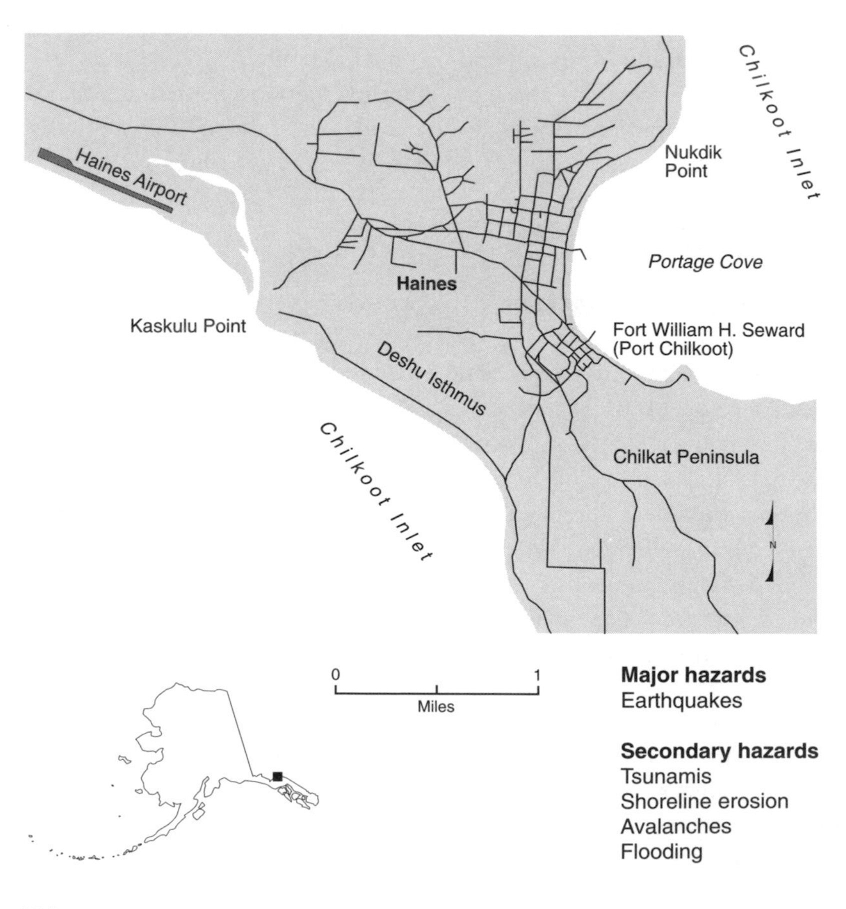

strip, its facilities, and the cargo dock adjacent to the ferry terminal. Floodplain and deltaic deposits on the Chilkat River will be affected in inverse proportion to grain size; the smaller the sediment, the greater the effect. Don't build on clay, especially in parts of Haines underlain by fine-grained marine deposits! In areas with thick muskeg and high water tables, seismic shaking might prove disastrous. Coarser shore and delta deposits should prove considerably safer than fine ones; such deposits underlie much of the main Haines business district. Consult the relevant geologic and planning maps before building or buying in Haines. Landslides and slumps are a possible complication of seismic activity in the Haines area. Once again, the most likely areas of concern are the Chilkat River, its delta, and any fine-grained deposits in town.

Tsunamis and seiches may present a significant hazard if local in origin and if coupled with high tide (the tidal range at Haines is 17 feet), because much of the city sits below the 100-foot contour. However, a 60–100-foot tsunami is considered highly unlikely. Distant earthquakes probably present little tsunami threat because of the sheltered configuration of Lynn Canal, although some damage in the lower areas adjacent to the waterfront is possible.

Subsistence fish camps, recreational areas, and riverside development are subject to flood hazards at the margins of the Klukwan River and Tsirku River fans upriver from Haines. In September 1967, a heavy rain (6.5 inches in 5 days) either flooded or damaged 35 miles of highway. Storm surge flooding may be complicated by wind-driven waves and high tides, especially if coupled with southeasterly winds, which can reach 46 mph. The 100-year storm surge water level is estimated at 25 feet above mean low low water.

Landslide and shoreline erosion are comparatively minor hazards in the Haines townsite. Residents adjacent to slopes should consider the likelihood of avalanches and debris slides, which have been reported on steep mountains north and northwest of Haines. Two areas of hazardous slopes have been identified by the Haines Coastal Zone Management Plan: (1) the lower slopes of Mount Ripinski, about 1 mile north of downtown; and (2) the Lutak Highway to the ferry terminal. Increased logging may exacerbate slope-related slides or rock falls.

To evaluate any property, contact the Haines Planning Commission or the Haines City Council and ask to see the flood and geophysical hazards map prepared in 1989.

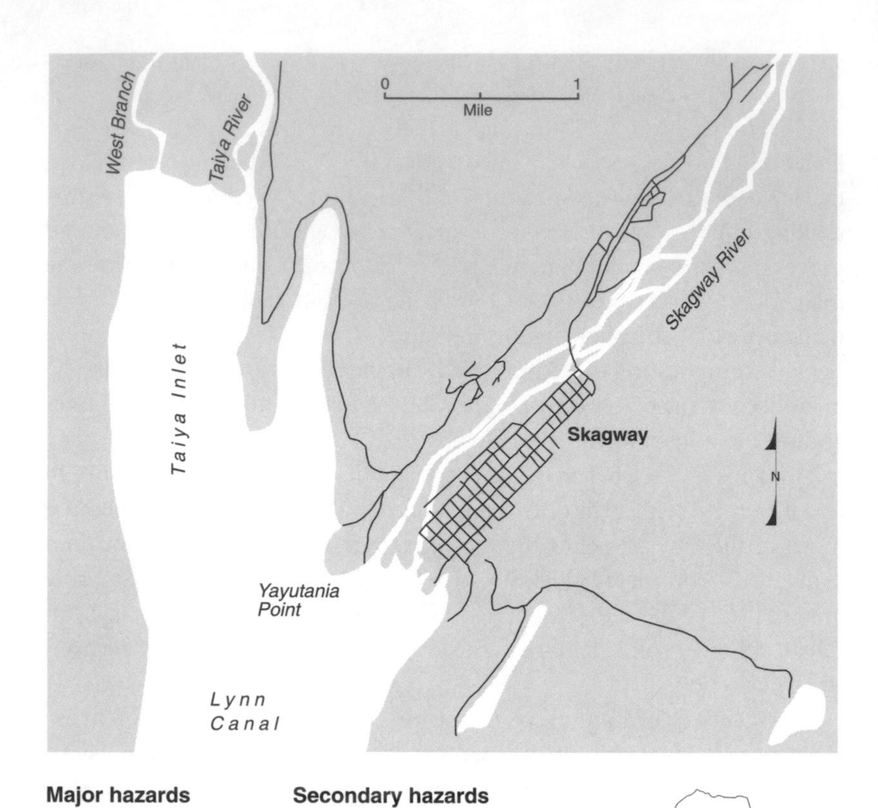

RM 12.3 Skagway

Skagway (RMs 12.1, 12.3)

Located at the north extremity of Lynn Canal, Skagway (pop. 754) served as the jumping-off point (via the Chilkoot Trail) to the Klondike gold fields and witnessed a spurt of prosperity and notoriety during the 1890s. Later, as the southern terminus for the White Horse and Yukon Railroad, tons of lead ore traveled to port via the Skagway docks. The surviving frame buildings, with their spires and bay windows, fuel Skagway's current prosperity as a theatrical backdrop for the annual onslaught of summer visitors (more than 200,000) in search of the gold rush "experience," replete with gold panning, train rides, and barroom skits celebrating local denizen Soapy Smith.

Circumscribed by steep slopes and the delta of the Skagway River, the Skagway townsite is limited to a level area within 1.5 miles of tidewater; only a handful of houses are on slopes (fig. 12.1). The Skagway River is a braided glacial stream whose average annual discharge pattern reflects snowmelt in the higher elevations. Of course, as on any river, overbank floods may follow heavy rains or snow, especially in autumn, when early snows melt readily and add to rain at lower elevations. Flooding may also follow the breakout of ice- or debris-dammed lakes. Nine major autumn floods have overtopped the Skagway River in the last century (1901, 1919, 1927, 1936, 1944, 1949, 1967, and 1990); the largest floods reached up to 6 feet above mean water level.

The city is protected by a series of earthen dikes built in 1940. Subject to daily erosion and modification by the river, most of the dike system is considerably past the 25-year life span that the U.S. Army Corps of Engineers anticipated. The Corps initiated a new survey and planning study in the early 1990s that will lead to the renewal of the dike system. At present, the entire townsite is within the impact zone of a 100-year flood. A much larger area of floodplain is within the city limits but is unprotected by dikes and is unrated for flood impacts. Like many areas in Alaska, Skagway could be subject to an evacuation hazard in the case of widespread flooding. Fortunately, many people could probably reach higher ground on the hillsides.

Like most southeast Alaska communities, the Skagway region is subject to earthquakes, landslides, and isostatic uplift. Uplift from isostatic rebound and tectonic forces is about 0.5 inch per year. Skagway harbor is becoming

Figure 12.1 Aerial view of Skagway showing its location at the head of Lynn Canal. The community is particularly susceptible to tsunamis. The river flooding hazard here is reduced by a series of dikes, some of which are in poor shape.

shallower by this incremental uplift as well as from the deposition of river sediment. Dredging maintains the harbor at a navigable depth.

Potential earthquake effects are related to several active fault systems: the Fairweather, 100 miles to the southwest, and the Chugach–Saint Elias and Denali faults to the north. Although long-term records are unavailable, planners believe that a magnitude 6 quake is likely close to town, and magnitude 8 quakes could occur farther from town. Skagway's location on unconsolidated deltaic sediments presents a serious threat to the livelihood of the town. In one scenario, the outermost delta could slide seaward, which would threaten the waterfront, the docks, and bulk fuel storage areas.

In November 1994, a submarine slide off the delta front produced a localized tsunami that destroyed the ferry terminal and adjacent docks (*Anchorage Daily News*, November 4, 1994). Fortunately, no ships and or crowds of people were at the docks; a similar event during the tourist season would have been disastrous. The economic impact of the loss of the ferry terminal was substantial. Earthquakes might trigger landslides and debris flows above town, but these would have little effect on the town itself.

Most of Skagway is located on a floodplain and thus is not at immediate risk from landslides. However, high rainfall can destabilize rocks and sediments and produce landslides that rupture gas lines or damage other facilities, as occurred in fall 1994 when a landslide-induced break caused more than 2,000 gallons of gas to spill (*Fairbanks Daily News Miner*, October 5, 1994). Similarly, landslides may cause tsunamis capable of destroying dock facilities.

The narrowness of Taiya Inlet and the steepness of its valley walls make tsunamis and seiches a considerable threat to Skagway. The wave's height and timing with respect to tides will determine its effects. A 25-foot tsunami hit Skagway at low tide following the 1958 quake in Lituya Bay. The most severe threats to Skagway derive from local tsunamis; open-water tsunamis would issue from 160 miles or more away, allowing adequate warning time. Hazard maps indicate that a 100-foot tsunami would cover the entire townsite; a 60-foot wave would reach only the lower one-third.

Taku winds can be a significant hazard because north winds often attain high velocities in winter. One unique wind-related hazard is spilled lead powder from ore shipped through the dock and railroad areas, which poses a serious health risk for children. For most of town, the avalanche hazard is minimal.

Juneau (RMs 12.1, 12.4, 12.5, 12.6)

Even in 1880, when Richard Harris and Joe Juneau discovered gold along Gastineau Channel, the future of Juneau was uncertain. As in most gold rush towns, planning for future development was not a top priority. The

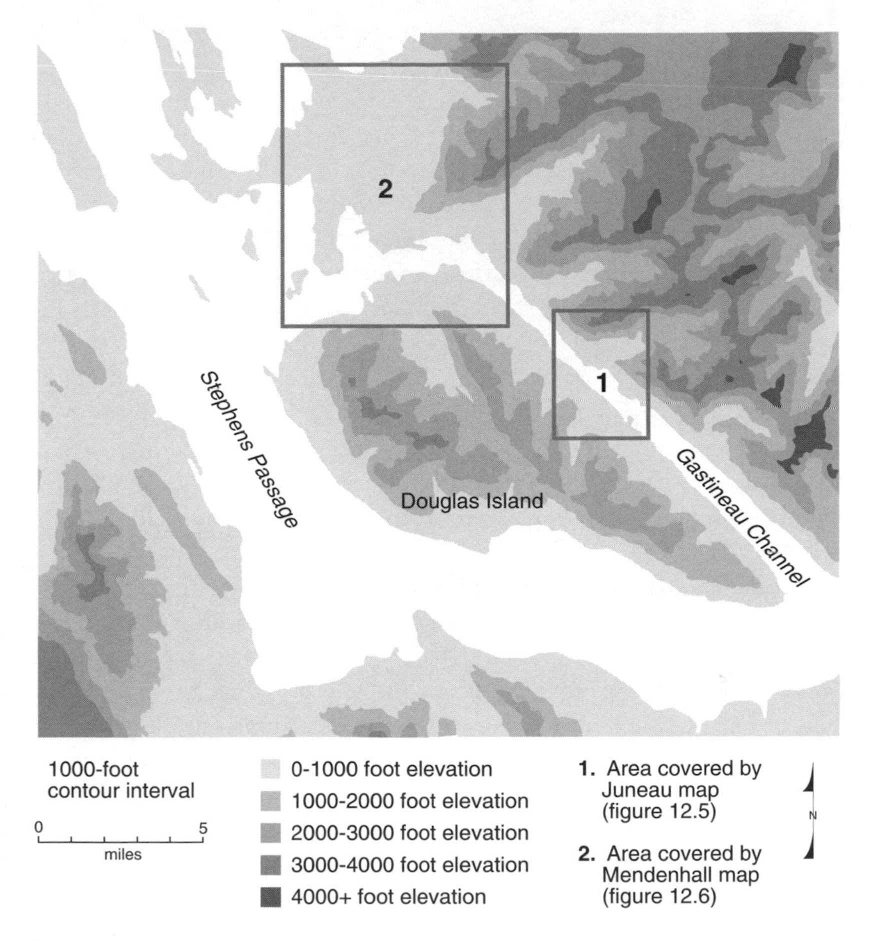

RM 12.4 Juneau area

survival of Juneau beyond the gold rush was not assured until the town assumed the status of territorial capital in 1906, and then state capital in 1959. Today, Juneau faces stiff competition from south-central Alaska business interests that have tried three times, without success, to move the capital to the Anchorage metropolitan area. After the defeat of the ballot initiative in 1994, the future of Juneau as state capital looks secure for the first time in a generation.

Commercial development in Juneau may now face pressure to accelerate as outsiders discover the town's picturesque setting. Developers and future residents, however, will be wise to consider the multifarious hazards that surround Juneau and build accordingly. Juneau sits at the base of Mount Juneau beneath a series of steep but relatively low mountains (4,000 feet above sea level) capped by an extensive ice field. The city is renowned for the reliability of its summer and fall rains, and its winter snow and wind.

The sheer, glacier-carved mountain walls act as a constricting ring around the city and virtually force builders to take risks.

Unlike the situation in Fairbanks and Nome, where gold is interspersed in valley alluvium or beach gravels, in sediments eroded from nearby mountains, the Juneau gold remained in the mother lode. Thus, the city became a hard-rock mining center with a system of deep tunnels running down into the gold deposits. Gold production peaked in the early 1900s, and although two mines closed in the early 1920s, production at the Alaska Juneau (AJ) Mine continued until 1944. In recent decades, government has been the prime industry of Juneau, although gold mining has returned on a smaller scale and tourism has expanded at a feverish pace. On warm, and not so warm, summer days it is not unusual for several large luxury liners to be docked in downtown Juneau, with buses spreading out in all directions to take cruise passengers to the well-provisioned tourist galleries and T-shirt shops while other day trippers crowd the steep and winding downtown streets.

The unified city and borough of Juneau has a population of 27,000 strung out along 30 miles from Auke Bay to the east side of Gastineau Channel and on the north coast of Admiralty Island. Although restricted to a single "roadway," the lifeline of many Juneau residents is the sea. Large boat harbors at Auke Bay and north of downtown shelter numerous pleasure craft and provide seasonal (and even year-round) housing for hundreds of people.

As mentioned above, the Juneau townsite was not selected on the basis of its suitability for development. Its location was purely a function of its proximity to gold deposits. By any measure, Juneau is a potentially dangerous area to live from the standpoint of natural hazards. An impressive number of people have died from debris flows (rock slides) and snow avalanches. That's the bad news. The good news is that the prudent developer or homeowner can take a number of steps to mitigate the hazards. First, you can avoid the dangerous areas by consulting borough planning documents and maps. In addition, you can use construction practices that will minimize damage from natural hazards.

Major differences in weather conditions are possible within the confines of Juneau as a result of the mountainous topography surrounding the town. For example, while the official average annual rainfall is 80 inches at the airport, nearly double that—140 inches—can fall in parts of Douglas to the south (RM 12.5). Douglas is subject to severe taku winds, but West Juneau, just to the north, is generally not.

One very important threat affects everyone in Juneau, no matter what their location: earthquakes. But the potential for earthquake-related hazards such as landslides, avalanches, soil liquefaction, and tsunamis to occur varies dramatically from place to place.

Snow avalanches are probably the most obvious of all the natural hazards in the Juneau area; their signs are unmistakable (fig. 12.2, also see figs. 4.11, 4.12). On the sides of most mountains, especially on the northeast, or mainland, side of Gastineau Channel, are swaths of young trees of uniform size bordered by taller forests. As everyone in Juneau knows or should know, these elongated patches are the paths down which avalanches thundered into

RM 12.5 Juneau, West Juneau, and Douglas

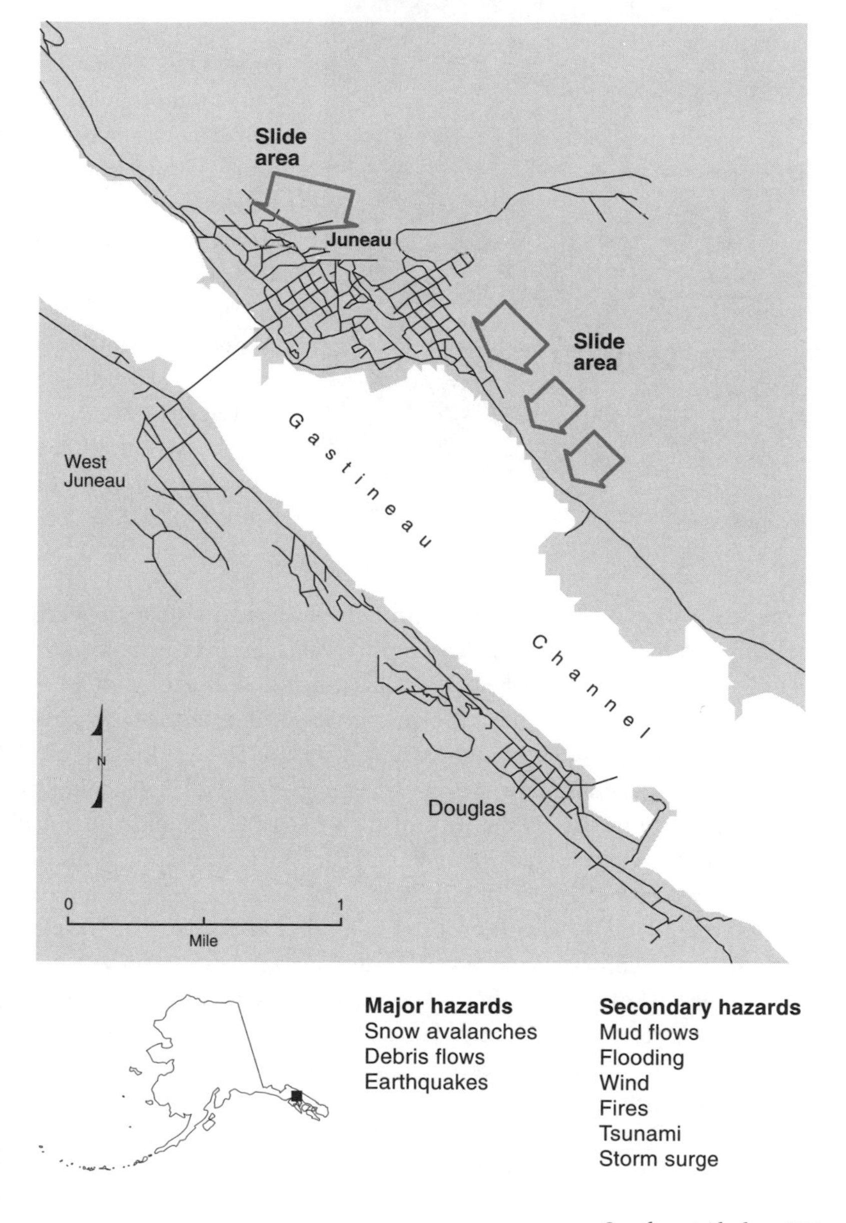

Figure 12.2 A house in Juneau with an avalanche-scarred mountainside behind it. When the next avalanche comes down the hillside, this house will be damaged. Numerous earth and snow avalanche scars exist on the mountainsides around Juneau, including one behind the high school. The Behrends Avenue snow avalanche track is visible in fig. 4.12.

the nearby valley. Avalanche paths with only alder trees visible are recent, probably less than 10 years old. The most spectacular avalanche in recent memory occurred in 1972, when a thick accumulation of snow slid off the top of Mount Juneau, became airborne, and crashed onto the Gold Creek Valley floor, damaging the city's water supply system. The resulting cloud of snow engulfed the entire downtown, enveloping even the tallest buildings. Fortunately, no one was in the path of the avalanche. The most frequent avalanches in recent years have been along Thane Road south of downtown. When snow accumulations are large and conditions are ripe for an avalanche, artillery shells have been used to initiate slides so that travelers along the road will have no surprises. In 1993 a large catch basin was bulldozed at the base of the slope in the hope of containing future avalanches.

Where an avalanche has occurred once, there is a likelihood that another will occur in the future. The Behrends Avenue slide in the Highland development just north of downtown may be the most dangerous development in America in terms of its potential for destruction by avalanche. As such, it has been featured twice in *National Geographic* magazine. Thirty-four buildings, including the local high school and the Breakwater Motel, rest within the clearly visible path of past avalanches. Studies show that avalanches can be expected in the Highland development every 14 years on the average! In some years, multiple small avalanches have occurred: there were 4 in 1971, and 7 in 1966. In 1962 the most destructive of all Juneau avalanches hit the subdivision and continued on to the boat harbor, damaging virtually everything in its path. In 1982 a dry avalanche in this area produced a damaging wind (see chapter 4). Some individuals who own homes in the old avalanche scar zone move their families away to temporary quarters when the National Weather Service says the time is ripe for an avalanche (which can happen any time from January through April). That's a good idea! Local forecasters also try to predict avalanche conditions. Listen to them. Anyone in close proximity to steep slopes should be concerned about avalanches and debris flows. Several areas are less susceptible to avalanches, especially the flats of Mendenhall Valley (RM 12.6) and the more gentle slopes of Douglas and West Juneau.

Rainfall is generally responsible for initiating debris flows in the Juneau area, although a strong earthquake might also be expected to initiate a lot of flows. Five inches or more of rain in a 24-hour period can produce debris flows. This rainfall amount occurs approximately every 5 years in Juneau. In between debris flows, weathered materials gradually fill in the upslope gullies, a process that takes centuries. In the old mining days, debris flows often occurred on Sundays, the day set aside by some mines for blasting. A debris flow in 1936 destroyed a boarding house and killed 36 people. Gaps between the otherwise very crowded houses along Gastineau Avenue are sites where houses were destroyed by earlier debris flows. At least two buildings in Juneau were built especially to avoid the rock slide hazard. The Marine View Building, a high-rise in downtown Juneau, has sacrificial, or breakaway, walls on the lower floor designed to let a debris flow pass right through. Far up the hill, another building is built well into the hillside in the hope that the rock flow will pass by overhead.

Contact the community development department to find out if your site is in a high-hazard zone with respect to snow avalanches, debris flows, and floods. They can also tell you whether you are in a V-zone or A-zone with respect to flooding. This office also handles building permits. See appendix B for the address and telephone number.

Development often involves extensive destruction of vegetation and alters preexisting drainage patterns. In either or both cases, runoff, and especially snowmelt, can mobilize the soil into mudflows that can be damaging to adjacent properties. Mudflows occurred during the 1984 development of Blueberry Hill in Douglas.

Several kinds of floods affect Juneau. On very rare occasions in the fall, the early snowfalls on Mendenhall and other glaciers melt rapidly, causing flash floods. Floods are also possible during fall and spring rains when the ground absorbs nothing because it is frozen, and during the spring melt of the accumulated winter snow. The highest discharges for local streams, however, usually occur during the rains of late summer.

Virtually all of the numerous small stream channels on the hill slopes of Juneau are capable of flooding, but the most dangerous sites in the borough may be along the banks of the Mendenhall River, especially south of Egan

RM 12.6 Mendenhall Valley

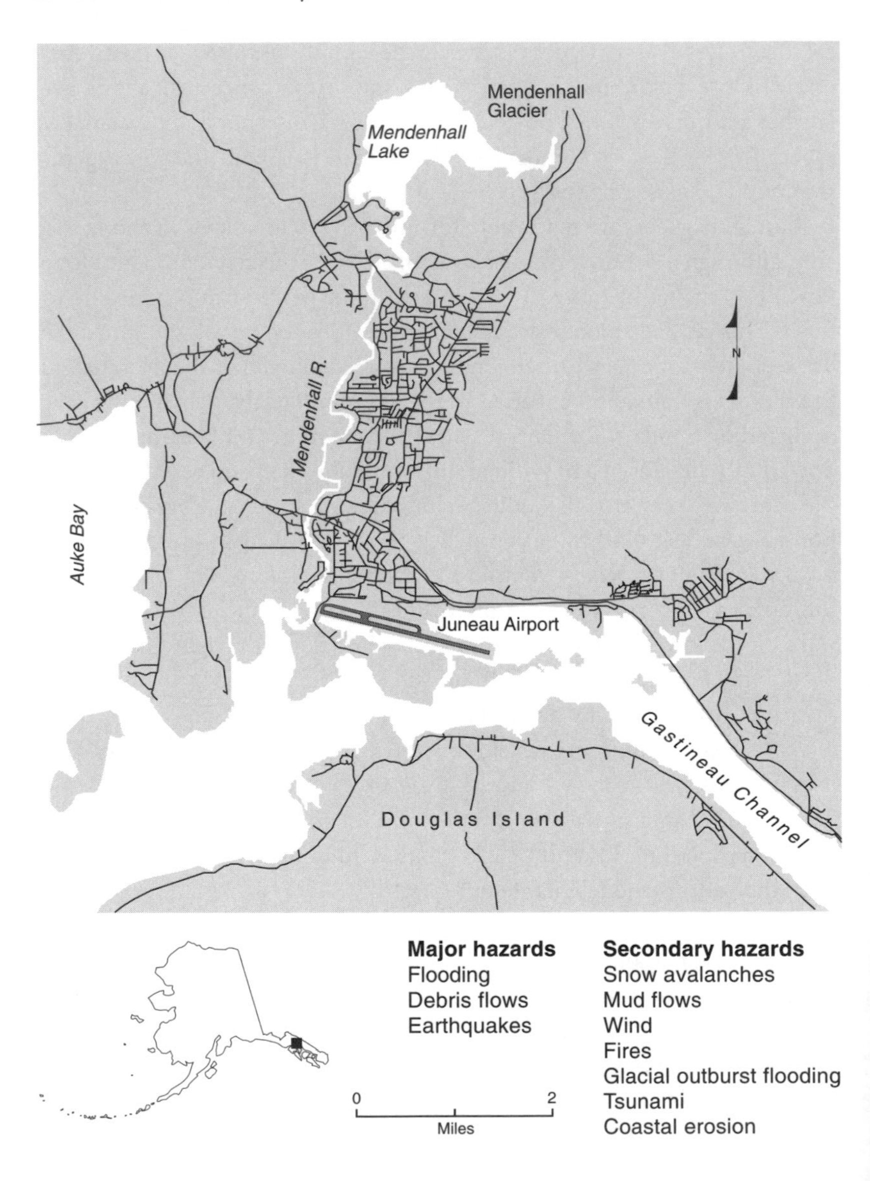

Drive. On the lower Mendenhall River, flooding may be exacerbated by high tide in Gastineau Channel. Rapid glacial retreat might also contribute to flooding on the Mendenhall River. Because Juneau participates in the Federal Insurance Administration program, it has defined various levels above river stage as probable 100- and 500-year flood elevations. Residents should view the elevations with caution in view of the limited data available to hydrologists working in Alaska. As discussed elsewhere, the term *500-year flood* does not mean that two such floods cannot occur in rapid succession. Bridges and culverts can complicate and enhance threats from flooding. The comprehensive plan of the Juneau Borough flatly states that numerous culverts and bridges in Duck and Jordan Creeks are inadequate and will contribute to backwater flows. Ice and debris can also block constricted drainages and culverts and contribute to flooding. Flood Insurance Rate Maps are available for the Juneau area and should be consulted by all concerned.

The Salmon Creek Reservoir, once a source of power for mining companies, now provides hydroelectric power to the city. The reservoir's dam is a very old, debris-filled, concrete-arch gravity structure. It is in need of repair, and as a safety measure the water level in the reservoir is kept quite low. In a worst-case scenario the dam could fail. A number of buildings are downstream from the dam although not necessarily in the flood path. The lower Salmon Creek area includes apartments, medical clinics, and a bus parking area. The hospital is also in the valley, but in a sheltered location. It might be cut off and isolated, but probably would not be flooded.

Shoreline erosion is not a problem for most of Juneau because much of the shoreline is naturally armored by large rocks and bedrock outcrops and because the area is experiencing a steady relative drop in sea level. Tide gauge measurements show that the land is rising nearly 1–2 feet per century as the land rebounds after glacial retreat, and probably as a result of tectonic activity as well. A sizable portion of the shoreline is seawalled. Of particular note in this regard is the long rock revetment protecting Egan Drive. The potential for shoreline erosion is probably greatest in Auke Bay because of the large fetch available to generate large waves during storms. At present the Auke Bay shoreline is stable, but the sand beach has recently been disappearing. No one has studied the cause of the beach loss, but it is probably a natural phenomenon.

A damaging earthquake could occur in the Juneau area, but most experts consider the area to be at moderate risk compared with the Anchorage–Cook Inlet region, which is in the high-risk category. Bear in mind that earthquakes can also unleash debris slides, avalanches, and mudflows. In fact, these subsidiary hazards may considerably outweigh the direct effects of earthquake motion. Much of the flat area of downtown is built on man-made fill that could be disrupted during a large quake. Similar circumstances pertain to poorly consolidated gravels deposited by glaciers or alluvial fans. Much of Juneau is underlain by fine-grained sediments created during a series of glacial expansions and subsequent flooding by the sea. These could be mobilized by earthquake motions, especially during summer, when the moisture-sensitive clays would be most susceptible to disruption. It is significant that a considerable portion of the Juneau Borough is mapped as having poor or marginal foundation materials; builders and homeowners should consult borough planning documents for details.

The likelihood of a significant tsunami in the Juneau area is reduced by its location within the innermost of the southeast labyrinth of estuaries and channels, far from the open ocean and the seismically active Queen Charlotte fault system. Of course, a dangerous tsunami could be generated by an earthquake close to town. Auke Bay, with its western exposure to open water, is at greater risk. In any event, the impact of a tsunami will depend largely on the tidal stage at the time of impact. According to some reports, the maximum tidal wave that may have affected parts of Juneau after the 1964 earthquake was on the order of 8 feet (see appendix C, *The Tsunami of the Alaska Earthquake 1964*).

The most recent major storm surge struck Juneau on Thanksgiving Day 1984. The storm spun off from a huge low-pressure zone in the Gulf of Alaska squeezed against a high-pressure area on the mainland. Gusts of more than 90 mph were measured at the Federal Building in downtown Juneau, and meteorologists believe that winds exceeded 100 mph in Gastineau Channel. Some wave and water damage occurred in Juneau, but the same storm caused much more damage in Tenakee Springs, which faces a longer expanse of open water. The Thanksgiving Day storm killed 2 people and did $2 million in damage in southeast Alaska. Although this was an unusually intense storm, dangerous fall windstorms are fairly common. Commonly such storms approach from the southeast, resulting in winds blowing straight up Gastineau Channel, providing maximum opportunity to generate both waves and storm surge in Juneau.

The damage from storm surge depends on the tide stage. The Thanksgiving Day storm struck during spring tides, which greatly increased the amount of flooding in low-lying buildings along Gastineau Channel.

The nature of storm winds in this area is greatly affected by the shape of the land. Winds tend to be funneled into and concentrated in the valleys. High gusts are common at the valley openings, and damage potential is high because of venturi effects and interactions with winds from adjacent valleys (see the "High Winds" section in chapter 5).

Taku winds have been responsible for deaths and property damage in Juneau. In the old days, corrugated tin roofing was a particular hazard for pedestrians who ventured out during windstorms. Occasionally during the winter, Juneau experiences taku winds that flow downslope from nearby ridge tops, sometimes achieving velocities in excess of 100 mph. These dangerous storms are produced by a special set of atmospheric conditions that include temperature inversions at or near the ridge tops. The winds that blow off ice fields may last for a day or two and are localized. Douglas is particularly hard hit by such winds. Take special precautions to ensure that your building construction follows high-wind building codes. Tie down your roof if you live in Douglas!

Any home or community on a wooded hillside is susceptible to fire from spreading forest fires. Many areas of Juneau have densely packed houses through which fire could spread rapidly, especially at times of high wind. The 1993 fire that whipped through the nearby community of Tenakee Springs holds an important lesson for all of southeast Alaska. Local firemen express concern that fuel storage facilities are stored on the fringes of downtown Juneau.

Sitka (RMs 12.1, 12.7)

The old Russian capital of Alaska, Sitka (pop. 8,400), remains a fishing and tourist center—and the lumber mill east of town may yet operate again. The fishery off Sitka owes its existence to a complex interplay between nutrient-rich cold water from the North Pacific and warmer fresh water along the shallow and rocky shores of Sitka Sound. Sitka is situated on a low bench 60 feet above sea level beneath spruce-timbered hills. Numerous islands extend offshore, forming a natural breakwater between the town and the open ocean to the southwest. The sheared-off cone of Mount Edgecombe, a dormant volcano, looms on the western horizon about 15 miles away. Sitka's hub is its harbor and the adjacent rock knobs where the Russians built fortresses and encampments in 1799. Mount Edgecombe School and the airport are on Japonski Island, which is connected to the old town by a suspension bridge (fig. 12.3).

Sitka's two surges in population growth came in the early 1960s and the mid-1970s—the first when the pulp mill was built, and the second when the pipeline boom enlarged government services and brought added tourism. The 1960s boom left a legacy of crowded trailer parks, many in low areas along the coast. Hurried building by inexperienced contractors is also evident in the cedar bark fill under some home foundations. An engineer's evaluation of a homesite and design is essential in any sloping area, especially on Halibut Point Road.

Island-studded Sitka Sound opens west into the Pacific Ocean. This setting, while spiritually inspiring, leaves the town of Sitka vulnerable to ocean waves and high water. The single most immediate and deadly threat is from tsunami waves generated in the nearby Queen Charlotte fault system and in far-off earthquake centers in the North Pacific. So far, only numerous small

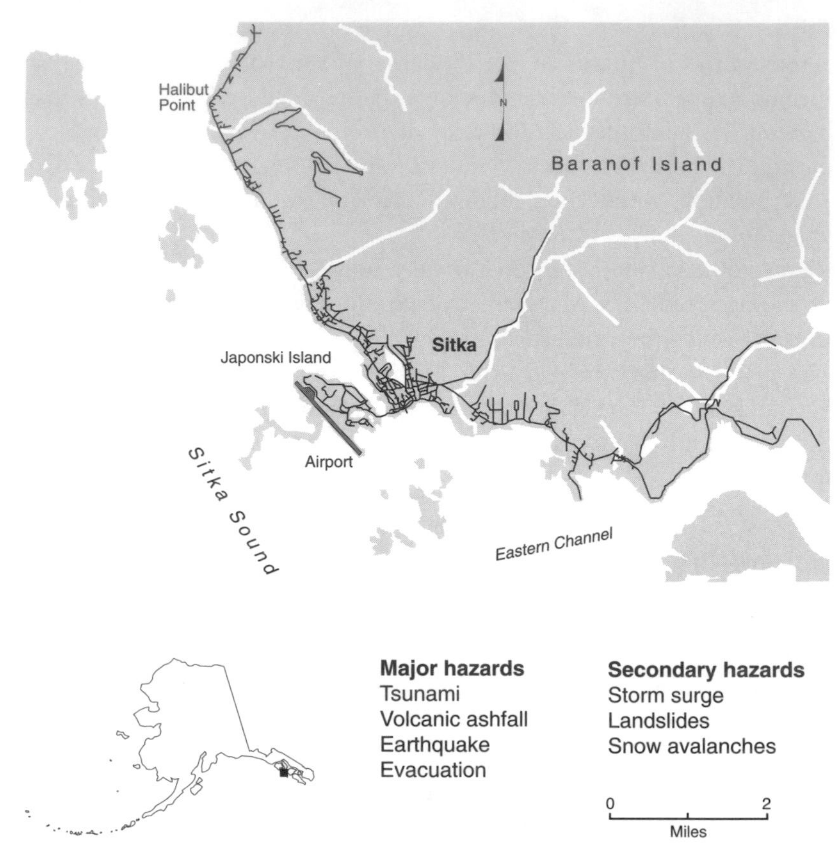

RM 12.7 Sitka

tsunamis have reached Sitka's shores; the largest, in 1964, produced only an 8-foot wave and did comparatively little damage. Nonetheless, the potential for a catastrophic tsunami is high here, particularly if one arrived at high tide. All land at elevations less than 50 feet and within 1 mile of the shore is in danger. This includes nearly half the populated area in town.

The presence of the offshore islands complicates the tsunami hazard. During the passage of a small tsunami wave, the passes between islands might act as constrictions and increase water levels, amplifying the wave by feet. By the same token, the islands might serve as a breakwater for a larger tsunami and lessen its effects. Significantly, most of the town's fuel storage tanks and hazardous chemicals are in locations threatened by tsunamis. The boat harbors vary in their degree of exposure.

Although the town is served by the Tsunami Warning Center in Palmer, people do not always react appropriately to warnings. In one case, parents crossed the bridge to Mount Edgecombe school to secure their children when a warning was issued. The resulting traffic jam could have produced numerous fatalities if a large tsunami had struck. The city now plans to move the school off Japonski Island. The airport is also seriously threatened by tsunamis but cannot be relocated. The threat of tsunamis is so great in Sitka that residents project a certain fatalism. "The whole town would be gone, puff," said one city official.

Earthquakes also pose a serious threat to Sitka; the town is less than 50 miles from the Queen Charlotte Islands fault, a major north-south lineament comparable in seismic activity and magnitude to the famed San Andreas fault in California. Several shocks of magnitudes 7–8 have been recorded since 1899. A 7.4-level quake in 1972 was only 25 miles from Sitka but did little damage. A severe quake could produce significant damage, and its

Figure 12.3 A view of downtown Sitka showing its proximity to the sea, the source of significant winter storms. Mt. Edgecombe rises in the background.

effects could be magnified in structures that sit on slopes or on unstable surfaces. Two settings are particularly unstable in Sitka: those underlain by muskeg, or bog, sediments of clay and silts, and those atop clays of volcanic ash, which present readily deformed slip planes. Problems with unstable ash sediments may continue offshore, where pilings may be destabilized by an earthquake.

The precise consequences of earthquakes in Sitka are difficult to anticipate because the extent, limits, and prevailing direction of movements are unmapped. Outside of structural damage or accidental deaths, the greatest long-term damage could be from tectonic subsidence or uplift. The former would disrupt dock facilities, and the latter would produce shoaling and render port facilities unusable.

Winter storms with a southwesterly fetch cause perennial problems to the Sitka waterfront, much of which is surfaced with artificial fill and rubble to augment bedrock outcrops. Every year storm waves push boulder-sized rocks onto the airport runway, forcing its closure for part of the day. High waves also undercut banks and remove gravel foundation fill from shorefront property, as well as goods stored on the beach. Structures may suffer damage if storm waves are coupled with high tides, as they were during the Thanksgiving Day storm of 1984, when part of the Seamart parking lot was washed out and several houses seaward from the Centennial Building were damaged by waves. Driftwood carried by waves could be hazardous, but most of it is removed from beaches by firewood entrepreneurs.

The prevailing longshore drift is northward here. Coastal erosion is a minimal hazard on bedrock cliffs, although the 3–5-foot-thick glacial sediments atop the bedrock may be subject to erosion.

Figure 12.4 Gravel beach in Sitka. Mount Edgecombe, in the distance, last erupted 4,400 years ago and is considered dormant but may pose a volcanic threat in the future.

Volcanoes could pose a serious threat to Sitka; eruptions bring acid rains and heavy ashfalls, and initiate earthquakes and tsunamis. The history of Mount Edgecombe's eruptions is poorly known, but present evidence classifies the peak as dormant (fig. 12.4). Nevertheless, inhabitants of Sitka should be prepared for volcanic eruptions.

Most of Sitka's residents live on the narrow coastal plain, and comparatively few are threatened by slope-related hazards. Two areas do experience recurrent debris or snow avalanches. Debris slides occur from an open scar on a slope just past the state vehicle shop at milepost 2 on Halibut Point Road, which has been closed by slides several times. Snow avalanches are common on slopes toward the mill about 3 miles east of town. Slope-related problems should be considered on newer homesites along Halibut Point Road to the ferry terminal.

Wrangell (RMs 12.1, 12.8)

Wrangell (pop. 2,481) is located on the northwest tip of Wrangell Island at the junction of Zimovia Strait with the inner Umner Strait and several channel arms and waterways, including the mouth of the Stikine River. Petroglyphs on the town's cobble beach attest to ancient residents, and historic Tlingit villages predate the present town's Russian America Company fort of Redoubt Saint Dionysius, built in 1834 to protect the fur trade. Over the years, Wrangell's natural deepwater port, one of the best in Alaska, has drawn fur traders, gold rush traffic on the Stikine, and modern timber and fishing commerce. Whether by chance or good choice, Wrangell is located in a relatively low-risk zone compared with several other southeast Alaska communities. Seated on an emergent marine terrace underlain by bedrock, the area is still rising at a nearly imperceptible rate of 0.25 inch or less per year.

Climatic hazards are the most frequent natural threats, but no more so, and usually less, than in other parts of Alaska. Strong winds funnel through the channels and down the Stikine Valley, usually in the equinox seasons of fall and spring. Flooding is not a threat in presently developed areas, except for a narrow zone along the shore on the north end of the development. Two or three houses in the vicinity of the petroglyph beach are at the edge of the sandy boulder beach, and homeowners to the south, adjacent to the beach, indicate that storm spray from waves occasionally reaches their windows. Shoreline erosion is not a problem on the resistant bedrock shores. Most of the houses along the shoreline are well elevated above the high-tide level (fig. 12.5). Storm surge is not reported or perceived as a threat. Concrete and rip-rap bulkheads are designed to hold the upland in place rather than to combat wave erosion.

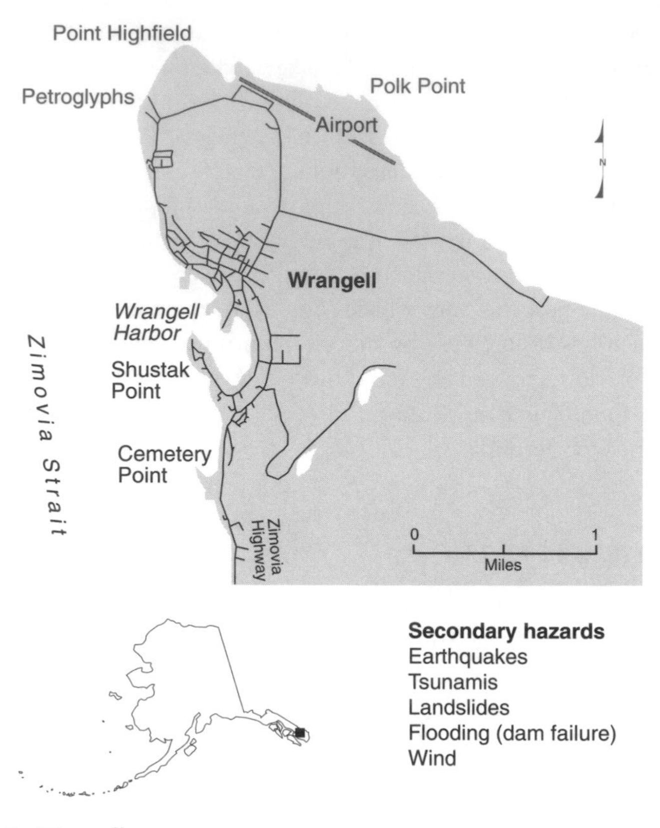

RM 12.8 Wrangell

Seismic hazards cannot be ruled out, but no serious damage or impact is reported in the 150 years of Wrangell's history. Great earthquakes have occurred along the Fairweather fault system, but the epicenters were far enough to the northwest to have little impact on Wrangell. Several modern earthquakes were felt in Wrangell (e.g., on July 10, 1958, and July 30, 1972), and earthquakes have broken submarine telephone cables in the region (e.g., August 22, 1949). Associated tsunamis are possible with earthquakes, and minor effects were observed in association with a 1945 earthquake and the 1964 quake. Evacuation was ordered after the 1964 Good Friday earthquake, but the tsunami came on a low tide. A record low water level was observed, revealing old boilers and other debris on the seafloor not seen before or since. Tsunami warnings should be taken seriously in the future in spite of the lack of past impact. Similarly, a great earthquake cannot be ruled out, and associated events of elevation changes, ground shaking, and ground (soil) failure are possible.

Human-caused hazards have done more damage in Wrangell than natural ones. The two greatest catastrophes in the town's history were the fires of

1906 and 1952, which burned down the waterfront. The March 1952 fire destroyed 22 properties at an estimated loss of $1,273,000 (in 1952 dollars). The town learned from these disasters, and the post-1952 reconstruction initiated greater reliance on artificial fill along the waterfront instead of wood pilings. Fill from past harbor dredging was used to create the land between Front Street and the harbor. The material was allowed to settle and was protected by rip-rap, and the area was later developed. The abundant high-quality rock in the area provides good fill material; however, seismic ground shaking could cause settling or failure in fill. Similarly, up from the dock, some housing sites were developed in areas of muskeg. Such sites have poor drainage and require pilings and suitable anchoring to withstand seismic ground shaking. The clustered buildings around the harbor, including the fuel storage and transportation facilities, remain a concern for the community from the standpoint of pollution, oil spill, and fire hazards.

Land slippage and slides are not common, but slides do occur on Wrangell Island. The most recent events were on steep slopes that were clear-cut and then subjected to heavy rain. In 1979 a major debris torrent damaged a trailer park and temporarily blocked the road, and in 1993 landslides closed the road along the north coast of Zarembo Island, west of Wrangell. Colluvium and glacial soil occur along the west side of Dewey Hill in Wrangell, and development there may induce slippage or suffer from poor drainage. The steep slopes above the town and south along Zimovia

Figure 12.5 Part of the Wrangell waterfront. Natural hazards are less important here than in other southeastern Alaska towns, although the slopes above town present some potential for slope failures or avalanches.

Highway should remain timbered and undeveloped. Concern has been expressed about the integrity of the old dam holding one of the town's two lake reservoirs. Failure of the dam would threaten downstream coastal development, including a trailer park.

Petersburg (RMs 12.1, 12.9)

Petersburg (pop. 3,207), Alaska's "Little Norway," was named after Peter Buschman, who left Norway with his family in 1891 and settled on the north end of Mitkof Island in 1897. The site was attractive not only for its natural harbor but also because of the abundant fish in nearby waters, the availabil-

RM 12.9 Petersburg

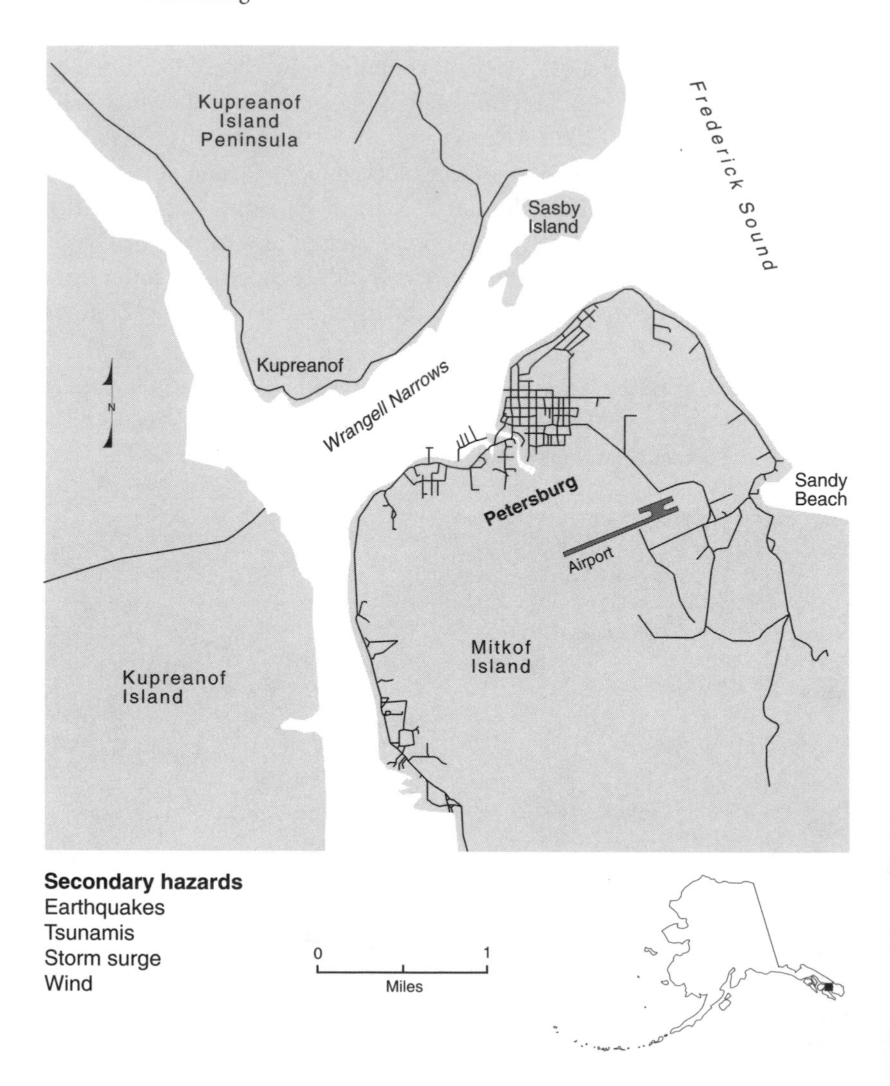

Figure 12.6 Aerial view of the Petersburg waterfront. This village is among the lowest-risk towns in Alaska from the standpoint of natural hazards.

ity of ice and timber, and the good building sites. Norwegian countrymen soon followed Buschman, and Petersburg was incorporated as a city in 1910.

Petersburg is located on gently sloping land, mostly less than 100 feet in elevation, covered by muskeg with occasional outcrops of bedrock and glacial sediments. Although most of Mitkof Island is rugged, there are no steep slopes to generate rock slides or avalanches into the town (fig. 12.6). Roads outside town may be subject to slides. The mean tidal amplitude at Petersburg is about 16 feet; spring tides are about 20 feet.

Occasionally icebergs from the LeConte glacier in LeConte Bay drift all the way across Frederick Sound and into Wrangell Narrows. Needless to say, these icebergs constitute a hazard to navigation, but they can also destroy piers and marina facilities. In the past, wayward bergs in Wrangell Narrows have been pushed away from the docks by tugboats.

In general, the muskeg and small streams around Petersburg can easily absorb the local rainfall and snowmelt. In 1991 the annual rainfall was a record 147 inches, and still no floods occurred. Nevertheless, stream and sheet flooding are possible in extreme 24-hour rainfalls. Storm surge flooding along any Petersburg shoreline is a stronger possibility, especially if the storm coincides with extra high tides. Buildings along Sandy Beach Road may be particularly susceptible to storm surge in some future major storm.

Petersburg participates in the National Flood Insurance Program, which means that 100-year flood levels have been calculated and are available in city hall for the asking. In order to obtain federal flood insurance it is necessary to build structures above the projected 100-year flood line. Unfortunately, this flood line does not include the height of storm waves above the still-water level. The prudent homeowner will elevate a few feet (4–5) above the law's requirements.

Shoreline erosion is not a serious problem in Petersburg, partly because the land is still rebounding after the retreat of the glaciers. Thus, sea level is, in a relative sense, dropping (at less than 0.5 inch per year), making the beaches wider. Destruction of rock revetments during storms has been fairly common along Sandy Beach Road. Seawalls and other forms of shoreline hardening are commonplace here.

One problem with hard structures is that they prevent shoreline erosion (which is why homeowners use them, of course), a process that furnishes fresh sediment to the beach on the upland during storms. A more immediate problem along Sandy Beach Road may be the recent tendency to build walls beyond the storm-tide line, presumably to get a better view of the sea. Having an irregular line of walls can eventually (in decades) create erosion and even storm surge problems and should be avoided.

Earthquakes pose a threat to Petersburg, but the likelihood of a big one occurring here is less than in communities farther west. The most active faults in southeastern Alaska are along the outer coast. A number of earthquakes have been felt in Petersburg (e.g., 1927, 1949, 1964, and 1972), but little damage has been reported in the local paper (the 1927 tremor broke a few windows).

Tsunamis are an ever-present possibility, as they are elsewhere in southeastern Alaska. Tsunamis can be generated locally or by earthquakes tens and even hundreds or thousands of miles away. Overall, the tsunami risk in Petersburg is low, but even a small tsunami could do moderate to severe damage if it happened to coincide with a high spring tide or a storm surge. During the 1964 earthquake a wave perhaps 3 feet high may have occurred in Wrangell Narrows, but this was probably a seiche (water sloshing back and forth as in a bathtub) rather than a true tsunami. According to the 1978 U.S. Geological Survey report by Lynn Yehle (see appendix C), an earthquake of magnitude 8 on the nearest of the major active faults (about 100 miles away) might produce a 3–7-foot tsunami.

On the Sukoi Islands (locally pronounced "sockeye"), people found a log 3.5 feet in diameter approximately 50 feet above the normal high tide line. The log had a lot of barnacles, and it could have been put there by a tsunami. An unknown but probably low potential exists for tsunami generation by underwater slope failure at the Stikine River delta and at the mouth of Thomas Bay. A large accumulation of muddy sediment, probably in an unstable configuration, lies at the river mouth adjacent to deep water. At the mouth of Thomas Bay, the old glacial moraine marking the most seaward extent of the local glaciers sits atop a steep slope leading to deep water. Shaking of either pile of sediment by an earthquake could lead to a large but localized tsunami. Such an underwater landslide caused the very damaging 1964 tsunami in Valdez. The possibility that Horn Cliff (east of Petersburg across Frederick Sound) could fail and cause a giant local tsunami is a hazard widely recognized by local citizens. During the 1970s a religious cult left Petersburg out of fear of a Horn Cliff tsunami. However, Paul Bowen, a Petersburg geologist who has studied local hazards, believes failure of the cliff is unlikely.

Ketchikan (RMs 12.1, 12.10)

Rising 3,000 feet from Tongass Narrows beyond the forested slopes of Revillagigedo Island, the southerly borough of Ketchikan (pop. 14,000), the "gateway of Alaska," remains foremost a fishing village and logging center. The steep stairways and colored frame houses on the labyrinthine streets lend a picturesque character to the historic downtown area, now the first Alaska port of entry for throngs of cruise ship passengers. Ketchikan continues to celebrate its pioneer past of thriving brothels and boasts a number of taverns that cater to a rowdy fishing and logging clientele. The town is defined by the steep topography of the island and the straightness of the coastline. Most residences and businesses are strung out along a two-lane road 20 miles long with only two stoplights. The airport is located on an island, requiring a ferry ride to town. A few substantial homes, cabins, and tent platforms are located on adjacent small islands approachable only from private docks. In summer, boat, kayak, and floatplane traffic is heavy on Tongass Narrows. The city of Ketchikan is only a small part of the inhabited area; the communities of Mud Bay and Ward Cove are to the north, and Saxman is to the south. Only the city of Ketchikan has piped-in water. Homeowners in the outlying areas rely either on delivery or rain catchment drums.

Ketchikan is affected by the same geological hazards that threaten most southeast Alaska communities, although the threats are not quite as dire as, for example, in Juneau. Glacially scoured metamorphic schist or phyllite bedrock lies only a short distance below the surface, and surficial deposits of any kind are comparatively thin. Some of the thickest sediments are on a prominent uplifted terrace of a 13,000-year-old sea level now about 35–40 feet above sea level. The present coastline is rocky and largely invulnerable to erosion. In addition, the strait between islands is narrow and prohibits the development of waves large enough to cause erosion or other damage. In some areas the shore fronts waters with greater fetch and may be threatened by storm surge; for example, newer development north of Mud Bay to

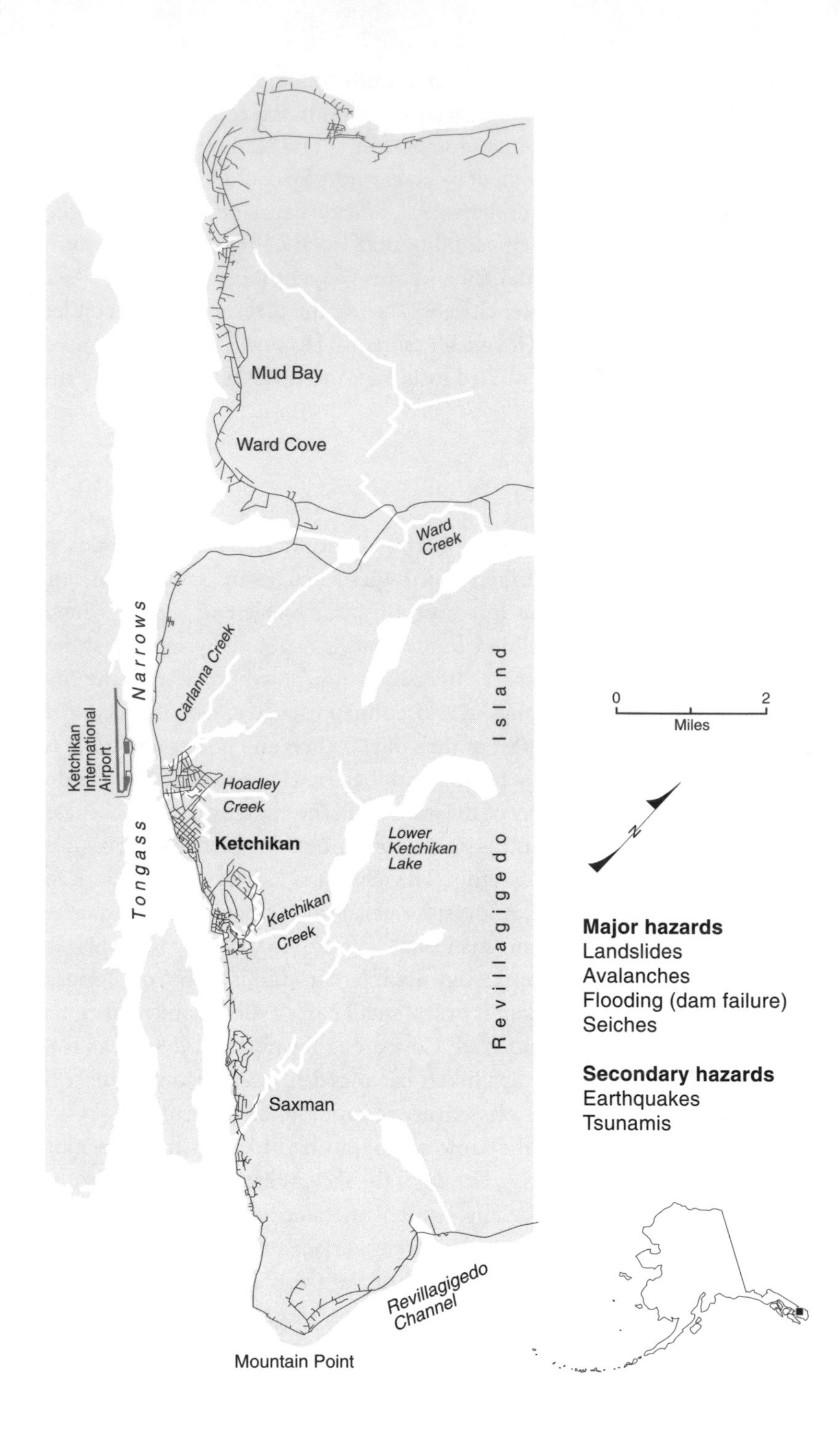

RM 12.10 Ketchikan

Clover Passage and south of Saxman to Mountain Point. Storm surges have only minimal effects unless coupled with high tides; only the northernmost part of the borough is susceptible to such surges. Few buildings are threatened by storms. Most of the shores in the city are artificial; seawalls or rock revetments underlie roads and enclose the boat harbor, docks, and most of downtown. The artificial protection differs little from the bedrock cliffs natural to the area.

A number of faults crisscross Ketchikan and its neighboring islands, but earthquakes are rare here. One large earthquake is documented in the last 200 years. Because much of Ketchikan is underlain by bedrock, earthquakes may present less of a threat than in other areas. Areas uphill from the waterfront can probably expect fewer earthquake effects. Residents living atop poorly consolidated materials (including man-made fill, fan deltas, streams, and modern beach deposits) should expect considerable shaking. Manmade fill underlies much of the waterfront, building pads, and parking lots. Deposits at the mouths of Ketchikan, Carlanna, and Hoadley Creeks are the most likely to amplify seismically induced shaking because of the combination of poorly consolidated sediments and a cap of man-made fill. Elevated marine deposits, if thick enough, may amplify seismic waves. Fan deltas and submarine fans offshore are susceptible to seismic-induced failure, and docks and piers may be undermined if the seismic activity is strong enough. Homeowners should follow earthquake-resistant building codes and use proper foundation materials to prevent liquefaction during a quake.

A potential tsunami threat does exist in the Ketchikan area, despite its sheltered location. Likely source regions include the Queen Charlotte fault, which produced a 0.3-foot tsunami in 1949, and the subducting Aleutian trench, which generated a 2-foot wave in 1964. Tsunamis occurring simultaneously with a high tide would cause the most damage. The Corps of Engineers estimates that a 10-foot tsunami is possible south of Saxman at Mountain Point along Revillagigedo Channel. Seiches generated in the reservoirs above Ketchikan could present serious problems if the dams failed under the pressure and water supplies were depleted. There are no masses of material offshore that could generate waves. Residents of "float homes" should be mindful of tsunami hazards.

The most threatening feature of the landscape arises from the hill slopes that lend Ketchikan its distinctive character. In general, the steeper the slope, the greater the hazard from loose rock or heavy snows. At higher elevations, avalanches can be a serious problem, but these follow known pathways that are plotted on official plats. Landslides also issue from steep upper slopes. Several areas may be subject to snow avalanches, landslides, or both, especially the steep slopes just south of downtown. Several avalanche tracks are documented on slopes adjacent to Ward Creek. As new development leads to deforestation, the possibility of landslides and avalanches will increase. Homeowners or buyers should consult the official plats, which are legally bound to delineate likely avalanche chutes, sites of recurring landslides, floodways, and V-notched drainages. The V-shaped character of some stream channels suggests that unusual or catastrophic rainfall will produce local flooding, a circumstance that can be avoided by building homes well above the 100-year flood level as shown on city hall or borough planning maps.

The entire downtown area lies downstream from earth-dammed Ketchikan Lakes, on the lower course of deeply incised Ketchikan Creek, which is influenced by tides at its lower end. The creek occasionally floods, as in 1963, although the prospect of flooding from the lake is much reduced because of the incised nature of the creek. Lake overflow during high tide could conceivably create a more serious problem. Failure of earthen dams, especially on Carlanna Creek as occurred in October 1974, is possible. The resulting flood destroyed a bridge and a mobile home park, cut off electricity and water, and severed the main highway.

Hydaburg (RM 12.1)

The fishing village of Hydaburg (pop. 388) is located on the southern part of Prince of Wales Island, 200 miles southeast of Juneau. Haida people have lived at the site for several centuries, certainly before Europeans arrived. The population of Hydaburg changes with the vicissitudes of the fishing industry. The village grew between 1910 and 1950, when fishing jobs were plentiful, and declined in the following decades until the next fishing boom in the late 1960s. The village is located on a narrow, low bench just above a small fan delta.

A variety of geological hazards may affect residents of Hydaburg. Erosion presents some hazard for those along the coastal strip, but the most severe hazard is reserved for those who live close to the slopes above the town to the north. Slope failure leading to debris flows and landslides is possible in several areas of town: (1) just above the church, (2) below the tennis courts, (3) up the road on the west to the marina, and, most seriously, (4) along the banks of the creek through the town.

Residents of Hydaburg could face major earthquake damage generated by seismic activity in the Queen Charlotte fault system, a north-south lineament that outlines the tectonic structure of southeast Alaska. This fault produced a 6.5 magnitude quake in 1956 from an epicenter only 85 miles southwest of Hydaburg. Hydaburg's reliance on floatplanes for transportation out of the area could make residents susceptible to an evacuation hazard in case of fire or excessive injuries incurred during other hazards.

Craig (RM 12.1)

Several Haida villages of unknown antiquity were located on the Craig townsite. Present-day Craig (pop. 1,260) owes its existence to the establishment of a cold storage facility and cannery at the site in 1908. Located on a small peninsula issuing from the west shore of Prince of Wales Island, Craig depends for its livelihood on logging, retailing, and fishing, activities known for pulses of growth followed by decline. The town rests on two 200-foot-tall hillocks covered by boulder-studded clay deposited by glaciers and an organic cover of forest debris formed over the last several thousand years.

As in most southeast Alaska communities, earthquake and tsunami hazards loom before Craig residents as unknown quantities. Depending on the percentage of sensitive clays underlying individual structures, residents should expect differing degrees of damage from large quakes. Half the town lies above the 100-foot contour and may be somewhat safe from a large tsunami. Storm damage and erosion are possibilities in view of the exposed nature of Craig to the southwest; more than 10 miles of open water extends across Bucarelli Strait. The steep slopes to the east of town, on the island proper, present a considerable hazard for debris flows, landslides, and small-scale avalanches. Slope-related problems will be heightened if logging continues to expose the underlying thin sediments. Craig's reliance on floatplanes for transportation outside the area could render residents susceptible to an evacuation hazard in case of extremely high seas, catastrophic fires, or tsunamis.

Metlakatla (RM 12.1)

Situated on Annette Island, Metlakatla (pop. 1,411) is the southernmost town in Alaska. Most of the town's residents are Tshimshian natives, and Metlakatla is one of the few Indian reservations in the state. Although Annette Island is as mountainous as the rest of southeast Alaska, the town of Metlakatla lies on a peninsula with low relief, less than 100 feet above sea level. Most of the town is built on 20-foot-thick uplifted pebble and sand beach deposits atop a metamorphic bedrock that outcrops along portions of the coast. The beaches represent the erosive action of rising seas on bluffs of glacial sediment deposited about 9,000 years ago. Glacial rebound since that time has upraised the beaches 50 feet. Glacial sediments still underlie a considerable part of the southwest portion of town. Some of the surface is a highly porous muskeg and decomposing organic debris peat several feet thick. Artificial fills lie beneath several areas in town and form part of the breakwater fronting the boat harbor.

The Metlakatla region is crisscrossed by a number of faults and is within striking distance of several of the major fault systems of southeast Alaska. The epicenter of the 1949 Queen Charlotte earthquake, of magnitude 8.1, was only 125 miles southwest of town, but no one in Metlakatla even noticed it. A 7.0 earthquake in 1929 on the closer Sandspit fault also caused no damage or concern. Even considering the likelihood of much stronger quakes from the two faults, the seismic hazard in the Metlakatla region is still ranked as moderate. The amount of risk is directly proportional to the type of materials underlying the buildings and other structures in town. The most dangerous setting is muskeg or artificial fill with a considerable amount of muskeg. The least dangerous (but still susceptible) setting is the emerged beach deposits and the metamorphic rocks.

At least a dozen tsunamis may have crested at Metlakatla during the last 100 years, but no precise records are available. The 1964 earthquake apparently did not produce a tsunami here. A tsunami with breaking waves arriving at high tide could inflict severe damage on shoreline facilities such as the harbor and fuel tanks. Similarly, the town could be seriously affected by any seiche that affected the outlets of the various lakes that supply power or water. Locally generated tsunamis are thought to be of little consequence for Metlakatla. Similarly, waves from storms usually dissipate before reaching the town, although the possibility exists that large waves from winter storms or landslides could affect northern Annette Island.

Part IV

Responding to Alaska's Hazards

Chapter 13
Mitigating Wind, Loading, and Permafrost Hazard Impacts through Construction

Deborah Pilkey

Design Requirements: High Wind

Strong winter winds with speeds of 100 mph and greater are common in many parts of Alaska and are among the most destructive forces to be reckoned with in the Alaskan coastal zone. Figure 13.1 illustrates the effects of high winds and other storm forces on structures.

Winds can be evaluated in terms of the pressure they exert. Wind pressure varies with the square of velocity, the height above the ground, and the shape of the object against which the wind is blowing. For example, a 100 mph wind exerts a pressure, or force, of about 40 pounds per square foot (psf) on a flat surface such as a sign. The effect on a curved surface such as a sphere or cylinder is only about half that. If the wind increases to 190 mph, the 40 psf force increases to 140 psf on the flat surface. Wind velocity increases with height above the ground, so a tall structure is subject to greater velocity at its top—and thereby greater pressure—than a low structure. The velocity and corresponding pressure at 100 feet above the ground may be almost double that at ground level.

A house or building designed for areas not subject to strong winds can be built primarily to resist vertical loads, assuming that the foundation and framing must support the load of the walls, floor, roof, and furniture against relatively insignificant wind forces. A well-built house in a high-wind area, however, must be constructed to withstand strong winds that may come from any direction. Although many people think that wind damage is caused by uniform horizontal pressure (lateral loads), most damage is actually caused by uplift (vertical), suctional (pressing outward from within the house), and twisting (torsional) forces (fig. 13.2). High horizontal pressure on the windward side is often accompanied by suction on the leeward side. The roof is subject to both downward pressure and, more important, to uplift. A roof may be sucked up by the uplift drag of the wind. Usually, the structural failure of a house results from deficiencies in the devices that

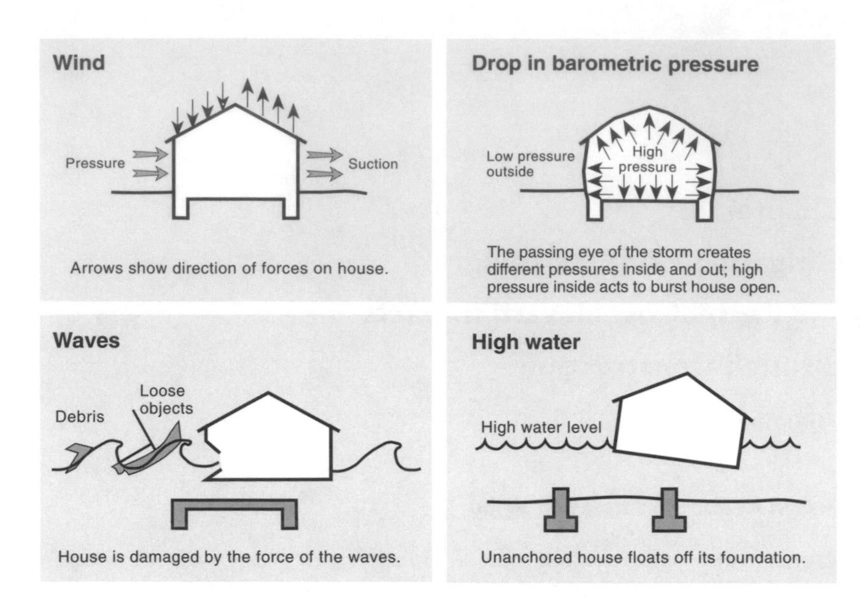

Figure 13.1 Natural forces to be reckoned with in the coastal zone.

tie the parts of the structure together. All structural members (beams, rafters, and columns) should be fastened together on the assumption that there will be strong winds and about 25 percent of the vertical load on any member may be a force coming from any direction (sideways or upward). Such structural integrity is especially important if it is likely that the structure may be moved at a later date to avoid destruction due to impending shoreline retreat or other hazards.

Exterior

A building's envelope is its protective shell. By *envelope* I mean the entire system by which the building resists wind penetration. A breach in the envelope occurs when an exterior enclosure fails, such as when the garage door or a window is open. When this happens during a strong wind, the pressure inside the structure builds up and roof uplift or wall suction may occur, eventually causing the entire system to fail.

It is necessary to secure all exterior openings in a house. The most susceptible parts of the house are the windows, garage doors, and double-wide entry doors. If a building's envelope is breached, major consequences are likely to result. Wind, rain, and snow can damage the interior contents of the structure. Internal pressure buildup can cause partial or complete blowout of the walls or the roof.

All doors in a home should be certified by the seller as to their strength under a given wind load. This is especially important for garage doors and double entry doors. The strength must be adequate to prevent damage from projectiles or flying objects such as tree limbs. Locking systems that use wind-resistant latches and dead bolts can be used to upgrade existing doors. The dead bolt will act as an additional rigid connection to the house frame. If the house has double entry doors, one door should be fixed at the

Figure 13.2 Modes of failure due to winds or waves. Modified from U.S. Civil Defense Preparedness Agency publication TR-83.

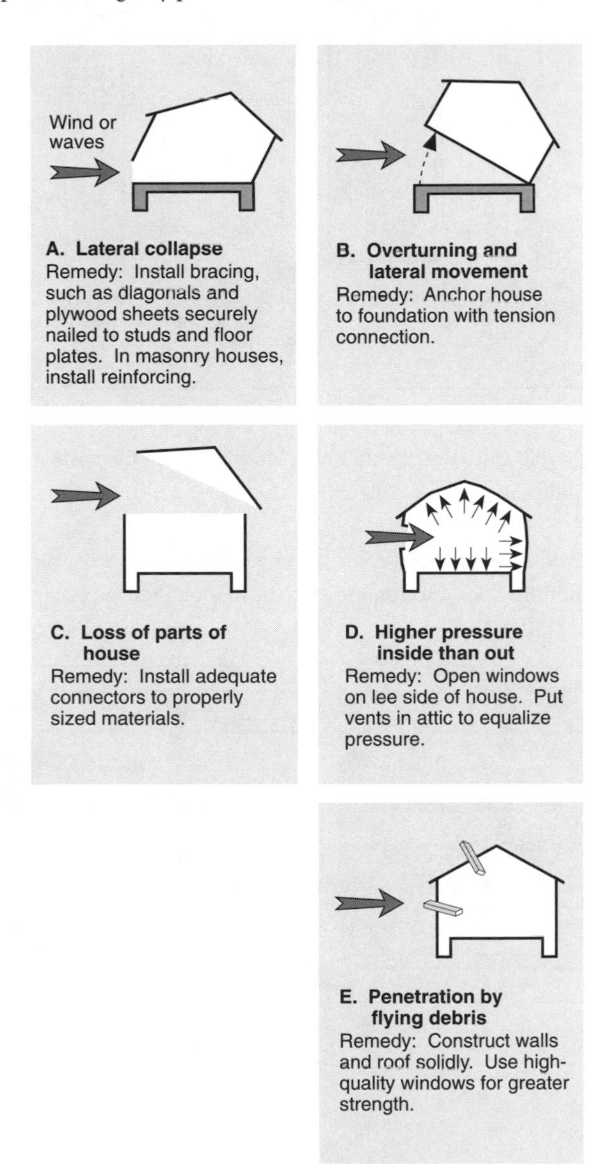

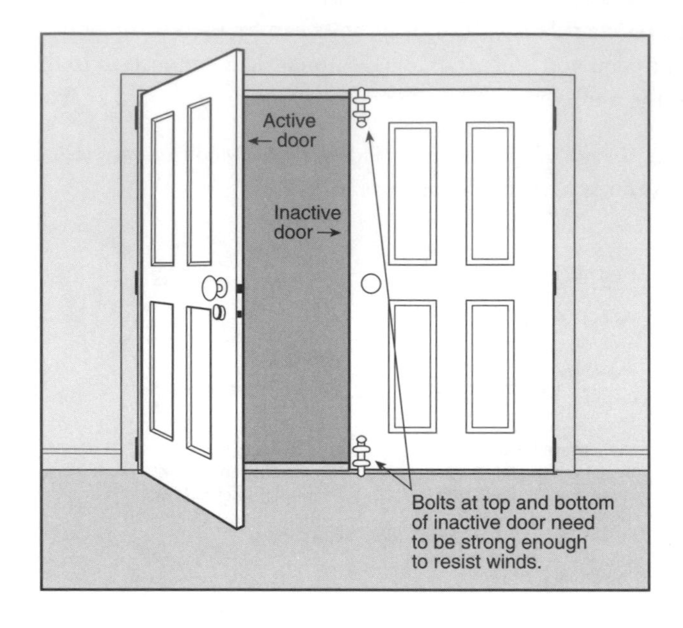

Figure 13.3 Double entry doors with bolts. Modified from the FEMA and American Red Cross publication *Against the Wind.*

Figure 13.4 Double-wide garage doors: wind-resistant features or where to retrofit. Modified from the FEMA and American Red Cross publication *Against the Wind.*

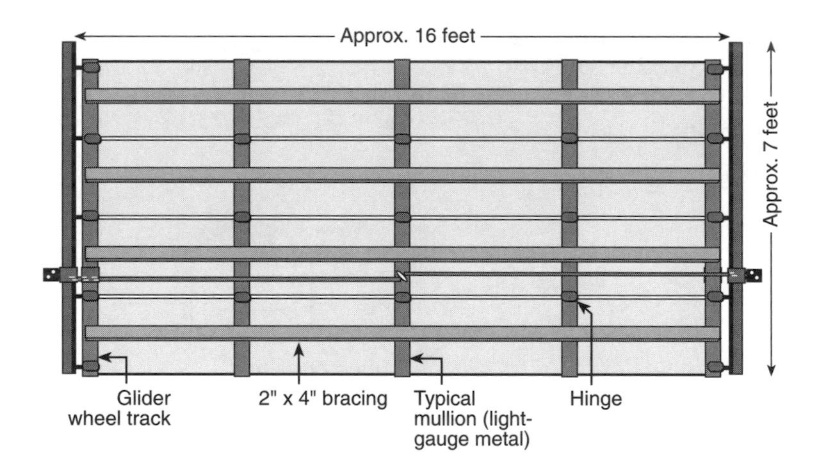

top and bottom with pins or bolts (fig. 13.3). Usually, the original pins are not strong enough to resist heavy wind forces. Homeowners should consider installing heavy-duty bolts (your local hardware store will be able to advise you regarding the strength of the bolts or pins).

Garage doors also pose a risk to the building envelope. Often, garage doors fail because of inadequate thickness. They tend to bend when subjected to strong winds. Double-wide, or two-car, garage doors are especially susceptible to forces of high winds. New and existing garage doors can be fixed to alleviate this problem. When purchasing a new garage door, study the manufacturer's certification of its strength to verify the adequacy of the system. Existing garage doors can be retrofitted by installing horizontal girts on each panel (fig. 13.4). A temporary strategy to reinforce your garage door during a windstorm is to back your car up against the inside of the door. This provides extra support against bending. Strengthen the track supports and glider tracks to prevent the garage door from falling off its tracks during high winds. It is also necessary to reduce the rotation of the door along its edges by chaining the door pin to the glider track connections.

The windows in your home—including skylights, sliding glass doors, and french doors—must be protected from missiles that could penetrate the building envelope. During a storm, missiles may be branches, roof pieces, or anything picked up by the wind. Windblown materials from poorly constructed buildings are a threat to pedestrians as well. Windows can be protected by using storm shutters. It is important to have strong windows to protect the structure against high winds. In commercial or office buildings where shutters are impractical, windows can be reglazed. Reglazing is a method of strengthening the windows that requires replacing normal glass with tempered laminated glass. This is an expensive option, however; it can cost up to 10 percent of the entire building cost. Finally, all windows and doors should be anchored to the wall frames to prevent them from being pulled out of the building. After missile impacts, this is the second most common mode of failure of the building envelope. Pay close attention to the manufacturer's specifications of wind resistance for shutters and windows. The town manager's office should be able to help you choose the right materials. In Alaska, local experience can be a useful guide in high-wind construction.

Roof

When a roof fails, either by losing its shingles or by flying off the house, it can spell disaster for the rest of the house and its contents. In high-wind areas, roofs can fail for a number of reasons, including inadequate tie-downs of the roof framing and poor connections of the roof to the wall compo-

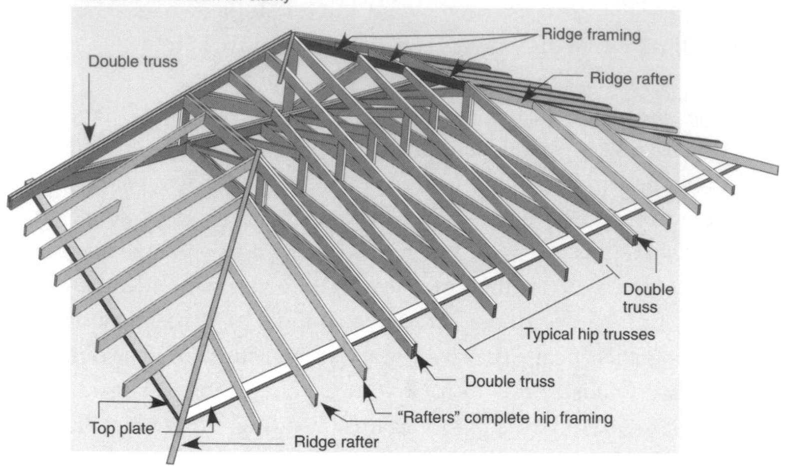

Figure 13.5 Wind-resistant hip roof framing.
Modified from FEMA publication FIA-22.

nents. Poorly attached roof sheathing and poorly placed asphalt on roof shingles can also cause failure.

The shape of the roof is an important consideration. A hip frame roof fares far better than a gable roof in high winds (figs. 13.5–13.8). The pitch angle of the roof also has an effect on its wind resistance. Steeply pitched roofs are better at resisting wind forces than low-pitch roofs because low-pitch roofs have higher uplift pressures exerted on their windward sides. A pitch angle greater than 40 degrees may prevent suction on the windward side of the structure.

To protect the home's contents, the roof's coverings must be correct and securely attached. Shingles can be rated by the manufacturer and recommended as satisfactory for high-wind areas. Metal is the least acceptable of all roof coverings. In Juneau, pedestrians have been killed by flying metal roof components. Asphalt shingles have also performed poorly because poor fastening techniques are often used. If your roof is covered with metal or asphalt, it might be wise to change to wood shingles, which have a history of performing well in high-wind areas. Unfortunately, wood shingles also increase the risk of fire.

Look around your neighborhood. What has worked in the past? Your town hall should have information that will be helpful in choosing proper roofing.

The roof trusses must be strong enough to withstand design wind loads. Structural rigidity can be created by using braces and connectors. Secondary bracing within the truss system can help resist lateral wind forces. An

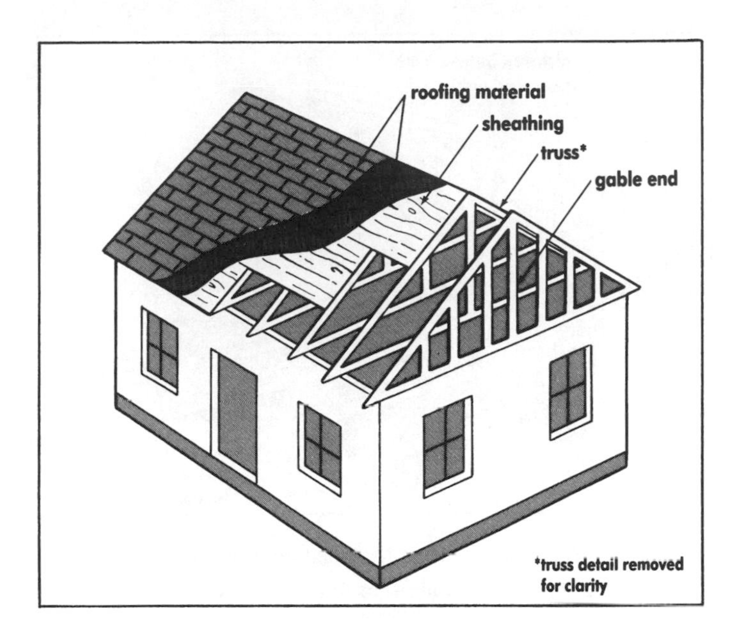

Figure 13.6 Wind-resistant gable roof design. Modified from the FEMA and American Red Cross publication *Against the Wind.*

Figure 13.7 Truss bracing for gabled roof. Modified from the FEMA and American Red Cross publication *Against the Wind.*

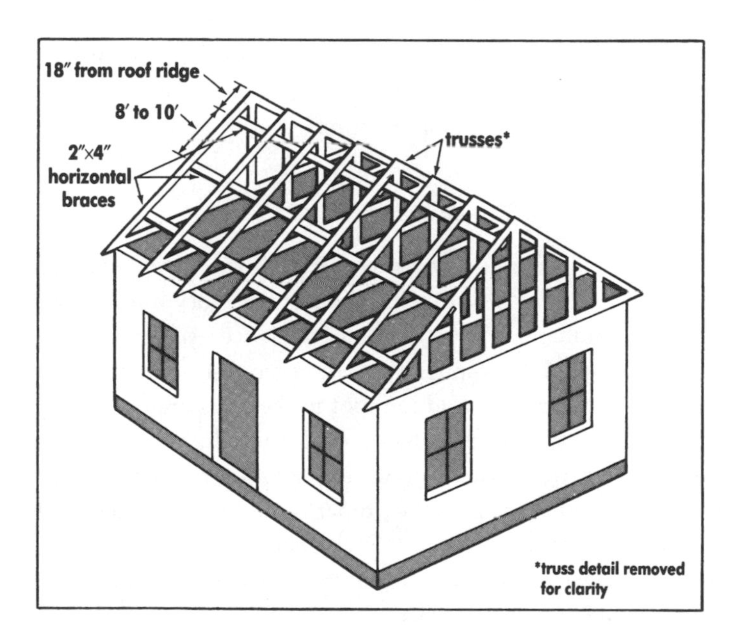

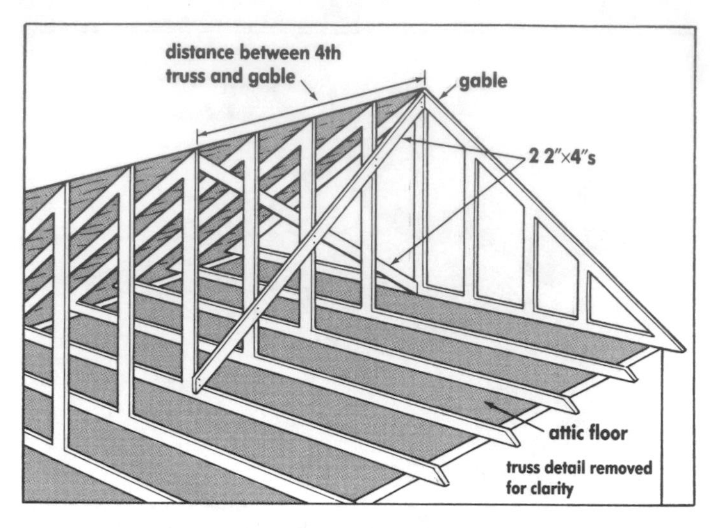

Figure 13.8 Gable end bracing. Modified from the FEMA and American Red Cross publication *Against the Wind.*

inherent method of bracing is accomplished by substituting hip roofs for gable roofs. Retrofitting gable roofs may be necessary to strengthen them. Figures 13.7 and 13.8 show how to place two-by-fours to strengthen your gable roof. In addition to strengthening the trusses, the overhang must be considered. To prevent roof failure there should be minimal overhang—only the distance required for proper drainage.

Finally, the roof must be vented to relieve internal pressures. Adequate openings can reduce induced pressures on the roof structure. Venting must be installed to exclude the entry of any uncontrolled air flow, which could result in a buildup of internal air pressure.

High-Wind Straps and Tie-downs

A continuous load transfer path is needed if a house is to remain structurally intact under extreme loads. This means that everything must be connected to the foundation. Structural integrity is achieved by using fasteners and connectors at all the joints, as shown in figure 13.9. In areas with high winds, the connectors that hold the roof to the walls are commonly called hurricane straps (fig. 13.10); tie-downs or anchor bolts hold the house to the foundation. High winds cause uplift and lateral forces on the structure's girders, trusses, and beams, but the proper connectors can transfer the load away from these vulnerable areas. Such reinforcement could reduce the structural damage or even save the home. It has been shown that wood structures reinforced with metal connectors perform better during windstorms than unreinforced structures.

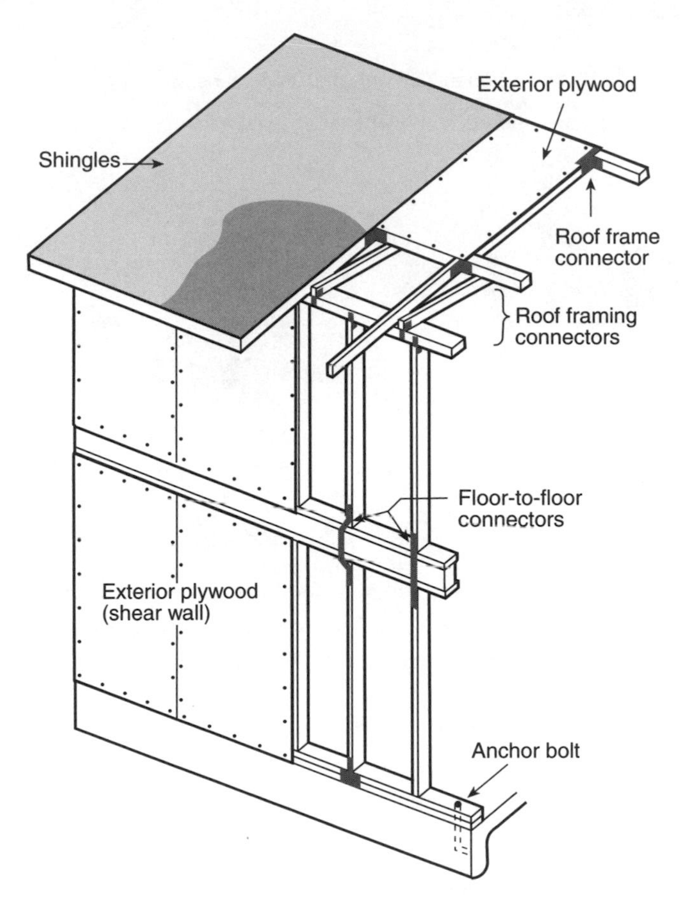

Figure 13.9 Recommended continuous load path design for wood-frame construction. Modified from FEMA publication FIA-22 and Simpson Strong-Tie Company, "Connectors for High Wind Resistant Structures," 1992.

A shear wall is an important component to help a house resist lateral loads. Plywood is an excellent shear wall when it is nailed to the building frame accurately and completely. The larger the nails and the closer together they are, the better the plywood will perform as a shear wall. The plywood should be attached at all levels of the building.

A multistory home must have floor-to-floor connectors to transfer the load path correctly. The first floor should be connected to the second floor with either nailed ties or bolted hold-downs to transfer the uplift forces from the upper stud to the lower stud. In addition, all homes must have connectors in the rafters and trusses.

The load transfer path includes a tie-down to the foundation of the building. This generally involves attaching a metal rod from the house frame into the concrete foundation (fig. 13.9). A professional should check

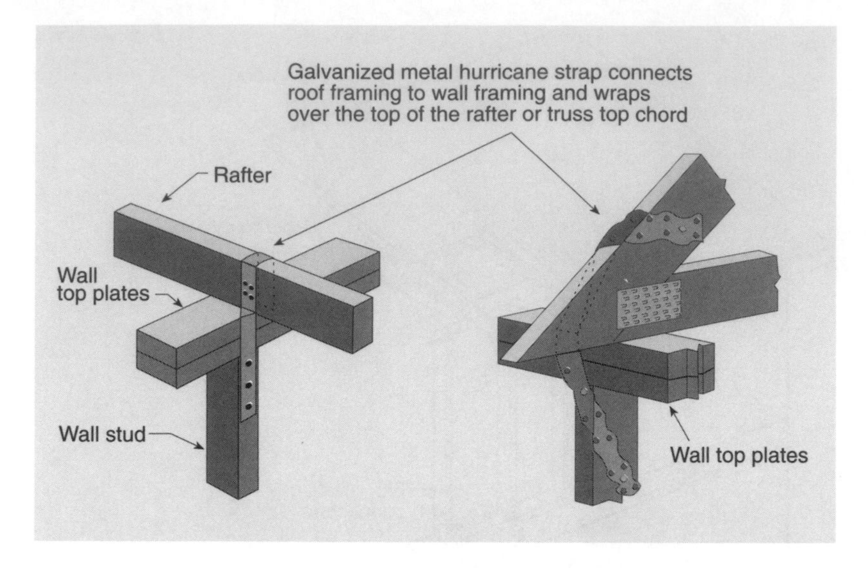

Figure 13.10 Wind-resistant strap-to-roof framing for rafter or prefabricated roof truss designs. Although called "hurricane straps," the connectors should be used in any construction subject to high winds. Modified from FEMA publication FIA-22.

the foundation of a house for termite infestation and dry rot damage and to see if it is compatible with the planned tie-down system. Remember, if any component of the house is not connected, the whole house could fail.

Mobile Homes: Limiting Their Mobility

Because of their light weight and flat sides, mobile homes are particularly vulnerable to high winds. Such winds can overturn unanchored mobile homes or smash them into neighboring homes and property. Millions of Americans live in mobile homes today, and the number is growing. Some 20–30 percent of single-family housing in the United States consists of mobile homes. High winds damage or destroy nearly 5,000 of these homes every year, and the number will surely rise unless greater protective measures are taken.

Several lessons can be learned from past storm experiences. First, mobile homes should be properly located. Placing mobile home parks in woods with the units close together will minimize wind damage. Mobile homes in unprotected areas have been overturned and even destroyed by the force of the wind. The protection afforded by trees is greater than the possible damage from falling limbs. Two or more rows of trees are better than a single row, and trees 30 feet or more in height give better protection than shorter ones. If possible, mobile homes should be positioned so that a narrow side faces the prevailing winds.

Locating a mobile home in a hilltop park greatly increases its vulnerability to the wind. A lower site screened by trees is safer from the wind, but it should be above storm surge flood levels. A location that is too low increases the likelihood of flooding in a flood-prone area. Slope gradient is also important in Alaska.

A second lesson taught by past experience is that the mobile home must be tied down or anchored to the ground so that it will not overturn in high winds (figs. 13.11 and 13.12, table 13.1). Simple prudence dictates the use of tie-downs. Many insurance companies, moreover, will not insure mobile homes unless they are adequately anchored with tie-downs. A mobile home may be tied down with cable or rope, or may be rigidly attached to the ground by connecting it to a simple wood-post foundation system. A conscientious mobile home park owner will provide permanent concrete anchors or piers to which hold-down ties can be fastened. In general, an entire tie-down system costs a nominal amount.

A mobile home should be properly anchored with ties to the frame and over-the-top straps; otherwise, the home may be damaged by sliding, overturning, or tossing. The most common cause of major damage is the tearing away of most or all of the roof. When this happens, the walls are no longer adequately supported at the top and are prone to collapse. Total destruction of a mobile home is more likely if the roof blows off, especially if the roof blows off first and then the home overturns. The need for anchoring cannot be overemphasized. Single mobile homes up to 14 feet wide should have over-the-top tie-downs to resist overturning and frame ties to resist sliding off the piers. Double-wides do not require over-the-top ties, but they do require frame ties. Although newer mobile homes are equipped with built-in straps to aid in tying down the system, it might be wise to add more straps if the home is in a particularly vulnerable location. Many older mobile homes are not equipped with built-in straps.

Also, keep in mind that most tie-down systems are designed to withstand 70–110 mph winds. If at all possible, evacuate the mobile home when high winds are expected.

For more information, obtain a copy of *Manufactured Home Installation in Flood Hazard Areas* from FEMA (see appendix C).

Snow and Ash Loading

Buildings in areas of high snowfall or potential volcanic ashfall are subject to above-normal loading. The bearing strength of the roof and walls must be adequate to tolerate such loads without failing. Most local building codes take this into account; however, property owners may wish to consult with the community building inspector or a knowledgeable contractor to be sure.

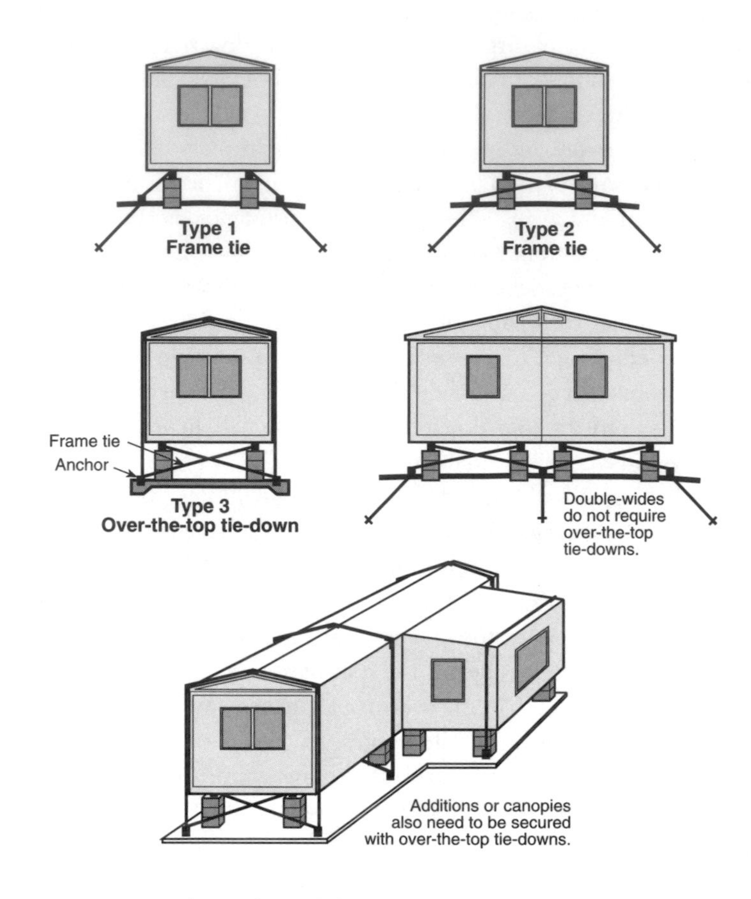

Figure 13.11 Tie-downs for mobile homes. Modified from *Protecting Mobile Homes from High Winds.*

Table 13.1 Tie-down Anchorage Requirements

	10- and 12-Foot-Wide Mobile Homes				12- and 14-Foot Wide Mobile Homes, 60–70 Feet Long	
	30–50 Feet Long		50–60 Feet Long			
Wind velocity (mph)	Number of frame ties	Number of over-the-top ties	Number of frame ties	Number of over-the-top ties	Number of frame ties	Number of over-the-top ties
70	3	2	4	2	4	2
80	4	3	5	3	5	3
90	5	4	6	4	7	4
100	6	5	7	5	8	6
110	7	6	9	6	10	7

In the case of severe snowfall or ashfall, remove the material as soon as possible to avoid structural damage (see appendix A).

Building on Permafrost

More than 85 percent of Alaska is underlain by permanently frozen ground, or permafrost (fig. 13.13). In the continuous permafrost zone, subsurface temperatures remain below -5°C, whereas the temperature ranges from -5° to 0°C in the discontinuous permafrost zone. The thickness of the permafrost layer varies from about 2,000 feet at Prudhoe Bay to 100–350 feet at Fairbanks. Although frozen ground can be as solid as rock, building on permafrost presents special problems because the ground near the surface can experience a thaw during the warm season. Depending on the location and the type of soil, the thaw may penetrate up to 15 feet. Heat transferred from the foundation to the ground can also cause thawing.

Buildings on permafrost should be designed to minimize ground thaw. The flow of heat from the building foundation to the ground must be controlled such that the ground remains frozen throughout the life of the

Figure 13.12 Hardware for mobile home tie-downs. Modified from *Protecting Mobile Homes from High Winds.*

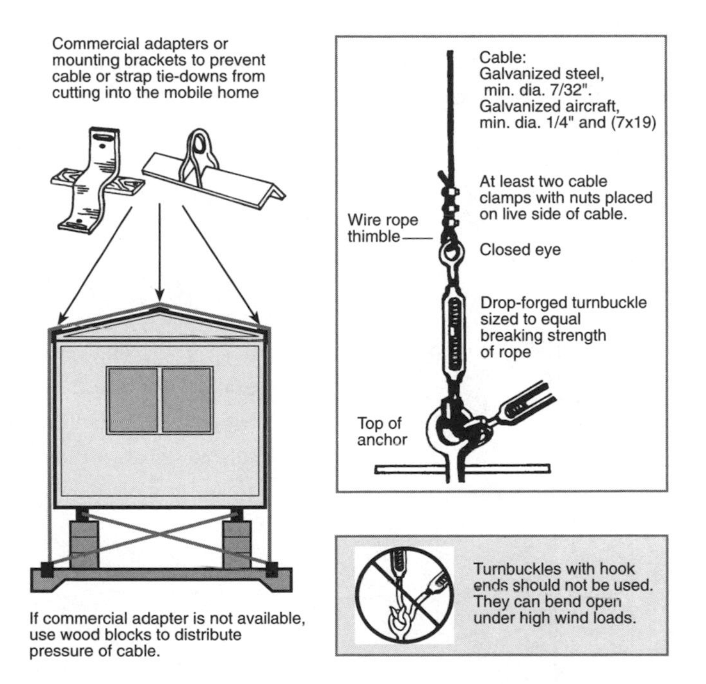

Figure 13.13 Regional distribution of permafrost within Alaska. Modified from *Permafrost and Related Engineering Problems in Alaska,* by O. J. Ferrians Jr., R. Kachadoorian, and G. W. Greene.

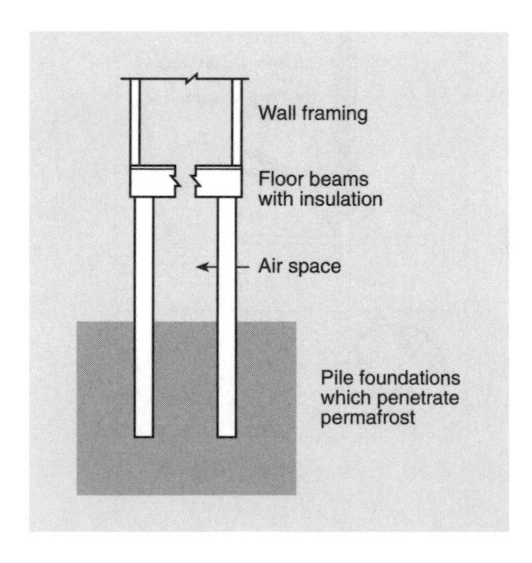

Figure 13.14 Foundation system using air space and pilings to avoid thawing permafrost. About 3 feet of vertical air space between the ground and the insulated structure floor is adequate to avoid thawing.

structure. Normally, the foundation is designed to minimize the contact area through which heat can flow from the foundation into the ground.

Some permafrost is classified as stable when thawed; the rest, which usually contains considerable quantities of ice, is identified as "thaw-unstable." In order to preserve the permafrost in thaw-unstable ground, the usual practice is to employ a pile foundation with an air space between the structure and the ground (fig. 13.14). If there must be a large area of contact between the ground and the structure, drained granular fills can be placed between an insulated slab and the ground surface. Sometimes ducts for circulating cold air are placed in the fill.

14 Earthquake-Resistant Design and Construction

Jane Bullock

Each new earthquake provides insights on how design and construction techniques influence the level of damage to houses and other buildings. For example, the 1994 Northridge, California, earthquake revealed that certain techniques for anchoring outside veneer to the framing of houses were inadequate, and that stucco applied over wall studs did not meet expectations for resisting seismic forces.

At the same time, every earthquake reinforces two fundamental concepts about seismic-resistant design and construction: (1) ground shaking causes more than 90 percent of all earthquake damage to homes, and (2) earthquake forces seek out structural weaknesses in buildings. These two principles were reconfirmed during the 1989 Loma Prieta earthquake. Significant damage occurred to houses built on soft soil with shallow groundwater as far as 60 miles from the earthquake's epicenter. In addition, most of the damage was to houses with unreinforced masonry foundations and those not anchored to their foundations.

Few people will claim that they can build an earthquake-proof house, but when certain fundamental principles of design and construction are applied and followed, the likelihood and extent of damages from earthquake forces are significantly reduced. The following outline provides a basic explanation of how structures react to earthquake forces and points out design and construction concepts that should be considered and understood by anyone purchasing or constructing a house in an earthquake-prone area. Many of these techniques are applicable to and complement design requirements for high winds and other hazards. In most cases, application of these techniques is not very costly, results in better-built structures, and provides a longer life for buildings.

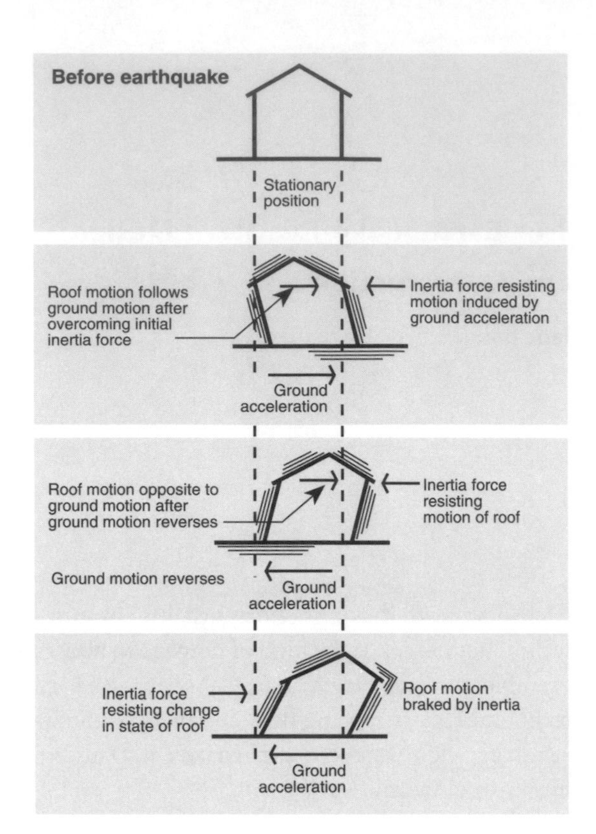

Figure 14.1 Inertial forces and ground motion cause shaking of structures.

Building Reaction to Earthquake Forces

Earthquakes create forces in structures because of inertia—the tendency for a body at rest (in this case the house or building) to resist motion, and for a body in motion to remain in motion. The forces acting on the structure depend on the direction of the ground motion caused by the earthquake and act side to side (horizontally), up and down (vertically), or both. During an earthquake, a structure resists the initial movement of the ground, but eventually, as the ground shaking continues, the structure begins to move back and forth with the ground until the shaking stops and the earthquake is over. The largest earthquake-related forces are usually horizontal or lateral forces acting parallel to the ground. In houses, the effects of vertical forces are normally considered to be offset by the weight of the building. Figure 14.1 shows how inertia acts on a building frame subjected to back-and-forth motion parallel to the ground. Similar effects would occur if the ground were stationary and a horizontal force was applied in a back-and-forth manner at the roof line.

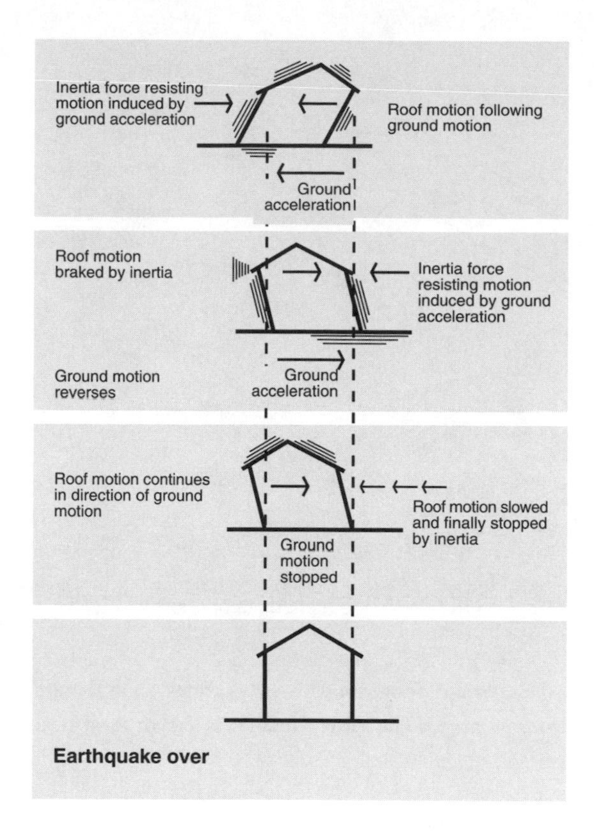

Elements of Earthquake-Resistant Design and Construction

Siting

The people of Alaska are subject to every form of risk connected with earthquakes. Besides the obvious damage caused by fault rupture and/or ground shaking caused by deep faults, Alaskans face the collateral problems of liquefaction, subsidence, and tsunamis. Liquefaction and subsidence caused significant loss and damage in the 1964 earthquake. Thus, it is crucial that the homeowner or buyer understand the nature of the ground under and around the house. Site evaluation often requires professional guidance and consultation, but it is well worth the cost and effort. Obvious areas to avoid are landslide-prone sites, marshes or other water-saturated soils, ocean bluffs, organic fills and alluvium, and loose, shifting soils at the base of hills.

The best sites to consider are stable and solid rock formations. Deep bedrock provides the least hazardous conditions and minimizes earthquake damages to dwellings. Natural sites that consist of firm, consolidated deposits of well-drained soil, either flat or sloping, and stable hillside slopes are

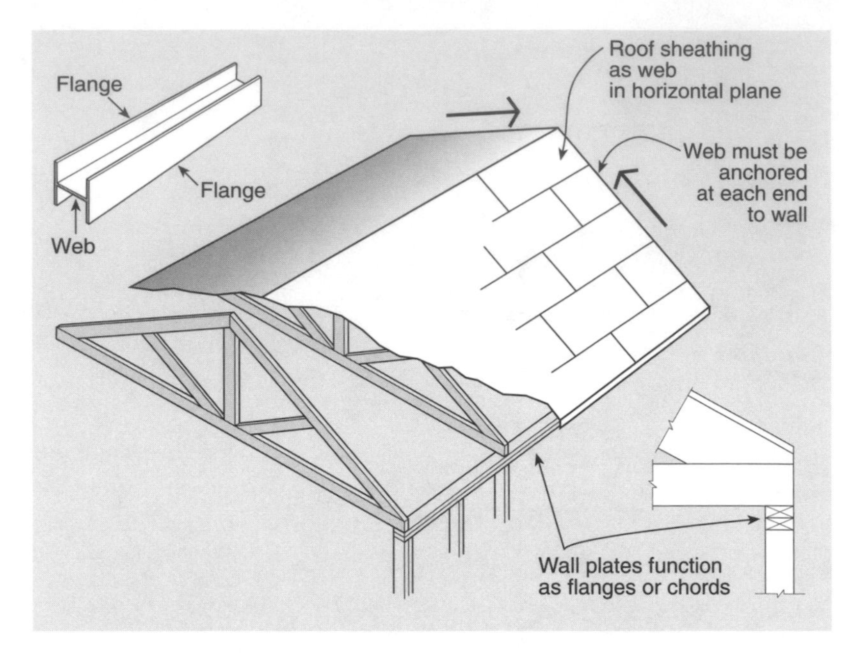

Figure 14.2 Simple earthquake-resistant design in which wall plates act as flanges and the roof sheathing functions as a web. The same function is shown for a steel beam.

acceptable. Less desirable sites may need to be improved by removing problem materials and replacing them with engineered fill, stripping and terracing sloped sites, providing additional drainage, or all of these. When constructing a building on bedrock, be careful to avoid areas with potential slippage planes.

Design and Construction

Dr. Karl Steinbrugge, the father of modern earthquake engineering, coined the phrase that has become the standard opening line for many an earthquake speech: "Earthquakes don't kill people, buildings do." Modern building codes and improved construction techniques and materials have spared Americans the enormous loss of life experienced several years ago in Armenia, and more recently in the 1995 Kobe, Japan, earthquake, when almost 5,000 people died in collapsed structures. Nevertheless, we should not become complacent. Attention to earthquake-resistant design and good construction techniques can reduce injury and property loss, as well as loss of life.

In general, one- and two-family dwellings, because of their design characteristics and materials, tend not to collapse in earthquakes. This statement applies particularly to areas in the western United States where wood-frame construction predominates. Wood-frame construction is known to perform better in earthquakes than other construction such as unreinforced masonry. This does not mean that there will be no significant damage to wood-frame structures. There often is, and it may make the structure uninhabitable and cause financial and emotional trauma to the occupants.

Most conventional homes are able to resist the horizontal forces of earthquakes (or wind) because they are built in a boxlike configuration. The plainer the design, the more stable the structure. The box configuration provides resistance by means of roof and floors (horizontal) and walls (vertical).

Roofs and floors are considered to be horizontal diaphragms. A horizontal diaphragm can be compared to a steel beam that has a top and bottom flange and web, with the web oriented in a horizontal plane. In house construction, the exterior wall's top plates act as flanges, and the roof sheathing functions as the web (fig. 14.2). For the roof to act most effectively as a horizontal diaphragm it should be of a simple, regular shape with unbroken planes. A flat, pitched, or gabled roof is the best configuration.

As with the roof, floor diaphragms are most effective when they are simple in shape and designed in a geometric pattern. Shape and design are particularly important in a first-floor diaphragm spanning a basement opening or on cripple wall foundations. Cripple walls are usually wooden stud walls set on top of an exterior foundation to support the house and create the crawl space. It is best to apply a conventional design combined with a corresponding symmetrical pattern for the shear walls or shear panels (see below).

Floors and roofs often have to be penetrated for practical purposes such as duct shafts, or for aesthetic purposes such as skylights. The size and location of such openings must be carefully considered because they can have a critical impact on the effectiveness of the diaphragm during an earthquake. Floors and roofs must be securely anchored to walls at the perimeter and at intermediate locations.

Walls that resist horizontal forces are known as shear walls or shear panels—the latter being resistant elements that might be part of a longer wall. These can be seen as upright beams on a fixed base comparable to a vertical cantilever beam (a beam supported at one end but not the other), with the end studs of the sheathed portion acting as flanges, and the sheathing between end studs acting as the web. During an earthquake, ground motions enter the structure and create inertial forces which move the floor diaphragms. This movement is resisted by the shear walls, and the forces are transmitted back down to the foundation.

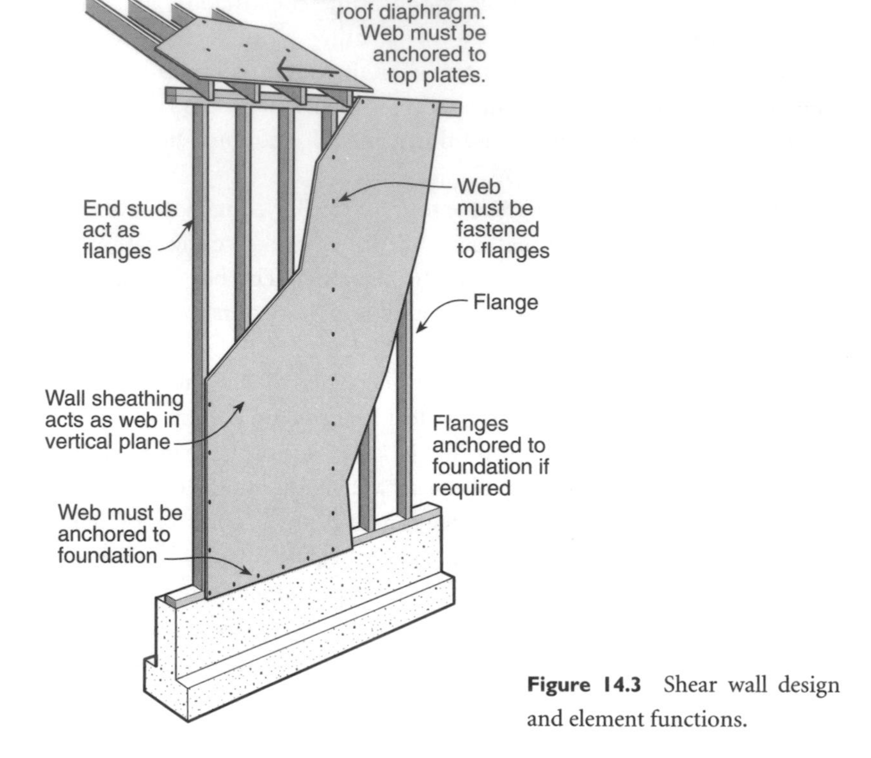

Figure 14.3 Shear wall design and element functions.

The shape and size of the shear walls are important design criteria. Walls that are too tall or too narrow tend to tip over before they slide. Because earthquakes can create forces in any direction, hold-downs must be placed at each end of a shear wall or panel. Shear wall patterns are configured so that opposite and parallel pairs of walls resist loads in a single direction, and the exterior walls help to prevent the house from twisting or racking (figs. 14.3 and 14.4). Shear walls must be located at or near the boundaries of the roof and floor diaphragms to be effective; that is, there should be shear walls at each exterior wall.

As in designing to resist the forces of winds, it is essential that a continuous load path be provided from roof to foundation in order to dissipate earthquake forces. Good connections between the resisting elements provide the continuity to ensure an uninterrupted path and tie the building's components together (figs. 14.5 and 13.9).

Among the most common earthquake damage to homes is an improperly anchored structure shifting off its foundation. This can result in a fire from a ruptured gas line, exterior damage to walls and windows, and inter-

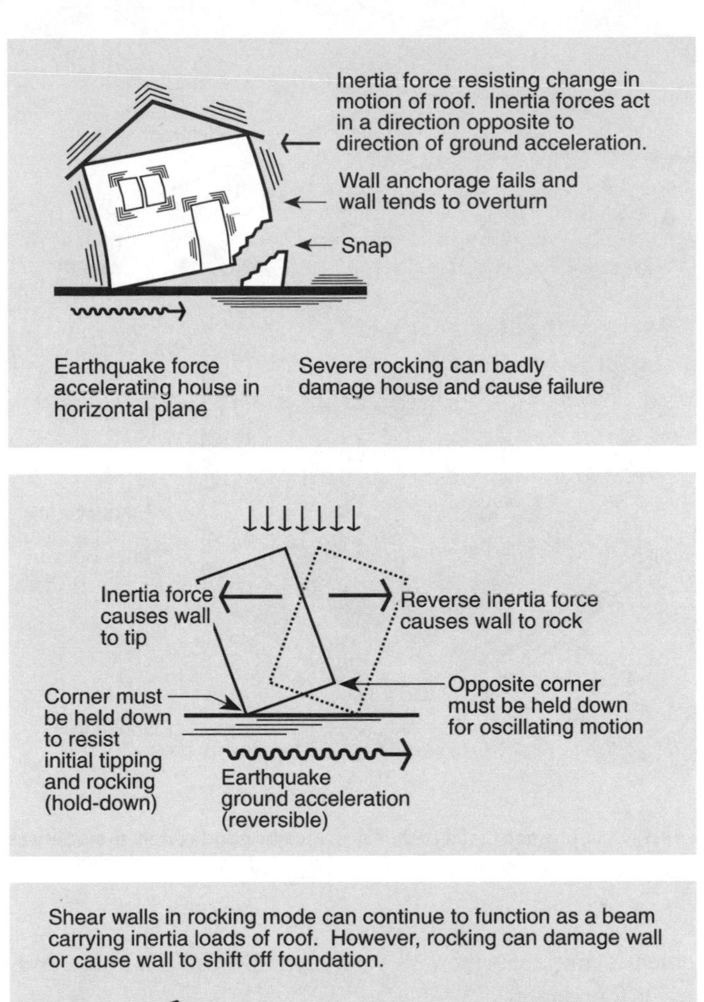

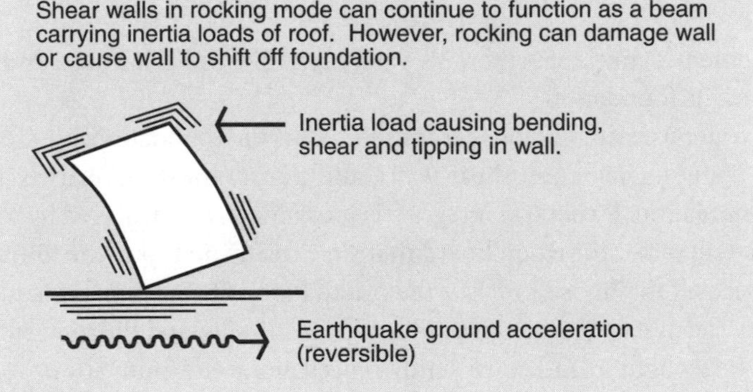

Figure 14.4 A secondary effect of earthquake forces is overturning of walls or shear panels. The structure's tendency to rock on its foundation must be resisted by anchors (hold-downs) that hold the wall to the foundation.

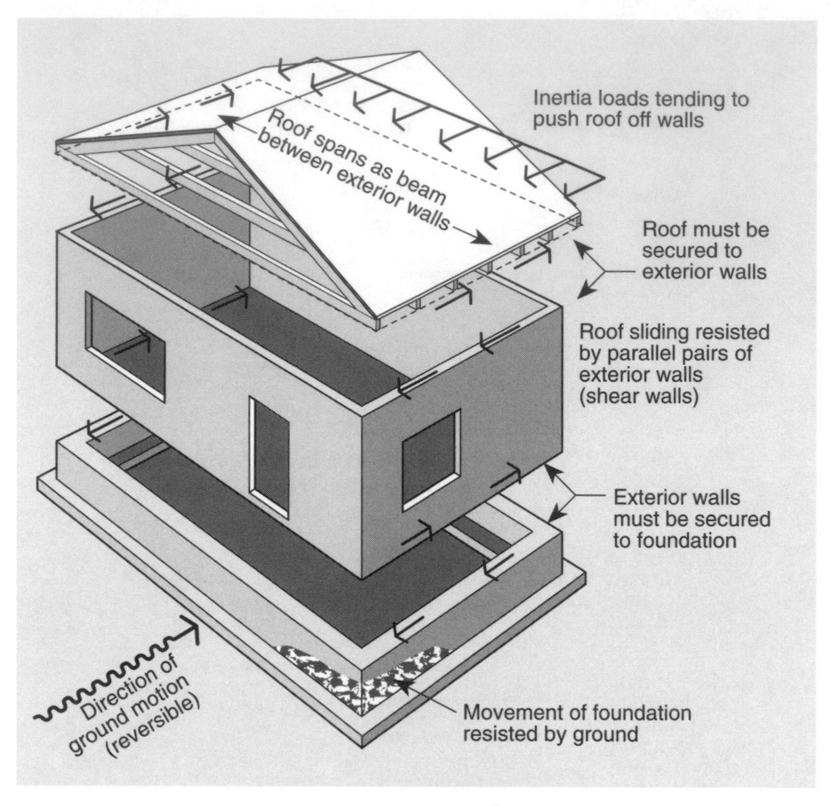

Figure 14.5 A continuous load path is provided by good connections between roof and walls, and walls and foundations (see also fig. 13.9).

nal content damage. Further, it's very expensive to lift a house and put it back on its foundation.

In order to keep the house from sliding off the foundation, all the structural components (roof, shear walls, and floors) must be securely tied to the foundation. Structural integrity is most effectively achieved by anchoring the sill plate (the wood board that sits directly on top of the foundation and secures the house to it) to the foundation. The anchor bolts must be located accurately on the center line of the mud sills, and the mud sill plates must be secured to and consistently spaced along the foundation. To check to see if a house is properly anchored, look in the crawl space for the heads of the anchor bolts. You should be able to see them every 4–6 feet along the sill plate. Other forms of connectors such as steel plates may be used instead of anchor bolts to connect the frame to the foundation.

Houses built on foundations constructed of unreinforced masonry such as brick or concrete block also tend to move off their foundations as a result of earthquake shaking. Brick foundations should have visible anchor bolts or steel reinforcing bars between the inner and outer faces of the brick foundation. Concrete block foundations should be reinforced with anchored steel bars embedded in the grout fill of the blocks. If the cells of the concrete blocks are hollow, the foundation is probably unreinforced and could fail, allowing the house to shift during an earthquake.

Improperly braced cripple walls do not perform well in earthquake-prone areas. Plywood panels and diagonal wood sheathing are the best braces for cripple walls.

Houses built on concrete slabs are generally considered to include a degree of seismic resistance because they do not have cripple walls and are bolted to the foundation. In most cases, such houses have foundation anchor bolts installed during construction. These anchor bolts are usually visible. If they are not, you can usually tell if they are there by examining the walls connecting the garage to the house.

Masonry chimneys are extremely vulnerable to earthquake forces. These heavy and brittle structures, usually constructed to be freestanding above the roof line, pose special problems when used with flexible buildings such as wood-frame homes. The chimney should be vertically supported on a reinforced concrete pad, and the chimney well should have a minimum of horizontal and vertical reinforcing. Separation of the chimney from the structure during ground shaking is the predominant damage pattern. To prevent this, the chimney should be anchored to the house frame by ties at every diaphragm level (floor, ceiling, and roof). The ties should be embedded in the masonry and strapped around the vertical reinforcing. Depending on the size and height of the chimney relative to the frame, additional ties may be needed. Consider replacing masonry chimneys with metal flues. Determining whether a chimney is susceptible to earthquakes is not easy. Tall, slender chimneys that dramatically exceed the plane of the roof tend to be the most vulnerable. Inspection by a professional often is necessary to determine the chimney's vulnerability.

Nonstructural Components

The contents of your home may pose a greater threat to your safety during an earthquake than the collapse of the structure. Few people realize that fixtures, appliances, and other objects can fall over and cause personal injury or a fire hazard. Several excellent documents are available to help you assess interior hazards around your house. Two in particular are the Bay Area Regional Earthquake Preparedness Program publication *Checklist of Nonstructural Earthquake Hazards* and the FEMA guide *Reducing the Risks of Nonstructural Earthquake Damage* (see appendix C). Both documents are available free of charge.

Two nonstructural components in particular need of bracing are water heaters and freestanding stoves. Water heaters not securely fastened to the wall tend to topple over during earthquakes and cause extensive water damage, but they also can ignite fires when connecting gas or electric lines break. Preventing this is relatively simple and inexpensive. Metal straps or braces should encircle the heater and be secured by bolts to the wall. These bolts should go into studs or concrete, but not drywall because it can't support the weight of the heater during intense ground shaking. Also, flexible pipes connecting the gas and water lines perform better during earthquake ground shaking than the more rigid aluminum or copper piping.

In most cases, fire codes set the requirements for the construction and siting of freestanding stoves. One of the requirements is significant clearance around the stove, which often means that the stove is unsupported on all four sides, making it extremely vulnerable to sliding or overturning during an earthquake. To prevent such a fire hazard, the stove should be anchored to the floor and the stovepipe sections secured to prevent separation. It is important that the materials used to do the anchoring and bracing be heat resistant.

Manufactured Homes

Manufactured homes are seldom destroyed by earthquakes. However, even moderate earthquakes can separate these homes from their foundations or piers and cause significant damage to awnings, decks, steps, skirting, and other accessory structures. Mobile homes often rotate on their foundations in an earthquake so that the doorway no longer opens onto the deck, porch, or steps. More than one mobile home resident has survived the quake only to step out the door into unexpected space and take a short fall to injury. Fires often result from the rupture of gas line connections to the structure or to the appliances within the home. Reinstallation of damaged mobile homes can be quite expensive.

Bracing systems are available to secure manufactured homes so they will not fall to the ground. To prevent earthquake-related fires, install flexible gas lines. All gas-burning appliances should be secured in place with straps to ensure that they won't be dislocated by ground shaking.

Codes and Current Practices

The foregoing concepts represent the very basics of earthquake-resistant design for residential buildings. The residential design and construction industry is one of the most rapidly changing and innovative industries in the world today. The numerous changes in construction technology for the housing market make it especially important for both buyers and residents to carefully review local building codes and consult reference materials (see appendix C). Incorporating earthquake resistance into the design and construction of a residence often requires specialized advice, and we encourage you to utilize the services of design and engineering professionals. The state-of-the-art home builder's guide to earthquake-resistant construction produced by FEMA (1992) in cooperation with the National Association of Home Builders is an excellent starting point to evaluate an existing structure or to plan your earthquake-resistant dream house. In addition, recent actions on the part of the model code groups indicate that they are considering adoption of the Council of American Building Officials' (CABO) One and Two Family Dwelling Code as the nationally accepted code for such construction. The seismic provisions of the CABO code would be substituted for the current provisions in the Uniform Building Code, the model building code most often applied in Alaska.

In addition to hearth and home, individuals and communities should be sensitive to the siting, design, construction, and interior safety of public buildings with high occupancy. How safe are local hospitals, schools, senior citizen facilities, and private buildings likely to hold large numbers of people (e.g., theaters)? Expressing your interest and concern will make everyone more sensitive to the seismic hazard and eager to take proper actions to reduce the associated risks (see appendixes A and C).

15 Natural Hazards and Coastal Zone Management in Alaska

In territorial days and before, Alaska's land and resources were under few legal constraints. The native people learned nature's laws the hard way and managed their lives in response to the natural environment. Coastal settlements were located by necessity and relocated when nature dictated.

The taming of the frontier brought a different philosophy. Resources were there for the taking, and nature could be overpowered, or at least outfoxed, through engineering. Hazards were either not recognized or ignored. While that attitude still remains to a certain extent, the pioneer period is essentially over. Today, resource management and mitigation of the impact of natural hazards through regulation are the marks of an orderly socioeconomic system. We now recognize the need to protect waters, critical environments such as wetlands and beaches, and fisheries and other resources. In doing so, we protect our own health, safety, and welfare.

Anyone buying, building on, or developing land—or taking any other action that will affect resources and environments—within Alaska's coastal zone must be aware of the regulations, codes, and ordinances that govern such actions. The following overview of federal, state, and local programs represents a starting point for coastal zone property owners and developers. Consult the appropriate state, borough, and community officials for up-to-date and complete information on coastal zone laws and regulations, as well as information on local hazards. Appendix B provides addresses for most of the agencies that furnish information, permit applications, and advice. Appendix C includes references that provide more details on specific regulations.

Federal Programs

National Flood Insurance Program

The National Flood Insurance Act of 1968, as amended by the Flood Disaster Protection Act of 1973, was passed to encourage prudent land-use planning and to minimize property damage in flood-prone areas, including the coastal zone. Local communities must adopt ordinances that reduce flood risks in order to qualify for the National Flood Insurance Program (NFIP). The NFIP offers property owners the opportunity to purchase flood insurance that would not be available from private insurance companies.

The Federal Emergency Management Agency (FEMA) Disaster Assistance Program Division serves as an advisory agency for the reduction of impacts caused by natural hazards and exerts some regulatory control to reduce future property damage. Under the authority of the federal Disaster Relief Act of 1974 (Public Law 92-288), FEMA evaluates potential hazards and determines plans to mitigate their effects. Reduction of loss due to flooding was specifically addressed under the authority of the Inter-Agency Agreement for Nonstructural Flood Damage Reduction, and Executive Order 11-988: Flood Plain Management, which designates FEMA as the lead agency to determine actions that will reduce the impact of flooding.

The initiative for qualifying for the NFIP rests with the community, which must contact FEMA. Any community may join the NFIP if that community requires permits for proposed construction and other development within the flood zone and ensures that appropriate construction materials and techniques are used to minimize potential flood damage. Initially, the community enters the "emergency phase" of the NFIP. The community is provided with a Flood Hazard Boundary Map (FHBM), which serves as a preliminary delineation of flood hazard areas. During this phase, the federal government makes a limited amount of flood insurance available, charging subsidized premium rates for all existing structures and their contents, regardless of the flood risk.

To enter the "regular program" of the NFIP, the community must adopt and enforce floodplain management ordinances that meet the minimum requirements for flood hazard reduction as set by FEMA. The advantage of entering the regular program is that more insurance coverage is available. All new structures are rated on an actual risk (actuarial) basis, which may mean higher insurance rates in potential high-hazard areas, but generally results in a savings for development within A-zones (areas flooded in a 100-year flood), as opposed to property within a V-zone, or velocity zone, such as a coastal site likely to be affected by turbulent wave action during the flood event. FEMA produces detailed Flood Insurance Rate Maps (FIRMS), which show flood elevations and the various categories of flood hazard zones. FIRMS identify base flood elevations (BFES), establish special flood hazard zones, and provide a basis for managing floodplains and establishing insurance rates.

FEMA maps commonly use the 100-year flood as the base flood elevation to establish regulatory requirements. Persons unfamiliar with hydrological data sometimes mistakenly take the term *100-year flood* to mean a flood that occurs once every 100 years, but it really refers to the flood level having a 1 percent statistical probability of occurring in a given year. While on average that level may occur only once in a century, the reality is that 100-year floods can occur in successive years, or more than once in a single year. Residents along the upper Mississippi River valley experienced a 100-year flood in 1973, but this event certainly did not preclude the even greater flood of 1993. The 100-year flood level provides an arbitrarily defined minimum risk for purposes of planning and attempting to avoid losses due to more frequent, smaller floods. The Dutch use 500-year and even 1,000-year floods for designing coastal dikes!

Flood elevations in coastal V-zones take into account the additional hazard of storm waves atop still-water flood levels. In V-zones, elevation requirements are adjusted—usually 3–6 feet above still-water flood levels—to minimize wave damage. When your insurance agent submits your application for a building within a V-zone, an elevation certificate must be attached that verifies the postconstruction elevation of the first floor of the building.

Existing FEMA regulations stipulate protection of dunes and vegetation in V-zones, but local communities are not always careful to monitor and implement this requirement. The existing requirements of the NFIP do not address other hazards of migrating shorelines, such as shoreline erosion; however, legislation was introduced in the early 1990s with the aims of strengthening the NFIP and establishing a new program to reduce coastal erosion along U.S. tidewaters and the Great Lakes shorelines. The proposal called for identifying erosion-prone communities, establishing setback requirements, requiring adoption of land-use restrictions in erosion-prone communities, and providing erosion mitigation assistance to owners of eligible structures. The number of times flood insurance could be collected on a structure in a high-risk zone would be limited, but the amount of insurance coverage would be increased. The great economic losses sustained in Hurricane Andrew in 1992 and the Mississippi River flood of 1993 will probably reinitiate consideration of revising the NFIP along the lines attempted in the previously mentioned legislation.

The insurance rate structure provides the incentive of lower rates if buildings are elevated above minimum federal requirements. Eligibility requirements are different for pole houses, mobile homes, and condominiums. Flood insurance covers both structural damage and contents. To determine if your community is in the NFIP, and for additional information, contact your local property agent, insurance agent, NFIP servicing contractor, or the Federal Insurance Administration office in Washington, D.C. (appendix B).

Most lending institutions and community planning, zoning, and building departments are aware of the flood insurance regulations and can provide assistance. It would be wise to confirm such information with the appropriate insurance representatives. Any authorized insurance agent can write and submit an NFIP policy application.

Before buying or building in the coastal zone or an inland floodplain, find out the answers to four basic questions:

1. Is the community I'm locating in covered by the emergency phase or regular program of the NFIP?
2. Is the building site above the 100-year flood level? Is the site located in a V-zone? (V-zones represent higher-hazard areas and can pose serious problems.)
3. What are the minimum elevation and structural requirements for my building?
4. What are the limits of my coverage?

Nature has a way of exceeding human expectations. Over time, the unusual event becomes the expected event: the year with the heavy rainfall and river flooding, or the storm waves that come from just the right direction at just the right time on a high spring tide to cause flood levels that exceed the predictions of the flood maps. Property owners should regard the requirements necessary to obtain flood insurance as minimal, and go beyond those requirements when elevating and flood-proofing their buildings. Investing in flood insurance is a prudent action, even in zones where flood risk is low.

Some flood insurance facts:

—Flood insurance offers the potential flood victim a less expensive and broader form of protection than can be gained through a postdisaster loan.

—Flood insurance is a separate policy from homeowners' insurance. Know the difference between the two. From the standpoint of water damage, homeowners' insurance often covers only structural damage from wind-driven rain.

—Flood insurance typically covers losses that result from the general and temporary flooding of normally dry land, the overflow of inland or tidal waters, and the unusual and rapid accumulation of runoff or surface water from any source.

Check the FIRM to see if your location is identified as flood prone. If you are located in a flood-prone area, you must purchase flood insurance in order to be eligible for most forms of federal or federally related financial, building, or acquisition assistance; that is, VA or FHA mortgages, Small Business Administration loans, and similar assistance programs. Keep three things in mind: (1) You need a separate policy for each structure. (2) If you own the building, you can insure the structure and contents, or the structure only, or the contents only. (3) If you rent the building, you need only insure the contents. A separate policy is required to insure the property of each tenant.

A condominium unit—that is, a traditional town house or row house—is considered a single-family dwelling and may be insured separately.

Mobile homes are eligible for coverage if they are on foundations, permanent or not, and regardless of whether the wheels are removed either at the time of purchase or while on the foundation.

Structures and other items that are not eligible for flood insurance include travel trailers and campers, fences, retaining walls, seawalls, septic tanks, outdoor swimming pools, gas and liquid storage tanks, wharves, piers, bulkheads, growing crops, shrubbery, land, livestock, roads, and motor vehicles.

One insurance broker cannot charge you more than another for the same flood insurance policy, because the rates are set by the federal government.

A five-day waiting period from the date of application is required before the coverage becomes effective.

Coastal Barrier Resources Act of 1982 and Coastal Barrier Improvement Act of 1990

Alaska does not participate in these federal programs.

Navigable Waterways and Wetlands

Federal jurisdiction over wetlands and navigable waterways, including tidewater areas and rivers in the coastal zone, is through the U.S. Army Corps of Engineers. The Corps' authority to regulate projects in "navigable waters of the United States" dates back to the Federal River and Harbors Act of 1899, which was principally aimed at maintaining the navigability of those waters. The Corps was given additional responsibilities under section 404 of the Federal Water Pollution Control Act of 1972, and its authority was later extended to all "waters of the United States," including wetlands, and was further defined by the Clean Water Act of 1977. Most projects affecting waterways and wetlands, including dredging, fill work, and construction (e.g., docks, piers, breakwaters, bulkheads, and similar structures), are subject to the Corps' approval and require a permit. Many types of projects may be exempt or covered by a general permit. Check with the Corps of Engineers district office (appendix B) for specific permit requirements before planning or initiating work in a wetland or waterway.

Federal Coastal Zone Management Act of 1972

The Coastal Zone Management Act of 1972 (CZMA), as amended through 1990 (Coastal Zone Act Reauthorization Amendments of 1990), officially recognizes the importance of the coastal zone—both its resource development potential and the need to protect the inherent value of habitats and environments that are fragile and dynamic. The CZMA declares that national policy is "to preserve, protect, develop, and where possible, to restore or enhance, the resources of the Nation's coastal zone for this and succeeding generations." With more than half of the nation's open-ocean coast, and 33,000 miles of general coastline, Alaska faces a heroic management responsibility under this law, which is explicitly designed to give coastal management authority to individual states.

The CZMA provides assistance in establishing state coastal zone management programs, both advisory and financial. Federal funding was made available to assist in the development and implementation of state programs. The act outlines a general framework for state coastal management whereby existing or new legislation can qualify a state for participation.

Alaska qualified under the CZMA with the adoption of the Alaska Coastal Management Act of 1977. As noted in the program's 1990 Annual Report (see appendix C): "The Alaska Coastal Management Act strives for orderly and balanced development of Alaska's coast, with an opportunity for coastal residents to take part in planning and decision-making. The Act provides for: 1) a coordinated program built on existing agency authorities and 2) the Alaska Coastal Policy Council to govern development and implementation of Alaska's coastal program." Specifics of Alaska's program are discussed below.

The federal law specifically calls for "the protection of natural resources, including wetlands, floodplains, estuaries, beaches, dunes, barrier islands, . . . fish and wildlife and their habitat, within the coastal zone" and requires states to minimize potential loss of life and property due to improper development in high-hazard areas such as floodplains, storm surge zones, erosion-prone areas, and other geological hazards—in the case of Alaska, this latter implies seismic zones. Recognition of specific coastal dynamics (e.g., vulnerability to sea level rise, land subsidence, saltwater intrusion, and loss of natural protection provided by beaches, dunes, barrier islands, and wetlands) clearly implies that management strategies must be long term in scope. The law anticipated the need for multihazard analysis in the coastal zone, a concept now extended throughout continental areas.

The CZMA also calls for improving coastal water quality, preserving habitats and aesthetics, supporting fisheries, providing public access and recreational opportunities, rehabilitating waterfronts, planning proper siting of new coastal facilities, and maintaining historical and cultural sites. These policy goals are not always mutually compatible, and managers often face a complex set of mixed legal requirements along with conflicting demands for coastal zone land use.

State and Local Laws

Alaska Coastal Management Program and Alaska Coastal Policy Council

Before receiving a Coastal Zone Management grant in 1974, the state had no active coastal management policy. The Alaska Coastal Management Program (ACMP) of the Department of Environmental Conservation (DEC) was first established during Governor William Egan's administration. The program sought to develop a scientific basis for decisions that affected the coastal zone. After the election of Governor Jay Hammond, the ACMP was shifted to the Office of the Governor, its present location. The citizen who wishes to find the office of the ACMP will need persistence and a good taxi driver; it is tucked away in a small office many blocks uphill from the state office building in Juneau. The ACMP also maintains an office in Anchorage.

For several years, the ACMP remained in a "developmental" phase of workshops and meetings, with the staff assisting the governor in drafting legislation. Official sanction for coastal zone planning did not arrive until the passage of the Alaska Coastal Management Act in 1977. Since 1978, state agencies are required to implement ACMP regulations in the course of their activities. Coastal areas are divided into a number of Coastal Resource Service Areas (CRSA) that largely correspond to local communities (Haines, Cordova, etc.), boroughs (Northwest Arctic, North Slope, Juneau, Anchorage, Sitka, etc.), or native corporations (Bering Straits, Calista-Ceñaliulriit CRSA on the Yukon-Kuskokwim Delta). Special CRSA districts are defined in some areas, such as the Bristol Bay and Aleutians West CRSAS.

The planning process involves several steps. Consider, for example, a request for shoreline erosion control. First, districts and state agencies must identify "known geophysical hazard areas and areas of high development potential in which there is a substantial possibility that geophysical hazards may occur" (6 AAC 80.050 [a]). Hazards are defined as threats to life or property. Second, areas of erosion are assessed by coastal zone management districts. Third, development in hazardous areas may not be approved unless siting, design, and construction measures are taken to minimize property damage and protect against loss of life. This process sounds straightforward, but it involves state permits and, of course, lobbying efforts by private contractors interested in offering erosion control "solutions." Federal lands are, of course, excluded from the ACMP.

The Alaska Coastal Zone Management program requires each district to produce a planning document that places resources and hazards within a framework for protection and permitting. The sophistication of these documents reflects the commitment and intentions of the local communities. The documents are available from the ACMP in Juneau and should be perused by citizens who are planning construction or want to influence the government.

The first step in undertaking construction or development in the coastal zone is to contact one of the two Coastal Zone Management offices to obtain a Coastal Project Questionnaire. This questionnaire will determine the number of permits required and evaluate the proposed project's consistency with the goals of coastal zone management. To meet federal requirements, it is necessary to sign the Certification of Consistency that is included in the questionnaire. The state must complete review of the questionnaire within 30 or 50 days, depending on whether a 30-day public review process is required.

Each applicant must include a detailed description of the project, including a timeline, site plan, topographic map, and information on the site's legal status (broadly stated: state, federal, private, municipal). The application provides a guide for securing other required permissions, including that of the Bureau of Land Management, the U.S. Forest Service, the Corps of Engineers (if dredging is involved), the U.S. Coast Guard (if a bridge is involved), and related agencies. Applicants must also indicate whether sewage or other waste will be discharged (U.S. Environmental Protection Agency) and whether an airport is within 5 miles of the site (Federal Aviation Administration).

The type of questions asked by the state are specific, but typically involve yes or no answers: "Is the rate of discharge [of wastewater] greater than 500 gallons per day?" "How deep is the bottom of the system to the top of the subsurface water table?" "Will your facility burn more than the following [units of fuel] per year in stationary equipment?" "Will you be harvesting timber?" "Is the proposed project within a known geophysical hazard?" "Will you be investigating or removing historical or archaeological resources on State-owned land?" Three different state agencies are involved in securing project approval: Department of Environmental Conservation, Department of Natural Resources, and Department of Fish and Game. The questionnaire has about 50 questions but should require no more than an hour's work. Applicants should be aware of the reason for the large number of questions. The state must consider the long-term and often irreversible effects of any construction or development in the coastal zone.

Numerous items pass across the desks of the coastal zone reviewers during a typical year. In 1990, for example, the 450 applications reflected such diverse projects as float homes 40 miles from Ketchikan; offshore oil drilling in the Beaufort Sea, where Kaktovik residents fear adverse impacts on whale populations; offshore placer mining near Nome; hard-rock mining chemical discharge into southeast Alaska streams; marine debris discharged by ships and mariculture plans; and aquatic farming of oysters and finfish. Of those 450 applications, only 2 were found to be inconsistent with coastal zone plans—a rejection rate of less than 0.5 percent.

The ACMP process is not perfect, as examination of a single project from start to finish will show. In 1993, the Alaska Department of Transportation (DOT) augmented a rubble seawall along the Safety Sound spit, east of Nome, at milepost 16 of the Nome-Council Road. The road crosses two sandy barrier bars that enclose Safety Sound. This low-lying section of coast is subject to storm surges every 10–20 years. On October 5–6, 1992, sustained southerly winds of 35 mph forced sea level up nearly 6 feet, flooding several areas of the road. The governor declared the Nome area a disaster emergency on October 12, and by October 27 the federal highway administrator in Portland, Oregon, had approved $8,791,000 for storm recovery projects. The federal authorities based their damage assessment on a video provided by the state DOT office in Nome. The design of the repair project was conducted by the Fairbanks DOT office and specified that the road be returned to its prestorm condition. As the project engineer put it: "Reconstruction of the road on the prestorm alignment and grade with the addition of rip-rap to prevent further damage from annual storms is the preferred alternative."

Rip-rap is an imprecise term in engineering circles; in this case it appeared to include boulders of considerable size. In other words, the DOT authorized emplacing a seawall on a barrier island. Because the road was a preexisting feature, no specific permits were required. The DOT contacted the governor's office using a questionnaire. The city of Nome found the project to be in compliance with the Alaska Coastal Management review criteria. The DOT placed an advertisement inviting public comments in the local newspaper, the *Nome Nugget,* and received no response; apparently, no public meetings were held. The whole process proceeded without any outside review. Only local officials actually observed the damage and authorized changes, and only engineers judged the reliability of the protection strategy proposed. The local government, which had a considerable financial stake in the project, acted as the only environmental oversight.

Alaska Water Use Act (Alaska Statute 46.15)

Alaska landowners do not have automatic rights to surface water or groundwater, and use of such water without a permit does not establish a legal right to the water. The Water Use Act protects established water right holders, helps prevent overappropriation of water sources, and guides management of the state's water resources. Established water rights usually transfer with the sale of land, but not always. Check with the Department of Natural Resources (see appendix B for regional offices) to make sure water rights are included before you purchase land.

If the water rights are not attached to the land, you need to apply for water rights and a permit to drill a well or divert water. The type of permit and the legal requirements depend on the amount of water to be used and the source of the water. Similarly, if a dam is required, a separate certificate of approval is necessary. For more information, obtain the *Water Users Handbook* and related water fact sheets from the Department of Natural Resources.

Hazards, Economics, and Politics

Our survey of Alaska's coastal communities has revealed a greater diversity of hazards, attitudes, and responses than is found in any other state in the union. Nevertheless, we see enough similarities to offer a few generalizations on living with the coast of Alaska.

1. Townsites came into existence without hazard planning. Port facilities to support the fur trade, gold mines, fishing, and transportation evolved haphazardly. Towns built on the shore, at the mouths of streams, on deltas or floodplains, in seismic zones, or at the foot of steep slopes are often extremely hazardous places to live. In some cases, such as upper Wrangell, towns were platted in traditional grids over terrain that was difficult or impossible to develop rather than being planned with site stability and low risk in mind.

2. Where settlers learned from experience, low-risk sites tended to be developed first, leaving high-risk sites to accommodate later growth. Newer developments on steep slopes (e.g., Juneau, Kodiak), floodplains (e.g., Juneau, West Valdez), weak soils, and along shorelines are at high risk. Such development, when threatened, often opts for engineering "solutions" rather than relocation. These projects continually increase in expense and are often ineffective.

3. Politicians usually give priority to promoting economic development and management, *not* to protecting the electorate.

4. "Protective" regulations to reduce natural hazards (e.g., prohibitions on fill and development of wetlands) for property owners are often threatening to developers. Developers often resist regulations designed to protect property owners.

5. Politicians are usually drawn from the economic community. They are often owners of undeveloped acreage, developers, suppliers of materials,

lawyers, businesspeople, and professionals who benefit from growth and development. Even if no conflict of interest is intended, they have a stake in development. So the new school site selection may be influenced more by a site's availability from a board member than by an evaluation of its hazard risk.

6. Politicians are the employers while the day-to-day work is carried out by the employees: the hired town manager, planner, and community development personnel. By and large, these employees do an excellent job in most communities. They are knowledgeable, realistic, and committed public servants—but they answer to the elected politicians.

7. When disasters do strike, firefighters and the police are our first line of defense. The people with the greatest responsibility for public safety sometimes seem to be the least appreciated. We rely on volunteer firefighters rather than a salaried staff. Media comments suggest that police are overpaid, but their high salaries are actually the result of overtime required by understaffing.

8. Collective community attitudes are variable. Shishmaref, for example, has a high percentage of permanent residents. Their perceived planning needs differ from those of more transient communities. Newcomers in the latter communities have no experience with Alaska's surprises. Their attitude is different with respect to planning, and they are more likely to locate in newer, higher-risk developments.

9. Developers are in business to make money, not to protect the public. Their emphasis is on build and sell, not site-specific hazard risk.

10. Banks and other lenders have a stake in property mortgages. In some cases, lenders can be a source of information on risk. If your credit is good, however, you can get the loan in spite of the risk. A property owner in Kodiak is now paying off a mortgage on air and a vacant lot after his house was destroyed by a debris flow.

11. Catastrophes are followed by several responses that affect both the local community and communities at large. Postcatastrophe recovery is a time of shock and haste to put things right again. Disposing of debris to reconstruct harbor land may create a new hazard. Big catastrophes bring new and/or stronger regulations; for example, upgraded building codes, prohibitions or restrictions on development, and mitigation against recurrence of the hazard. Expect it.

12. The quality of management and politics in Alaska is as diverse as the communities themselves and the hazards they face. There are first- and second-class towns, cities, boroughs, native lands, and state and federal regulatory agencies.

You, the individual citizen and property owner, are the final decision maker. The foregoing chapters provide a starting point for you to identify and evaluate hazards in terms of their potential for both property damage and risks to human health and safety. The appendixes that follow this chapter offer hazard checklists, contact agencies, and recommended reading so that you can continue to broaden your Alaskan horizons.

Appendix A
Responding to Hazards: Checklists

The best precaution against hazards is to avoid them. Locate in an area outside the hazard zone. This solution is reasonable for local hazards but may be impossible with respect to processes that are regional in extent such as earthquakes, volcanic ashfalls, and tsunamis (table A.1). When a hazard is recognized after an area has already been developed, relocation away from the hazard is the most prudent remedy. The messages are: Don't build on the floodplain or shoreline. Set structures well back from the bluff edge. Do not site buildings at the toe of steep unstable slopes, in rock avalanche chutes, or at the base of potential snow avalanche runs. Consider site safety for all structures, not just occupied buildings; for example, don't locate oil and fuel tanks in hazard zones.

Local hazards can be mitigated in part by proper site selection, site maintenance, and construction techniques. Build on high ground. Avoid overwash passes and zones of erosion or ice override. Preserve vegetation cover, and in the case of beaches, preserve the sand supply. Be skeptical of plans that rely on engineering schemes to hold back the sea, rivers, and the effects of gravity. If you are in a restricted or low-lying area such as an island, spit, or floodplain, be sure to have an evacuation plan and heed warnings to prevent becoming trapped.

The ultimate rule is to expect the unexpected and be prepared. After following the precautions enumerated in the foregoing chapters, consider how you can best prepare for the big events of regional scale. The following checklists provide a guide.

A-1 Earthquake Checklists

The following lists and information are slightly modified from the U.S. Geological Survey publication *The Next Big Earthquake* (1994). Most people in Alaska will survive the next big earthquake with little loss; some will be severely affected. Actions you take now can help determine how much you and your family will lose.

—Protect yourself
- —Practice "duck, cover, and hold" drills at home with your family and at work.
- —Injuries and deaths during earthquakes are caused by falling objects and collapsing structures. Knowing how to protect yourself when the shaking starts may save your life. Show children a safe area in which to duck and cover.
- —Practice counting 60 seconds. Most quakes do not last that long, and it will help you to keep calm when a real earthquake strikes.

—Develop an earthquake plan at home, in your neighborhood, at school, and at work.
  —Determine the safest places in your home and at work. These should be away from heavy furniture or appliances, woodstoves, fireplaces, open shelves and bookcases, and large panes of glass, pictures, or mirrors.
  —If the earthquake hits during the day, family members may be separated for several hours to several days. Plan ahead and select a safe place where you can reunite after the earthquake. Consider your family's possible needs, and also select alternate meeting places near work or schools.
  —Designate an out-of-the-area telephone contact. Select a relative or friend to act as a clearinghouse for information about your family. Family members should call this contact to report their condition and location. Make sure family members carry this number with them at all times, and that the number is known by other friends and relatives.
  —Learn how to fight fires, how to rescue people trapped under debris, how to provide first aid, how to find help for dire emergencies, and how to assist others, especially the elderly, immobile, or handicapped. Ask your local American Red Cross Office for more information.
  —The most common cause of earthquake-related fires is broken gas lines. Everyone should know how to turn off the gas supply at the meter in case they smell gas after a large earthquake. Buy a special wrench that fits your gas shut-off valve and fasten it next to the valve.
  —Find out the policy of your local school concerning the release of children after an earthquake. Arrange with neighbors to watch out for your family and property in case you are not at home.
—Store emergency supplies
  —After a major earthquake or other natural disaster, medical aid, transportation, water, electricity, and communication may be unavailable or severely restricted for

Table A.1 General Hazards List and Scale of Events in Terms of Impact

Hazard	Local impact	Regional impact	Reference Checklist
Earthquakes	Yes	Yes	A-1
Volcanoes	Yes	Yes	A-2
Tsunamis	Yes	Yes	A-3
Floods	Yes	Yes	A-4
Storms	Yes	Usually	A-5
Storm surge	Yes	Possibly	
Coastal erosion	Yes	Possibly	
Ice override	Yes		
Wind	Yes		
Mass wasting	Yes	Yes, if induced by	
Avalanches	Yes	seismic activity or	
Landslides/flows	Yes	regional precipitation	
Human-induced	Yes	Possibly	
Oil spills	Yes	Possibly	
Fire	Yes	Forest fires	
Site-specific	Yes		

several days to weeks. Be prepared to take care of yourself, your family, and your neighbors for at least three days, longer if you live in a remote area.
—At home, at work, and in your car, store flashlights, batteries, an A-B-C rated fire extinguisher, a battery-operated radio, a first-aid kit and handbook, at least one gallon of water per person per day, canned and dried food, can opener, warm clothes, sturdy shoes, gloves, matches and candles, and a fresh supply of any medications you and your family members may need.
—Make sure smoke detectors are properly installed and functional.
—Know how to turn off utilities.
—Have an adjustable wrench on hand for turning off gas and water.
—Consider what you will need if an earthquake takes place in the winter. Have warm clothes and sleeping bags and pads for all members of your family.
—Make sure emergency supplies are located in a safe and readily available place.
—Make sure everyone in your family knows where these supplies are and how to use them. Keep a list of emergency phone numbers (i.e., police, fire department, doctors) with the supplies.
—Include pets in your planning. Plan for their food and water supplies for at least three days. Make arrangements with a neighbor to care for your pets in the event you are unable to get home.

During an earthquake

—If you are indoors, duck or drop down to the floor, or stand in a framed doorway. Take cover under a sturdy desk, table, or other furniture. Hold on to it and be prepared to move with it. Hold the position until the ground stops shaking and it is safe to move. Stay clear of windows, fireplaces, woodstoves, and heavy furniture or appliances that may fall over. Stay inside to avoid being injured by falling glass or building parts. If you are in a crowded area, take cover where you are. Stay calm and encourage others to do likewise.
—If you are outside, get into the open, away from buildings, power lines, and trees.
—If you are driving, stop if it is safe, but stay inside your car. Stay away from bridges, underpasses, overpasses, and tunnels. Move your car as far out of the normal traffic pattern as possible. If possible, avoid stopping under trees, light posts, power lines, or signs.
—If you are in a mountainous area or near unstable slopes or cliffs, be alert for falling rocks and other debris that could be loosened by the earthquake.
—If you are at the beach, move quickly to higher ground or several hundred yards inland.

After an earthquake

—Check for injuries. Provide first aid. Do not move a seriously injured person unless that person is in immediate danger of further injuries.
—Safety check. Check for the following hazards:
— Fire or fire hazards.
—Gas leaks. Shut off the main gas valve only if a leak is suspected or identified by the odor of natural gas. Wait for the gas company to turn it back on after the damage is repaired.
—Damaged electrical wiring. Shut off power at the control box.
—Downed or damaged utility lines. Stay away from downed lines even if power appears to be off.
—Broken water lines. Shut off supply valves.

—Fallen objects in closets and cupboards. Displaced objects may fall when you open the door.
—Downed or damaged chimneys. Approach chimneys with caution. They may be weakened and could topple during an aftershock.
—Check your telephone. Make sure each phone is on its receiver. Telephones that are off the hook tie up the telephone network unnecessarily.

—Clean up. Clean up potentially harmful materials and spilled medicines.
—Tsunami hazard. If you live along the coast, be alert for tsunami warnings issued by the Alaska Tsunami Warning Center. If you experience a strong earthquake, there may not be time for the center to issue a warning. Move to higher ground as soon as you are able, and stay there until the authorities issue an all clear.
—Aftershocks. Expect aftershocks. Most of these are smaller than the main earthquake. Some may be large enough to do additional damage to weakened structures.
—Use flashlights or battery-powered lanterns; do not use lighters, matches, candles, or lanterns until you are sure there are no gas leaks.
—Use your telephone only in the event of life-threatening emergencies.
—Turn on a battery-powered radio for information, damage reports, and for information on volunteering your assistance.
—Keep streets clear for emergency vehicles. Cooperate with public safety officials.

Emergency Public Information Following a Disaster

News dissemination after a disaster takes time and is frustrating when we want, and have come to expect, immediate and complete information. Search the radio and television dials to find stations that are able to give information. Remember that initial reports may be inaccurate. Don't believe everything you hear. Pay particular attention to information from governmental sources.

The first information about a large local earthquake may come from the Alaska Tsunami Warning Center or from the Alaska Earthquake Information Center. The initial estimates of location and magnitude are likely to be revised as more information becomes available.

Initial reports of damage based primarily on eyewitness accounts may be misleading and cause speculation. Local news-gathering capabilities may be severely hindered by the disaster because the station's power may be off or the news staff may be unable to communicate with the station. Below are telephone numbers useful in emergencies.

American Red Cross
Anchorage: (907) 277-1538
Fairbanks: (907) 456-5937
Juneau: (907) 789-0920
Kodiak: (907) 486-4040

Local emergency management offices
Fairbanks: (907) 459-1481; 24-hour number: 474-7721
Kenai: (907) 262-4910
Mat-Su Valley: (907) 373-8800

Children and Earthquakes

Earthquakes are traumatic for everyone, but they are especially frightening for children, who may have to leave their homes and all that is familiar to them. A child does not usually understand such events and feels anxious, confused, and frightened. Fear is a normal reaction to any danger that threatens life or well-being. After an earthquake, a child's fears are those of reoccurrence, injury, death, or being separated from the rest of the family. Aftershocks can increase this fear. Parents sometimes ignore the emotional needs of their children once assured of their physical safety. A child's persistent fears may generate disruptive behavior, surprising and frustrating a parent who is trying to continue the family's daily routine.

How a parent can help:

—Keep the family together. This provides immediate reassurance to children and alleviates fears of being abandoned and unprotected.

—Reassure children by words as well as actions. Emphasize the positive: "We are all together and nothing has happened to us." "You don't have to worry, we will look after you."

—Encourage the child to talk. It may be helpful to include other family members, neighbors, and their children in a discussion of reactions to the earthquake.

—Include the child in family activities. There will be important concerns and things to do after an earthquake: checking on the damage, cleaning up broken glass, picking up fallen objects. Whenever possible, include children in these activities.

—Children may have difficulty falling asleep or may wake up during the night, and may have nightmares for weeks or months after the earthquake. Allow the child to move into a room with another child or to sleep on a mattress in the parents' room, or simply spend a little extra time in the child's room giving reassurance.

Protect Your Belongings

Falling objects and toppling furniture present both the greatest danger and the biggest potential financial loss for most people. Imagine all of the contents of your kitchen cabinets falling to the floor or on your head! At home, at work, and in schools, building contents should be secured.

—Be sure that no heavy items, such as pictures or mirrors, can fall on your bed, where you typically spend a third of each day.

—Secure tall furniture and bookcases to the wall. Add lips to shelves to prevent costly items from sliding off. Be sure adjustable shelves cannot slide off their supports.

—Put strong latches on cabinet doors, especially at home in your kitchen and at work in laboratories. Fasten heavy or precious items to shelves or tables. Secure file cabinets, computers, televisions, and machinery that may overturn during an earthquake.

—Store potentially hazardous materials such as cleaners, fertilizers, chemicals, and petroleum products in appropriate containers in sturdy cabinets fastened to the wall or floor.

—In your office, be sure heavy objects are fastened to the building structure and not to a movable wall. Ask a carpenter or an electrician to check light fixtures and modular ceiling systems.

—Be sure your water heater is fastened to the wall studs and that all gas heaters and appliances are connected to the gas pipe with flexible tubing. If you use propane gas, be sure the storage tank is secured against overturning and sliding.

—Secure your woodstove to wall or floor studs. Make sure you have a fire extinguisher close at hand.

—Check with school officials to be sure they have taken similar precautions.

Wood-burning stoves

Freestanding wood-burning stoves pose an additional risk to many in Alaska—especially in bush communities. Heavy objects such as stoves are actually more likely than lighter objects to move during strong ground shaking. Fire codes dictate that stoves be unsupported on all four sides, but that makes them more vulnerable to sliding or overturning during an earthquake. If the stove were to tip and/or separate from the stovepipe, cinders or sparks might easily cause a fire in the home.

In order to reduce the potential fire hazard following an earthquake, the stove should be anchored to the floor and stovepipe sections secured. It is important that the seismic anchors or braces do not conduct heat from the stove. Although there are many types of stoves, the following recommendations apply to most installations:

—Stoves resting on a brick hearth can be anchored using bricks and mortar.
—Woodstoves resting on a concrete slab on grade can be anchored directly to the concrete.
—Stovepipes should be anchored to the flue exit, and each of the stovepipe segments should be securely fastened together.
—Mobile home–approved units come with predrilled holes in the pedestals or legs and can be safely anchored to the underlying floor framing.

Propane tanks

Many residents in rural areas of Alaska use above-ground propane tanks. These tanks may move, slide, or topple over during strong ground shaking. Gas leaks are frequently the cause of earthquake-related fires. The following recommendations will help you reduce the postearthquake fire hazard associated with propane tanks.

—Mount the tank on a continuous concrete pad and bolt the four legs to the pad.
—Install a flexible hose connection between the tank, the supply line, and the entrance to your home or business.
—Clear the area of tall or heavy objects that can fall and rupture the tank or supply line.
—Keep a wrench tied on a cord near the shut-off valve and make sure family members or employees know how to use it.
—Seismic shut-off valves are available for large tanks.

A-2 Volcanic Ash Checklists

Volcanic ashfalls are a regional hazard and require special preparations. Ash is gritty, abrasive, and acidic. Heavy ashfalls reduce visibility and may create sudden electrical power demands resulting in brownouts. Ash may cause machine failure and contaminate or clog water supplies. Accumulation of ash may cause structural collapse (a 1-inch layer of ash weighs 10 pounds per square foot), and fine ash is slippery, making both walking and driving hazardous. Motorists may become stranded. Ash can cause respiratory irritation, and in extreme cases respiratory failure, especially among infants, the aged, and those with respiratory ailments.

The following precautions are slightly modified from *Ash Alert,* a pamphlet published by the Alaska Division of Emergency Services.

Home supply list

—Obtain NIOSH-approved dust/mist respirators (coded TC-21C-XXX). Some will not fit children; check fit before buying.

—Obtain sufficient nonperishable food for two weeks.
—Store water (1.5 gallons per day per person) in clean, airtight plastic containers.
—Have required medicines and a first-aid kit on hand.
—A battery-operated radio with extra batteries will be useful to monitor ashfall warnings and all-clear messages.
—Store extra pet food if applicable.
—Keep an A-B-C fire extinguisher handy.
—Make sure family members know the location and operation of these items.

Auto supply list
—NIOSH-approved dust/mist respirators (coded TC-21C-XXX)
—Eyeglasses, to replace contact lenses
—Blankets
—Fire extinguisher
—Extra clothing
—Emergency food ration
—First-aid kit and critical medication
—Flashlight, extra batteries, and bulbs
—Basic tool kit
—Portable radio and extra batteries
—Shovel
—Matches, candles, emergency flares
—Heavy rope or tow cable
—Extra air filter
—Extra windshield washer fluid
—Extra windshield wiper blades

Workplace supplies (see also home supply list)
—Large plastic bags to protect office equipment
—Critical personal medication

Prepare Your Home

Store additional water if your water supply is vulnerable to power outages or contamination. Maintain a home emergency kit (see home supply list above).

During an ashfall
—Stay indoors if possible.
—Close doors, windows, and dampers.
—Avoid lighting woodstoves or fireplaces.
—Eliminate draft sources.
—Do not run exhaust fans or clothes dryers.
—Listen to your radio.
—Vacuum furniture, carpets, and other surfaces. Try not to wipe surfaces, as ash will scratch.
—Brush, shake, and presoak clothes. Use plenty of water and detergent. Do not use soap; it tends to gum up.
—If you have been working in an ashfall, have your work clothes laundered at work or outside the home.

—Cover and don't use personal computers, stereos, and other sensitive electronic and mechanical equipment.

After an ashfall

—Wear a respirator during ash cleanup.
—Consider using goggles for eye protection.
—Remove heavy accumulations of ash from flat or low-pitched roofs and from rain gutters. This may have to be done repeatedly.
—When sweeping, dampen or sprinkle ash with Industrial Sweep™ to avoid raising unnecessary dust.
—Mow lawn when damp and bag lawn clippings to reduce dust.
—Replace items used from the emergency kit.

Prepare Your Family

Instruct your family on emergency and safety procedures. Have quiet games and activities available. Consider organizing a community day-care center to help working parents during ashfall cleanup and keep children in a cleaner environment. Plan for sheltering pets and livestock.

During an ashfall

—Stay indoors if possible.
—Keep children indoors.
—Minimize exertion to reduce inhaling ash.
—Do not attempt to pick up your children at school unless directed to do so. Schools will be notified of emergency procedures.
—As much as possible, maintain normal routines for children.
—Use a respirator if you must go outdoors.
—If ashfall is prolonged, take children outside as weather conditions permit (use dust/mist respirators).
—People with heart or lung disease should consult a physician at once if they experience symptoms.
—Use the telephone only for emergencies.
—Keep pets indoors as much as possible.
—Tightly restrict the outdoor movement of pets and livestock to reduce inhaling ash; likewise prevent livestock from grazing ash-covered plants.
—Get clean water to livestock as soon as possible.
—If pets go out, brush or vacuum them before letting them back inside. Do not let them get wet, and don't try to wash them.
—Keep extra dry, clean pet food on hand.

After an ashfall

—Limit outdoor activities of children and pets until ash dust is no longer evident.

Prepare Your Automobile

Maintain an auto emergency kit (see auto supply list above). Make sure that windshield wiper blades are in good condition.

During an ashfall

—Do not drive unless it is absolutely necessary.
—If you must drive, drive slowly. Do not follow the car ahead too closely. Ash is slippery.
—Use your windshield washer system anytime you must use your wipers.
—Do not drive without an air filter.
—Change your air filter if you notice a loss of power in your car's engine.
—If the car stalls, push it off the road to avoid collisions, and then stay inside it.

After an ashfall

—Change the oil and oil filter.
—Change the air filter.
—Wash your car thoroughly with water to remove all ash.
—Replace any item used from the auto emergency kit.

Prepare Your Workplace

Become familiar with your employer's emergency plans. Encourage employers to have an early release policy to allow employees to get home before an ashfall occurs. Maintain a workplace emergency kit.

During an ashfall

—Don't operate nonessential equipment.
—Protect office equipment such as copiers, fax machines, and personal computers as recommended by the manufacturer.
—Store computer diskettes inside sealed bags or containers.
—Go home, if possible, before ash begins to fall.
—If ash is already falling heavily, stay at work. If you work outside, go indoors until the ash has stopped and settled.
—If it is not possible to go indoors, get into your car and go directly home.

After an ashfall

—Clean up all ash before removing protective covers from office equipment.
—Replace items used from the emergency kit.

Information

For additional information, contact:

Alaska Division of Emergency Services
P.O. Box 5750
Ft. Richardson, AK 99505-5750

A-3 Tsunami Checklists

Earthquakes and volcanic events both locally and far away may generate tsunamis. All coastal residents should be familiar with the tsunami warning system and know what to do when a tsunami threatens. The following rules are derived from the U.S. Geological Survey publication *The Next Big Earthquake* and the National Oceanic and Atmospheric Administration (NOAA) brochure *Tsunami Watch and Warning.*

Tsunami Safety Rules

—Tsunamis follow no discernible pattern of occurrence. When you receive a tsunami warning, you must assume that a dangerous wave is on its way. History shows that when the great waves finally strike, they claim those who have ignored the warning.

—The warning and evacuation of personnel in endangered areas is the job of designated agents participating in the Tsunami Warning System. The agent in your location will know what measures to take and will take them, with your cooperation. You can help by remembering several facts.

—All earthquakes do not cause tsunamis, but many do. When you hear that an earthquake has occurred, prepare for a tsunami emergency.

—An earthquake in your area is a natural tsunami warning. Do not stay in low-lying coastal areas after a local earthquake. Quickly move to high ground and away from the coast.

—A tsunami is not a single wave but a series of waves. The first wave is not necessarily the largest. Stay out of danger areas until an all-clear is issued by a competent authority.

—Approaching tsunamis are sometimes heralded by a noticeable rise or fall of coastal water. This is nature's tsunami warning. Heed it. Move to high ground quickly, and away from the coast.

—A small tsunami at one beach can be a giant a few miles away. Don't let the modest size of one make you lose respect for all.

—Local subsea topography and coastal configuration modify the tsunami. The fact that you are far up the arm of a fjord or embayment, away from the open sea, may amplify the effect of a tsunami and make it even more dangerous.

—The Tsunami Warning System does not issue false alarms. When an *ocean-wide warning* is issued, a tsunami exists. When a *regional warning* is issued, a tsunami probably exists. The tsunami of May 1960 killed 61 people in Hilo, Hawaii, who thought it was "just another false alarm."

—All tsunamis—like hurricanes—are potentially dangerous, even though they may not damage every coastline they strike.

—Never go down to the beach to watch for a tsunami. When you can see the wave, you are too close to escape it.

—Sooner or later, tsunamis visit every coastline in the Pacific. Warnings apply to *you* if you live in any Pacific coastal area.

—During a tsunami emergency, your local Civil Defense, police, and other emergency organizations will try to save your life. Give them your fullest cooperation.

—Know your safety refuge zone. Unless it has been determined otherwise by competent scientists, potential danger areas are less than 50 feet above sea level and within 1 mile of the coast for tsunamis of distant origin, and less than 100 feet above sea level and within 1 mile of the coast for tsunamis of local origin. When seeking refuge, move well above these elevations and beyond these minimum distances.

—Stay tuned to your radio or television during a tsunami emergency—bulletins issued through emergency agencies and NOAA offices can help you save your life!

—As in the case of other hazards, develop and discuss a tsunami disaster plan with your family, particularly how children should respond (e.g., follow the instructions of school officials when at school) and where to meet or contact each other after the threat has passed.

A-4 Flood Checklists

The following information on planning for and responding to flood hazards is slightly modified from the NOAA pamphlet *Floods, Flash Floods and Warnings.*

Flood Checklist

Before the flood

—Keep materials on hand that are useful in emergencies, such as sandbags, plywood, plastic sheeting, and lumber.
—Install check valves in building sewer traps to prevent floodwater from backing up in sewer drains.
—Hospitals and other operations critically affected by power failure should arrange for auxiliary electrical supplies.
—Keep first-aid supplies on hand.
—Keep your automobile gas tank near full; if electric power is cut off, filling stations may not be able to operate pumps for several days.
—Keep a stock of food that requires little cooking and no refrigeration; electrical power may be interrupted.
—Keep a portable radio, emergency cooking equipment, lights, and flashlights in working order.
—Know your elevation above flood stage.
—Know your evacuation route.
—Make a photographic record of your house and furnishings for insurance purposes, and store it where it will not be affected by a flood.
—Keep important records, including flood insurance policies, stored somewhere safe from flooding; e.g., your bank safety deposit box.

When you receive a flood warning

—Store drinking water in clean bathtubs and containers. Water service may be interrupted.
—If forced to leave your home, and time permits, move essential items to safe ground; fill fuel tanks to keep them from floating away; grease immovable machinery (a coat of grease will protect against rust and corrosion).
—If time permits, some flood-proofing may be attempted, such as sandbagging and sealing around doors and other openings.
—Move to a safe area before access is cut off by flood water.

During the flood

—Avoid areas subject to sudden flooding.
—Do not attempt to cross a flowing stream if the water is above your knees.
—Do not attempt to drive over a flooded road. You could be stranded or trapped. The depth or velocity of water is not always obvious.
—Do not return to your flooded property until the flood has abated and the all clear has been given by public officials.

After the flood

—Do not use fresh food that has come in contact with floodwaters.
—Boil drinking water before using it. Wells should be pumped out and the water tested for purity before drinking.

—Seek necessary medical care at the nearest hospital. Food, clothing, shelter, and first aid are available at Red Cross shelters.
—Do not visit disaster areas; your presence might hamper rescue and other emergency operations.
—Do not handle live electrical equipment in wet areas; electrical equipment should be checked and dried before being returned to service.
—Use flashlights, not lanterns or torches, to examine buildings; flammable material may be inside.
—Report broken utility lines to the appropriate authorities.

Flash Flood Checklist

Flash flood waves, moving at incredible speeds, can carry boulders, tear out trees, destroy buildings and bridges, and scour out new channels. Killing walls of water can reach heights of 10–20 feet. You won't always have warning that these deadly sudden floods are coming.

When a flash flood warning is issued for your area, or the moment you first realize that a flash flood is imminent, act quickly to save yourself. You may have only seconds.

—Move to high ground. If in a narrow valley, move up the valley walls rather than downstream.
—Get out of areas subject to flooding. This includes dips, low spots, canyons, washes, and underpasses, and usually the entire valley floor.
—Avoid already flooded and high-velocity flow areas. Do not attempt to cross a flowing stream on foot if the water is above your knees.
—If driving, know the depth of water in a dip before crossing it. Remember, the roadbed may not be intact under the water.
—If the vehicle stalls, abandon it immediately and seek higher ground—rapidly rising water can engulf a vehicle and its occupants and sweep them away.
—Be especially cautious at night when it is harder to recognize flood dangers.
—Do not camp or park your vehicle along streams and washes, particularly during threatening conditions.

During any flood emergency, stay tuned to your NOAA weather radio, commercial radio, or television station. Information from NOAA and Civil Defense emergency forces may save your life.

A-5 Winter Storm Checklists

Experienced Alaskans are generally knowledgeable about winter storms and their extremes, but people sometimes become complacent or careless, and novice pioneers may not know how to prepare for winter. Blinding blizzards, heavy snows, wind chills to -90°F , plus local effects such as ice override, avalanches, and ice jam flooding create serious hazards. To survive Alaska's multiple hazards only to fall victim to frostbite, hypothermia, or heart attack while shoveling snow would be ironic. Yet winter storms claim such victims annually.

The same advice and preparedness for earthquakes, volcanic eruptions, tsunamis, and floods applies to winter storms. Know the hazard and its secondary effects (e.g., coastal erosion, snow avalanches), develop a disaster plan, and implement the plan (i.e., prepare a standard 3-day water and food supply and materials disaster aid kit; have safety equipment on hand, including emergency heating; and know safety procedures). Be weather-aware and know the meaning of the various watches, warnings, and advisories. Avoid overexertion and follow the suggestions given in the NOAA pamphlet *Winter Storms . . . the Deceptive Killers: A Guide to Survival.*

When Caught in a Winter Storm

Outside

—Find shelter.
—If no shelter is available:
—Try to stay dry.
—Cover all exposed parts of the body.
—Prepare a lean-to, windbreak, or snow cave for protection from the wind.
—Build a fire for heat and to attract attention.
—Place rocks around the fire to absorb and reflect heat.
—*Do not eat snow.* It will lower your body temperature. Melt it first.

In a vehicle

—Stay in your car or truck. Disorientation occurs quickly in wind-driven snow and cold.
—Run the motor about 10 minutes each hour for heat:
—Open the window a little for fresh air to avoid carbon monoxide poisoning.
—Make sure the exhaust pipe is not blocked.
—Exercise from time to time by vigorously moving arms, legs, fingers, and toes to keep blood circulating and to keep warm.
—Make yourself visible to rescuers:
—Turn on the dome light at night when the engine is running.
—Tie a colored cloth (preferably red) to your antenna or door.
—Raise the hood, indicating trouble, after snow stops falling.

At home or in a building

—*Stay inside.* When using alternative heat from a fireplace, woodstove, or space heater of any type:
—Use fire safeguards.
—Properly ventilate.
—If no heat:
—Close off unneeded rooms.
—Stuff towels or rags in cracks under doors.
—Cover windows at night.
—Eat and drink. Food provides energy to maintain body heat. Fluids prevent dehydration.
—Wear layers of loose-fitting, lightweight, warm clothing. Remove layers to avoid overheating, perspiration, and subsequent chill.

Appendix B
A Guide to Federal and State Agencies and Sources of Hazard Information

Frequently Used Addresses and Phone Numbers
United States Government

Federal Government Information Center
Phone: (800) 326-2996

Building Structures in the Coastal Zone
U.S. Army Corps of Engineers
P.O. Box 898
Anchorage, AK 99506-0898
Design: (907) 753-2800
Emergency management: (907) 753-2513
Engineering: (907) 753-2662
Geotechnical: (907) 753-2680
Wetlands: (907) 753-2512

Bureau of Land Management
Alaska Regional Office
222 West 7th Avenue
Anchorage, AK 99513
Alaska Resources Library: (907) 271-5025
Applications: (907) 271-3796
Land information: (907) 271-5960
Resources management: (907) 271-5057

U.S. Coast Guard
Phone: (800) 478-5555

U.S. Environmental Protection Agency (EPA)
Suite 100, 410 Willoughby
Juneau, AK 99801
Phone: (907) 586-7619

Federal Emergency Management Agency (FEMA)
Region 10 (includes Alaska)
130 228th Street, SW
Bothell, WA 98021-9796
Phone: (206) 487-4678

FEMA documents should be ordered from:
P.O. Box 70274
Washington, DC 20024

U.S. Fish and Wildlife Service
Alaska Regional Office
1011 East Tudor Road
Anchorage, AK 99503
Phone: (907) 786-3487
Ecological services: 786-3544
Marine mammal management: 271-2394
Toll free: (800) 362-5148

U.S. Forest Service
Alaska Regional Office
709 West 9th Street
Juneau, AK 99801
Environmental coordinator: (907) 586-8887
Planning, projects: (907) 586-8884
Public affairs:(907) 586-8806
Resource management: (907) 586-8861

U.S. Geological Survey (USGS)
4230 University Drive
Anchorage, AK 99508
Maps, information, library: (907) 786-7011

Water Resources Division: (907) 786-7100

U.S. Minerals Management Service
Alaska Outer Continental Shelf (OCS) Region
949 East 36th Avenue
Anchorage, AK 99508

National Park Service
Alaska Regional Office
2525 Gambell Street
Anchorage, AK 99503
Coastal programs: (907) 257-2526
Environmental quality: (907) 257-2648
Planning: (907) 257-2647

State of Alaska

Archaeological Site Impacts
State Historic Preservation Office
Department of Natural Resources
P.O. Box 107001
Anchorage, AK 99510-7001
Phone: (907) 762-2626

Coastal Zone Management Information and Reports
Division of Government Coordination
P.O. Box 110030
240 Main, Suite 500
Juneau, AK 99811-0030
Phone: (907) 465-3562
Fax: (907) 465-3075

Coastal Zone Management Permits
Division of Government Coordination
3601 C Street, Suite 370
Anchorage, AK 99503-5930
Phone: (907) 561-6131
Fax: (907) 561-6134

Department of Fish and Game
333 Raspberry Road
Anchorage, AK 99518-1599
Phone: (907) 344-0541

Juneau office:
1255 West 8th Street
Juneau, AK 99802-5526
Phone: (907) 465-5331

Department of Natural Resources
Division of Land
Frontier Building
Southcentral District Office
P.O. Box 107005
Anchorage, AK 99510-7005
Phone: (907) 762-2253

Juneau office:
Division of Land
400 Willoughby Avenue
Juneau, AK 99801-1796
Phone: (907) 465-3400

Fire Marshal, Department of Public Safety
5700 East Tudor Road
Anchorage, AK 99507
Phone: (907) 269-5000
Fire prevention, building codes: (907) 269-5604

Specific Information Sources

Alaska Division of Geological and Geophysical Surveys (ADGGS)
794 University Avenue, Suite 200
Fairbanks, AK 99709
Phone: (907) 474-7147
Technical publications and maps about hazards and water resources.

Alaska Earthquake Information Center (AEIC)
U.S. Geological Survey/Geophysical Institute, University of Alaska Fairbanks
903 Koyukuk Drive
P.O. Box 757320
Fairbanks, AK 99775-7320
Phone: (907) 474-7320
Provides seismic monitoring for the state of Alaska and has compilations of location, magnitude, and depth of Alaskan earthquakes.

Alaska Tsunami Warning Center (ATWC)
Palmer, AK
Phone: (907) 745-4212
Provides tsunami watches and warnings for the northwestern Pacific Ocean by rapidly determining the location and size of tsunami-causing earthquakes in Alaska.

Applied Technology Council (ATC)
3 Twin Dolphin Drive
Redwood City, CA 94065

Phone: (415) 595-1542
Provides technical publications for engineers, architects, and others interested in designing to reduce earthquake damage to buildings and their contents.

Department of Military and Veteran Affairs (DMVA)
Division of Emergency Services
P.O. Box 5750
Fort Richardson, AK 99505-5750
Phone: (907) 428-7000, (800) 478-2337
Conducts preparedness and mitigation programs and workshops. Materials available on request.

Earthquake Engineering Research Institute (EERI)
6431 Fairmount Avenue, Suite 7
El Cerrito, CA 94530-3624
Phone: (415) 525-3668
Offers technical information for engineers, researchers, and practicing professionals, including videotapes, annotated slide sets, and reconnaissance reports on earthquake hazard mitigation and the response of buildings and bridges during major earthquakes around the world. Free catalog.

Aerial Photographs and Topographic Maps

AeroMap U.S., Inc.
2014 Merrill Field Drive
Anchorage, AK 99501
Phone: (907) 272-4495
Fax: (907) 274-3265
Photogrammetric consultants; the firm makes aerial photographs on contract basis, photos available to public.

Geo-Data Center, U.S. Geological Survey map distribution for central Alaska
Geophysical Institute, University of Alaska Fairbanks
903 Koyukuk Drive
Fairbanks, AK 99775
Phone: (907) 474-7487
Aerial photographs, USGS topographic maps.

USGS Earth Science Information Center
4230 University Drive
Anchorage, AK 99567-4667
Phone: (907) 786-7011
Publications and maps concerning earthquake hazards, faults, volcanoes, and permafrost.

Mail orders to:
USGS Books and Report Sales
P.O. Box 25425
Denver, CO 80225
For orders less than $10.00, include $1.00 for postage and handling.

Conservation Agencies

Sierra Club
241 East 5th Avenue, Suite 205
Anchorage, AK 99501
Phone: (907) 276-4048
Fax: (907) 258-6807

Mailing address:
P.O. Box 103441
Anchorage, AK 99510

Disaster Assistance

DMVA/Division of Emergency Services (*see above*)

FEMA (*see above*)

Municipality of Anchorage Emergency Management and Civil Defense
P.O. Box 196650
Anchorage, AK 99519
Phone: (907) 267-4904

State of Alaska Emergency Services
Phone: (800) 478-2337

Flooding

USGS, Water Resources Division
4230 University Drive, Suite 201
Anchorage, AK 99508-4664
Phone: (907) 786-7100

Insurance

FEMA
Federal Insurance Administration
500 C Street, SW
Washington, DC 20472
Phone: (800) 638-6831

FEMA
Flood Map Distribution Center
6930 (A-F) San Tomas Road
Baltimore, MD 21227-6227
Phone: (800) 333-1363

Libraries

Alaska Resources Library, Bureau of Land Management
222 West 7th Avenue
Anchorage, AK 99513
Phone: (907) 271-5025

Z. J. Loussac Public Library
3600 Denali
Anchorage, AK 99503
Phone: (907) 261-2975

USGS Library
4230 University Drive
Anchorage, AK 99508
Phone: (907) 786-7011

Oil Spills and Other Emergencies

To report toxic spills:
(800) 424-8802

Department of Environmental Conservation
Oil/Hazardous Spills
555 Cordova Street
Anchorage, AK 99501
Phone: (907) 563-6529

Anchorage office:
800 East Dimond Boulevard
Anchorage, AK 99515
Phone: (907) 349-7555

Oil Spill Public Information Center
645 G Street
Anchorage, AK 99501
Phone: (907) 278-8008

Pipeline Corridor Regional Office
411 West 4th Avenue
Anchorage, AK 99501
Phone: (907) 258-5400

Spill Response Center
National Guard Armory, Anchorage
Phone: (907) 428-7080

Permits

Municipality of Anchorage
Project Management and Engineering
Department of Public Works
P.O. Box 196650
Anchorage, AK 99519-6650
Phone: (907) 786-8251

Building Safety Division
3500 East Tudor Road
Anchorage, AK 99507
Phone: (907) 786-8301

Zoning and Platting
P.O. Box 196650
Anchorage, AK 99519-6650
Phone: (907) 343-4267

Inspections (Structural Code Compliance)
P.O. Box 196650
Anchorage, AK 99519-6650
Phone: (907) 786-8343

Planning Documents and Reports

Anchorage
Municipality of Anchorage
P.O. Box 196650
Anchorage, AK 99519-6650
Community Planning and Development: (907) 343-4309
Land-use plans: (907) 343-4224
Maps, publications: (907) 343-4283

Other Areas
State of Alaska
Community and Regional Affairs
Suite 220, 333 West 4th Avenue
Anchorage, AK 99501
Phone: (907) 269-4500

Weather Information

National Weather Service
Alaska Regional Headquarters
222 West 7th Avenue
Anchorage, AK 99513
Phone: (907) 271-5136
Anchorage weather: (907) 936-2525
Avalanche and mountain weather forecast: (907) 271-4500
General weather information: (907) 271-5105
Marine and boating information: (907) 271-2727
River information: (907) 271-3479

Coastal Zone Management Districts

Aleutians East Borough
Coastal Coordinator
P.O. Box 349
Sand Point, AK 99661
Phone: (907) 383-2699

Aleutians West CRSA
Program Coordinator
308 G Street, #225
Anchorage, AK 99501
(907) 272-6700

Municipality of Anchorage
Department of Community Planning and Development
P.O. Box 196650
Anchorage, AK 99519-6650
Phone: (907) 343-4261

City of Angoon
P.O. Box 189
Angoon, AK 99820
Phone: (907) 788-3653

Annette Islands Indian Reserve
Annette Island Metlakatla Indian Community
P.O. Box 360
Metlakatla, AK 99926-0360
Phone: (907) 886-4200

Bering Straits CRSA
Bering Straits Coastal Management Program
P.O. Box 190
Unalakleet, AK 99684
Phone (907) 624-3062

City of Bethel
Coastal Coordinator
P.O. Box 388
Bethel, AK 99559
Phone: (907) 543-4456

Bristol Bay Borough
P.O. Box 189
Naknek, AK 99633
Phone: (907) 246-4224

Bristol Bay CRSA
Nanvaq Building, Room 207
P.O. Box 849
Dillingham, AK 99576
Phone: (907) 842-2666

Ceñaliulriit CRSA
P.O. Box 357
St. Mary's, AK 99658
Phone: (907) 438-2638

City of Cordova
P.O. Box 1210
Cordova, AK 99574
Phone: (907) 424-6200

City of Craig
Coastal Coordinator
P.O. Box 725
Craig, AK 99921
Phone: (907) 826-3275

City of Haines
P.O. Box 1049
Haines, AK 99827
Phone: (907) 766-2231

City of Hoonah
P.O. Box 360
Hoonah, AK 99829
Phone: (907) 945-3663

City of Hydaburg
P.O. Box 49
Hydaburg, AK 99922
Phone: (907) 285-3761

City and Borough of Juneau
Department of Community Development
155 South Seward Street
Juneau, AK 99801
Phone: (907) 586-5230

City of Kake
P.O. Box 500
Kake, AK 99830
Phone: (907) 785-3804

Kenai Peninsula Borough
Coastal Coordinator
144 North Binkley Street
Soldotna, AK 99669-7599
Phone: (907) 262-4441

Ketchikan Gateway Borough
Coastal Coordinator
344 Front Street
Ketchikan, AK 99901
Phone: (907) 228-6610

City of Klawock
Coastal Coordinator
P.O. Box 113
Klawock, AK 99925
Phone: (907) 755-2261

Kodiak Island Borough
Department of Community Development
710 Mill Bay Road
Kodiak, AK 99615-6340
Phone: (907) 486-9360

Lake and Peninsula Borough
P.O. Box 495
King Salmon, AK 99613
Phone: (907) 246-3421

Matanuska-Susitna Borough
Coastal Management Coordinator
350 East Dahlia Avenue
Palmer, AK 99645-6488
Phone: (907) 745-9865

City of Nome
P.O. Box 281
Nome, AK 99762
Phone: (907) 443-5242

North Slope Borough
P.O. Box 69
Barrow, AK 99723
Phone: (907) 852-0440, x266

Northwest Arctic Borough
Coastal Coordinator
P.O. Box 1110
Kotzebue, AK 99752
Phone: (907) 442-2500

City of Pelican
Coastal Coordinator
P.O. Box 737
Pelican, AK 99832
Phone: (907) 735-2202

City of Petersburg
Department of Community Development
P.O. Box 329
Petersburg, AK 99833
Phone: (907) 772-4533

City and Borough of Sitka
100 Lincoln Street
Sitka, AK 99835-7540
Phone: (907) 747-1812

City of Skagway
Coastal Coordinator
P.O. Box 415
Skagway, AK 99840
Phone: (907) 983-2297

City of St. Paul
P.O. Box 901
St. Paul, AK 99660
Phone: (907) 546-2331

City of Thorne Bay
P.O. Box 19110
Thorne Bay, AK 99919
Phone: (907) 828-3380

City of Valdez
Department of Community Development
P.O. Box 307
Valdez, AK 99686
Phone: (907) 835-4313

City of Whittier
P.O. Box 608
Whittier, AK 99693
Phone: (907) 472-2327

City of Wrangell
P.O. Box 531
Wrangell, AK 99929
Phone: (907) 874-2381

City and Borough of Yakutat
P.O. Box 160
Yakutat, AK 99689
Phone: (907) 784-3323

Appendix C. References

This appendix includes technical background literature for each chapter as well as general works related to chapter topics that are recommended as additional reading. Some of the references are annotated to indicate content and explain how the publication relates to coastal zone hazards and related topics. Older publications may be found in your local library. The nearest college or university library is a good source of technical journal articles and the publications of government agencies. Recent publications may be ordered through your bookstore or from the publisher or agency that produced the text. Publications produced by state and federal agencies are often free or inexpensive, and we encourage readers to pursue their interests and take advantage of these resources.

To save space, we use the following agency abbreviations:

AIPG American Institute of Professional Geologists
APA American Plywood Association
ASCE American Society of Civil Engineers
BAREPP Bay Area Regional Earthquake Preparedness Project
CERC Coastal Engineering Research Center
DOE U.S. Department of Energy
FEMA Federal Emergency Management Agency
NOAA National Oceanic and Atmospheric Administration
OIES Office for Interdisciplinary Earth Studies
USGS U.S. Geological Survey

Chapter 1. The Lay of the Land and Its Occupation by Humans

Geology of Alaska

"Cenozoic Glaciation of Alaska," by Thomas D. Hamilton, 1994, pp. 813–44 in *The Geology of Alaska*, ed. George Plafker and Henry C. Berg. *Geology of North America*, volume G-1 (Geological Society of America, Boulder, Colo.). Technical but readable to an educated audience. Details the extent, stratigraphy, and timing of various ice advances through the last several million years, with focus on the last 25,000 years.

The Geology of Alaska, ed. George Plafker and Henry C. Berg, 1994, *Geology of North America*, volume G-1 (Geological Society of America, Boulder, Colo.), 1,055 pp. Includes articles on bedrock geology of the entire state, with a focus on tectonic evolution; well illustrated, technical. State-of-the-art synthesis of thousands of reports and surveys. Assumes familiarity with geologic concepts and theories. The chapters on glaciation (see below) and volcanism are fairly accessible to general readers.

"Growth of Western North America," by David L. Jones, Peter Coney, and Myrl Beck, 1982, *Scientific American* 247(5):70–84. This is a basic discussion of plate tectonics, exotic terranes, and geologic structures.

Roadside Geology of Alaska, by Cathy Connor and Daniel O'Haire, 1988 (Mountain Press, Missoula, Mont.), 250 pp. Although oriented toward the state's roads, the book does offer a brief geologic survey and extensive coverage of the islands of the marine "highway" via the state ferry system in southeast Alaska. Well illustrated, good figures, covers earthquake effects in some detail. Designed for use by both amateurs and professional geologists.

Bering Land Bridge

Paleoecology of Beringia, ed. David M. Hopkins et al., 1982 (Academic Press, New York), 489 pp. Results of a symposium on the past vegetation and fauna of the land bridge between Alaska and Siberia. Individual chapters include discussion of pollen evidence, large mammals, and humans. Discussion of sea level and geologic data is limited. A considerable body of more recent data supercedes this volume but is available only in journals. Earlier research (pre-1970s) is collected in *The Bering Land Bridge,* ed. David M. Hopkins, 1967 (Stanford University Press, Stanford, Calif.)

Regional Syntheses of Geology, Ecology, and Oceanography

The Eastern Bering Sea Shelf: Oceanography and Resources, by D. W. Hood and J. A. Calder, 1982, 2 volumes (NOAA, U.S. Department of Commerce, Washington, D.C.). The first volume is most informative for hazards-related topics: meteorology, tides, ice dynamics, distribution of fast ice, and interactions between oil and the ocean; technical in places. Much of the second volume is concerned with ecological issues, including marine mammals and fish. Available from Superintendent of Documents, Government Printing Office, Washington, DC 20402.

The Gulf of Alaska: Physical Environment and Biological Resources, ed. Donald W. Hood and Steven T. Zimmerman, 1986 (NOAA, U.S. Department of Commerce, Washington, D.C.), 656 pp. Scientific survey of numerous aspects of the gulf region, including meteorology, geologic setting, hazards, and biota. Detailed, assumes familiarity with basic concepts and terms; comprehensive glossary. Available from Superintendent of Documents, Government Printing Office, Washington, DC 20402.

Coastal Evolution and Processes

Coastal Environments: An Introduction to the Physical, Ecological and Cultural Systems of Coastlines, by R. W. G. Carter, 1988 (Academic Press, New York). Designed as a textbook for undergraduates; well illustrated with line drawings and maps; some technical descriptions of processes. Covers a wide range of material; focuses on sandy shores, although gravel and rock beaches are also discussed. The chapters on coastal management are comprehensive and international in scope but limited largely to developed countries. Included are oil spills, nuclear power plants, the effects of groins, and pros and cons of beach nourishment.

Coastal Evolution: Late Quaternary Shoreline Morphodynamics, ed. R. W. G. Carter and C. D. Woodroffe, 1994 (Cambridge University Press, Cambridge). Aimed at college students; technical in places. Several chapters provide scientific background on Arctic coastal processes, paraglacial coasts, and spits, barriers, and rocky shores.

The Geomorphology of Rock Coasts, by Alan S. Trenhaile, 1987 (Oxford University Press, Oxford). Textbook on coastal processes with some mathematics. Describes development of shore platforms, notches, and wave-cut terraces, and includes discussions of sea level changes and the physical basis of processes, rates, and lithological factors.

Sea Level History

"Heightened North Pacific Storminess during Synchronous Late Holocene Erosion of Northwest Alaska Beach Ridges," by Owen K. Mason and James Jordan, 1993, *Quaternary Research* 40:55–69. Derives a climatic history using geological and archaeological data from beach ridges and coastal deposits along the northwest Alaska coast.

"Paleoecology of Late Glacial Peats from the Bering Land Bridge, Chukchi Sea Shelf," by Scott Elias, S. K. Short, and R. L. Philips, 1992, *Quaternary Research* 37:371–78. Presents evidence that the Bering Strait was flooded less than 11,000 years ago; technical.

Sea Surface Studies: A Global View, ed. R. J. N. Devoy, 1987 (Croom Helm, London), 649 pp. Includes articles on topics ranging from evidence for sea level changes worldwide to methods of dating geologic deposits to and tectonic effects on sea levels. Discusses the response of the shore to rising sea level, human impacts on the coastal zone, and the greenhouse effect as it relates to sea levels.

Chapter 2. Settling Alaska: 11,000 Years of Human History

Archaeology and Peopling of the New World

Ancient Men of the Arctic, by James Louis Giddings, 1967 (Knopf, New York), 391 pp. A memoir by one of Alaska's pioneering post–World War II archaeologists. Although enjoyable reading, it should be viewed critically because of the author's tendency for flights of fancy beyond what the data support.

Eskimos and Aleuts, rev. ed., by Don E. Dumond, 1990 (Thames and Hudson, London), 200 pp. Introduction to archaeological evidence for most of Alaska's coastal regions; popular account, well illustrated. Further references in bibliography. Limited discussion of southeast Alaska.

The Great Journey: The Peopling of Ancient America, by Brian M. Fagan, 1987 (Thames and Hudson, London), 288 pp. Readable, popular account of the archaeological and geological evidence for the peopling of North America. Anecdotal, good set of basic references.

History of Alaska

Alaska: A History of the 49th State, 2d ed., by Claus-M. Naske and Herman E. Slotnick, 1987 (University of Oklahoma Press, Norman). Basic treatment of Alaska's history, predominantly after U.S. acquisition (1867). Brief introductory chapters on prehistory and on Russian America. Especially good on territorial days.

Buildings of Alaska, by Alison K. Hoagland, 1993 (Oxford University Press, Oxford), 338 pp. Architectural history of Alaska, with general introduction and town-by-town sec-

tions. Illustrated; maps for all major and many minor places. The detailed bibliography is especially good for historical data.
"Cyclic Formation of Debris Avalanches at Mount St. Augustine Volcano," by James Begét and Juergen Kienle, 1992, *Nature* 356:701–4. Establishes history of avalanches and, indirectly, of tsunamis from Augustine via radiocarbon dates from organic remains buried by volcanic products.
"A 500-Year-Long Record of Tephra Falls from Redoubt Volcano and Other Volcanoes in Upper Cook Inlet, Alaska," by James Begét, S. D. Stihler, and D. B. Stone, *Journal of Volcanology and Geothermal Research* 2:55–67. Uses a dated sequence of ash layers from Skilak Lake on the Kenai Peninsula to reconstruct the history of eruptions on the Alaska Peninsula. Technical in places; reports use of electron microprobe to source ashfalls.
"The Periodicity of Storm Surges in the Bering Sea from 1898 to 1993, Based on Newspaper Accounts," by Owen K. Mason, David K. Salmon, and Stefanie L. Ludwig, *Climatic Change* 34:109–23.
"Postglacial Eruption History of Redoubt Volcano, Alaska," by James Begét and Chris J. Nye, 1994, *Journal of Volcanology and Geothermal Research* 62:31–54. Fairly technical; good discussion of the effects of lahars and flooding on the Drift River terminal area.
Storm Surge Climatology and Forecasting in Alaska, by James L. Wise, Albert L. Comiskey, and Richard Becker Jr., 1981 (Arctic Environmental Information and Data Center, Anchorage). Maps and calculates recurrence intervals for all reaches of the coast. Includes compilation of storm surges from newspapers, 1900–1970s.
The Wake of the Unseen Object: Among the Native Cultures of Bush Alaska, by Tom Kizzia, 1991 (Holt, New York). *Anchorage Daily News* reporter Kizzia visited 11 native communities, most of them coastal. The aim of the book is to convey a sense of bush life. Very readable.

For information on the history of earthquakes, see chapter 3 and chapter 11 references.

Chapter 3. Earthquakes, Volcanoes, and Tsunamis: A Deadly Trio

Earthquakes and Volcanoes

The Citizen's Guide to Geologic Hazards, by Edward B. Nuhfer, Richard J. Proctor, and Paul H. Moser, 1993 (AIPG, Arvada, Colo.). Order by mail from AIPG, 7828 Vance Drive, Suite 103, Arvada, CO 80003, (303) 431-0831.
Earthquakes, by Bruce Bolt, 1988 (W. H. Freeman, New York), 282 pp.
Earthquakes, by Bruce Walker, 1982 (Time-Life Books, Alexandria, Va.), 76 pp. Out of print.
Earthquakes and Volcanoes, a bimonthly publication of the U.S. Geological Survey, is available yearly for $6.50 from the Superintendent of Documents, Government Printing Office, Washington, DC 20402, (202) 783-3238.
Facing Geologic and Hydrologic Hazards: Earth Science Considerations, ed. W. W. Hays, USGS Professional Paper 1240-B, 109 pp.
Living with Volcanoes, by T. L. Wright and T. C. Pierson, 1992, USGS Circular 1073, 57 pp. The USGS cooperative volcano hazards program is outlined in this illustrated booklet; topics include how scientists study volcanoes, historic eruptions, associated hazards, risk mapping, predicting problems, and avoiding disasters.
The Next Big Earthquake, by Peter Haeussler, 1994 (USGS, Washington, D.C.), 25 pp. An excellent summary of the earthquake hazard. Much of this booklet is reproduced in

this book in chapter 3 and appendix A. Available from Earthquakes Branch of Alaska Geology, USGS, 4200 University Drive, Anchorage, AK 99508.

On Shaky Ground: America's Earthquake Alert, by John J. Nance, 1989 (Avon Books, New York), 440 pp.

Terra Non Firma, by J. M. Gere and H. C. Shah, 1984 (W. H. Freeman, New York), 203 pp.

Earthquakes, Volcanoes, and Tsunamis in Alaska

The Alaska Earthquake, March 27, 1964: Regional Effects, by George Plafker, 1966, USGS Professional Paper 543-1. A classic; describes elevation changes associated with seismic effects of the Good Friday quake. Based on pioneering fieldwork conducted immediately after the event.

The Alaska Earthquake Series: Effects on Communities, USGS Professional Papers 542-A–F, 1966–68. This is a series of reports on Anchorage, Whittier, Valdez, Homer, Seward, Kodiak, and other communities (Chenega, Cordova). The reports are fascinating descriptions of the 1964 earthquake's effects on hydrology, submarine landslides, shoreline changes, and subsidence, plus observations of the quake's impacts on people and animals.

Alaska Volcano Observatory. Bimonthly report, 1987 to present. An informal discussion of recent volcanic activity in Alaska. Detailed accounts of earthquake activity as well. Provides information on time, depth, magnitude, and related data; plus information on gas emissions, ash, and volcanic activity. A typical issue is about 20 pages. Available from Alaska Volcano Observatory, 4200 University Drive, Anchorage, AK 99508, (907) 786-7497; and Department of Geological and Geophysical Surveys, 794 University Avenue, Fairbanks, AK 99709, (907) 474-7147.

Catalog of Tsunamis in Alaska, by D. C. Cox and G. Pararas-Carayannis (revised by J. P. Calebaugh), 1976, NOAA Report SE-1 (World Data Center A for Solid Earth Geophysics, Boulder, Colo.), 43 pp.

Earthquake Alaska: Are We Prepared? Proceedings of Conference on the Status of Knowledge and Preparedness for Earthquake Hazards in Alaska, November 19–20, 1992, ed. Rodney Cumbellick, Roger Head, and Randall Updike, 1994, USGS Open File Report 94-218, 192 pp. Distributed free by the USGS, this collection of transcripts provides clear and concise discussions of seismic and tsunami hazards, with special focus on Anchorage. Topics covered include engineering and infrastructure effects, emergency response, experiences in recent California quakes, and historical anecdotes on rebuilding after the 1964 earthquake and prequake preparations (or lack of). Highly recommended. Also available from Alaska Department of Geological and Geophysical Surveys, 794 University Avenue, Fairbanks, AK 99709, (907) 474-7147.

Effects of the March 1964 Earthquake on Various Communities, by George Plafker, Reuben Kachadoorian, Edwin Eckel, and L. R. Mayo, 1966, USGS Professional Paper 543-G. Detailed description of effects of the quake on buildings, people, and geomorphic features; well illustrated, maps.

The Eruption of Redoubt Volcano Alaska, December 14, 1989–August 31, 1990, ed. S. R. Brantley, 1990, USGS Circular 1061, 33 pp. A fascinating summary of the eruption events and their impact on surrounding areas. Observations of ashfalls and other impacts on the adjacent Kenai Peninsula are included.

The Great Alaska Earthquake of 1964, by National Research Council, 1972, 8 volumes (National Academy of Sciences, Washington, D.C.). This series is a monumental work supervised by the National Academy of Sciences. Volumes include *Geology, Seismology, Hydrology, Biology, Oceanography, Engineering, Human Ecology,* and *Summary.* Most volumes include well-illustrated technical papers. This is the most complete scientific study of any earthquake. Much of this material is also published in USGS Professional Papers 543-A–G.

Paleoseismicity of the Cook Inlet Region: Evidence from Peat Stratigraphy in Turnagain and Knik Arms, by R. A. Combellick, Department of Geological and Geophysical Surveys Professional Report 112, 1991 (Alaska Department of Natural Resources, Fairbanks). Establishes the likelihood of large-magnitude earthquakes in the Anchorage area using geological evidence and dating techniques; technical.

Redoubt Volcano, Southern Alaska: A Hazard Assessment Based on Eruptive Activity through 1968, by Alison B. Till, M. Elizabeth Yount, and J. R. Richle, 1993, USGS Bulletin 1996, 19 pp. Includes a very useful hazards map. Some of the report is technical, but the lay reader will find it interesting.

"Submarine Landslides: An Introduction," by H. J. Lee, W. C. Schwab, and J. S. Booth, 1993, pp. 1–13 in *Submarine Landslides: Selected Studies in the U.S. Exclusive Economic Zone,* ed. W. C. Schwab, H. J. Lee, and D. C. Twichell, USGS Bulletin 2002. Discusses full range of submarine slumps, including continental shelf margin area.

"Submarine Landslides That Had a Significant Impact on Man and His Activities: Seward and Valdez," by M. A. Hampton, R. W. Lemke, and H. W. Coulter, 1993, pp. 123–34 in *Submarine Landslides: Selected Studies in the U.S. Exclusive Economic Zone,* USGS Bulletin 2002. Straightforward account of slope failures, focuses principally on the 1964 earthquake; updates the 1966 reports.

Tsunami Hazard and Community Preparedness in Alaska, by G. W. Carte, 1981, National Weather Service Technical Memorandum AR-29, NOAA, Anchorage.

Tsunami Predictions for the Coast of Alaska: Kodiak Island to Ketchikan, by Peter L. Crawford, 1987, CERC Technical Report 87-7, prepared for FEMA. Uses mathematical models and historical data to predict 100-year and 500-year tsunami elevations at 1,249 locations. Cook Inlet, Seward, Valdez, and Kodiak Island are covered fairly well, but much of southeast Alaska is not included. Use with caution; a small distance can make a big difference. Text discussion is technical.

Volcanic Hazards from Future Eruptions of Augustine Volcano, Alaska, by Juergen Kienle and Samuel Swanson, 1985, UAGR-275 (Geophysical Institute, University of Alaska at Fairbanks), 125 pp. A review of the history, hazards, and future of this volcano located on an uninhabited island in Cook Inlet. The Augustine volcano is an important threat to the Kenai Peninsula and to low-lying towns such as Homer Spit.

Emergency Preparedness

Earthquake Survival Manual, by Lael Morgan, 1993 (Epicenter Press, Seattle), 168 pp. Order by mail from Epicenter Press, 18821 64th Avenue NE, Seattle, WA 98115, (206) 485-6822.

The Emergency Survival Handbook, American Red Cross, 1985, 63 pp. Order by mail from American Red Cross, Los Angeles Chapter, 2700 Wilshire Boulevard, Los Angeles, CA 90057.

Family Survival Guide, by the American Red Cross, 1990, 32 pp. Obtain from your local Red Cross office or order by mail from American Red Cross, Los Angeles Chapter, 2700 Wilshire Boulevard, Los Angeles, CA 90057.

Safety and Survival in an Earthquake, by the American Red Cross, 1989, 52 pp. Order by mail from American Red Cross, Los Angeles Chapter, 2700 Wilshire Boulevard, Los Angeles, CA 90057.

Surviving the Big One: How to Prepare for a Major Earthquake, 1990, revised. A very informative 1-hour video developed for public television. Order by mail from KCET Video, 4401 Sunset Boulevard, Los Angeles, CA 90027, (800) 228-5238, $19.95 + $3.50 postage and handling.

Regional Planning to Reduce Earthquake Risk

Geologic-Hazards Mitigation in Alaska: A Review of Federal, State, and Local Policies, by R. A. Combellick, 1985, Alaska Division of Geological and Geophysical Surveys Special Report 35, 70 pp.

Putting Seismic Safety Policies to Work, by M. Blair-Tyler and P. A. Gregory, 1988, 44 pp. Order by mail from ABAG, P.O. Box 2050, Oakland, CA 94064-2050, (415) 464-7900.

Seismic Hazard Mitigation: Planning and Implementation, the Alaska Case, by L. L. Selkregg and J. Pruess, 1984 (University of Alaska, Anchorage), 332 pp. Order by mail from University of Alaska, Anchorage Bookstore, 2905 Providence Drive, Anchorage, AK 99508.

Seismic Hazards and Land-Use Planning, by D. R. Nichols and J. M. Buchanan-Banks, 1974, USGS Circular 690, 33 pp. Oriented toward California but applicable to Alaska. Free from the USGS.

Seismic Hazards and Land-Use Planning: Selected Examples from California, by M. L. Blair and W. E. Spangle, 1979, USGS Professional Paper 941-B, 82 pp.

A Workshop on Evaluation of Regional and Urban Earthquake Hazards and Risk in Alaska, ed. W. Hayes and P. Gori, 1986 (USGS, Washington, D.C.), 386 pp.

Chapter 4. The Problem of Unstable Slopes

"Cold Regions Science and Engineering. Part III. Engineering, Section A3: Snow Technology: Avalanches," by Malcolm Mellor, 1968 (U.S. Army Material Command, Cold Regions Research and Engineering Laboratory, Hanover, N.H.), 215 pp.

"The Determination of Land Emergence from Sea Level Observations in Southeast Alaska," by Steacy W. Hicks and W. Shofnos, 1965, *Journal of Geophysical Research* 70(14):3315–20.

"Informal Cooperative Avalanche Warning System and Public Education Program for South-Central Alaska, U.S.A.," by Steve W. Hackett and Douglas Fesler, 1980, *Journal of Glaciology* 26(94):497–500.

"Mass-Wasting Hazards Inventory and Land Use Control for the City and Borough of Juneau, Alaska," by Douglas Swanston, 1972, pp. 17–51 in *Technical Supplement: Geophysical Hazards Investigation for the City and Borough of Juneau,* by Daniel, Mann, Johnson and Mendenhall, Architects and Engineers, Portland, Ore.

"Mass-Wasting in Coastal Alaska," by Douglas Swanston, USDA Forest Service Research Paper, PNW-3 (Pacific Northwest Forestry and Range Experimental Station, Portland, Ore.).

"Measurement of Creep in Shallow, Slide-Prone Till Soil," by D. J. Barr and D. N. Swanston, 1970, *American Journal of Science* 269:467–80.

"Probable Maximum Precipitation and Rainfall Frequency Data for Alaska," by John Miller, 1963, National Weather Service Technical Paper 47-6, 9 pp.

"Report of the Preliminary Evaluation of the Behrends Avenue Avalanche Path," by Keith Hart, 1967. Report for the City and Borough of Juneau.

"Slope Movement Processes and Characteristics," by Douglas Swanston and D. E. Howes, 1991, pp. 1–18 in *A Guide for Management of Landslide-Prone Terrain in the Pacific*

Northwest. Land Management Handbook 18 (British Columbia Ministry of Forests, Victoria, B.C., and USDA Forest Service Pacific Northwest Research Station, Juneau).

Chapter 5. Hazards in Alaska

Storm Surges and Ice Override

Alaska Marine Ice Atlas, by Joseph C. LaBelle et al., 1983 (Arctic Environmental Information and Data Center, Anchorage). Maps of ice limits by month and area. Weather data for localities; includes maps. Brief discussion of icing problems and storm surge.

"Another Ivu at Nome—This Time with a Storm," by Theodore F. Fathauer, March 20, 1989, Technical Attachment to the Alaska Region Staff Notes, National Weather Service, Fairbanks.

Climatology of the Ice Extent in the Bering Sea, by Bruce D. Webster, 1981, NOAA Technical Memorandum NWS AR-33.

"Coastal Floods on the Yukon Kuskokwim Delta of November 8–10, 1979," by Theodore F. Fathauer, December 1979, Technical Attachment to the Alaska Region Staff Notes, National Weather Service, Fairbanks.

Empirical Probability of the Ice Limit and Fifty Percent Ice Concentration Boundary in the Chuckchi and Beaufort Seas, by Bruce D. Webster, 1982, NOAA Technical Memorandum NWS 14-34.

"Fast Ice Sheet Deformation during Ice-Push and Shore Ice Ride-up," by Lewis H. Shapiro, Ronald C. Metzner, Arnold Hanson, and Jerome B. Johnson, 1984, pp. 137–58 in *The Alaskan Beaufort Sea,* ed. Peter W. Barnes, Donald M. Schell, and Erk Reimnitz (Academic Press, Orlando).

A Forecast Procedure for Coastal Floods in Alaska, by Theodore F. Fathauer, 1978. NOAA Technical Memorandum NWS AR-23.

"The Great Bering Sea Storms of 9–12 November 1974," by Theodore F. Fathauer, 1975, *Weatherwise,* April 1975.

Hunters of the Northern Ice, by Richard K. Nelson, 1989 (University of Chicago Press, Chicago).

Interagency Flood Hazard Mitigation Report, 1986. Response to the October 27, 1986, Disaster Declaration, State of Alaska, FEMA 781-DR. FEMA, Region 10, Interagency Hazard Mitigation Team.

Nome Nugget, December 17, 1987, February 2, 1989, and February 16, 1989.

"Onshore Ice Pile-up and Ride-up: Observations and Theoretical Assessment," by Austin Kovacs and Devinder S. Sodhi, 1988, pp. 108–42 in *Arctic Coastal Processes and Slope Protection Design* (American Society of Civil Engineers, New York). Technical in places, describes processes of formation using experimental data and some mathematical formulas. Good history of override across the Arctic.

"The Periodicity of Storm Surges in the Bering Sea from 1898 to 1993, Based on Newspaper Accounts," by Owen K. Mason, David K. Salmon, and Stefanie L. Ludwig, *Climatic Change* 34:109–23.

"Sealed in Time," *National Geographic,* June 1987. Journalistic account; somewhat sensationalistic. The continuing uncovering of well-preserved human "mummies" in the same embankment calls the ivu hypothesis into question.

Storm Surge Climatology and Forecasting in Alaska, by James L. Wise, Albert L. Comiskey, and Richard Becker Jr., 1981 (Arctic Environmental Information and Data Center, Anchorage). Maps; limited data. Recurrence intervals are calculated for all reaches of the coast; includes compilation of storm surges from newspapers. Incomplete.

Chapter 6. Shoreline Erosion

Erosion Control and Stabilization

Arctic Coastal Processes and Slope Protection Design, ed. Andrie T. Chen and Craig B. Leidersdorf, 1988 (American Society of Civil Engineers, New York). A useful book for understanding geomorphic processes and conventional engineering strategies for controlling nature. Good chapters on Arctic coastline, including bluff erosion and ice override. The engineering methods section is technical and has limited or no discussion of alternatives to engineering solutions.

Atmospheric Carbon Dioxide and the Greenhouse Effect, Bulletin DOE/ER-04II, 1989 (DOE, Office of Energy Research, Washington, D.C.), 36 pp. This easy-to-read pamphlet on the greenhouse effect is divided into six sections: "Introduction," "Increases in Atmospheric Carbon Dioxide," "Climate," "Plant," "Sea Level," and "Response to the Challenge." Summarizes current thought on greenhouse warming in the format of answers to 15 questions about the topic. Available from DOE, Office of Energy Research, Office of Basic Energy Sciences, Washington, DC 20545.

At the Sea's Edge: An Introduction to Coastal Oceanography for the Amateur Naturalist, by W. T. Fox, 1983 (Prentice-Hall, Englewood Cliffs, N.J.). Excellent nontechnical, richly illustrated introduction to coastal processes, meteorology, environments, and ecology.

The Beaches Are Moving: The Drowning of America's Shoreline, by Wallace Kaufman and Orrin Pilkey, 1983 (Duke University Press, Durham, N.C.). Highly readable account of the state of America's coastline that explains natural processes at the beach, provides a historical perspective of humans' relation to the shore, and offers practical advice on living in harmony with the coastal environment. Written from the barrier island point of view but applicable to sandy shorelines everywhere.

The Climate System, by Robert Dickinson and Richard Monastersky, 1991 (OIES and NOAA, Washington, D.C.). This report, one of the Reports to the Nation on Our Changing Planet series, discusses greenhouse gases, melting ice, the hydrologic cycle, and the global heat engine. Available from OIES, (303) 497-1682.

Coastal Environments: An Introduction to the Physical, Ecological and Cultural Systems of Coastlines, by R. W. G. Carter, 1988 (Academic Press, New York). Designed as a textbook for undergraduates; well illustrated with line drawings and maps; some technical descriptions of processes. Covers a wide range of material; focuses on sandy shores, although gravel and rock beaches are also discussed. The chapters on coastal management are comprehensive and international in scope but limited largely to developed countries. Included are oil spills, nuclear power plants, the effects of groins, and pros and cons of beach nourishment.

Coastal Land Loss, by Orrin H. Pilkey, Robert A. Morton, Joseph T. Kelley, and Shea Penland, 1989, *Short Course in Geology,* volume 2 (American Geophysical Union, Washington, D.C.). Notes for a short course presented at the 28th International Geological Congress, Washington, D.C. Covers social implications of land loss, factors affecting land loss, sea level rise, alternative responses to land loss, methods of quantifying land loss, predicting shoreline retreat, and regional land loss. Available from the American Geophysical Union, 2000 Florida Avenue, Washington, DC 20009, (202) 462-6900.

The Corps and the Shore, by O. H. Pilkey and K. Dixon, 1996 (Island Press, Washington, D.C.), 272 pp. A critical view of the shoreline activities of the U.S. Army Corps of Engineers.

The Effects of Seawalls on the Beach, ed. Nicholas Kraus and Orrin Pilkey, 1988, *Journal of Coastal Research,* special issue no. 4. The eight constituent papers discuss various aspects of seawalls. See especially Kraus's literature review and the paper by Orrin Pilkey and Howard Wright, "Seawalls versus Beaches," which includes a field study of beach width and stabilization on the East Coast.

Help Yourself, by the U.S. Army Corps of Engineers, is a small brochure that addresses erosion problems in the Great Lakes region. It may be of interest to all shore residents because it outlines shoreline processes and illustrates a variety of shoreline-engineering devices used to combat erosion. Free from the U.S. Army Corps of Engineers, North Central Division, 219 South Dearborn Street, Chicago, IL 60604.

Low Cost Shore Protection is a set of four reports written for the layperson by the U.S. Army Corps of Engineers, 1981. The set includes the introductory report, a property owner's guide, a guide for local government officials, and a guide for engineers and contractors. The reports suggest a wide range of engineering devices and techniques to stabilize shorelines, including beach nourishment and vegetation. The reports are available from the Section 54 Program, U.S. Army Corps of Engineers, USACE (DAEN-CWP-F), Washington, DC 20314.

Policy Implications of Greenhouse Warming: The Full Report, by the National Academies of Sciences and Engineering, 1992 (National Academy of Sciences Press, Washington, D.C.), 700 pp. A collection of four reports (natural science, mitigation, adaptation, and synthesis) reviewing the policy implications of global warming.

Responding to Changes in Sea Level, Engineering Implications, by Committee on Engineering Implications of Changes in Relative Mean Sea Level, Marine Board Commission on Engineering and Technical Systems, National Research Council, 1987 (National Academy of Sciences Press, Washington, D.C.). This book is of interest to community planners, officials, and legislators because of the implications of sea level rise for all coastal development. No specific solutions are provided, but the text does conclude with relevant general recommendations.

Sea Level Variations for the United States, 1855–1980, by S. D. Hicks, H. A. Debaugh Jr., and L. E. Hickman Jr., 1983 (NOAA, National Ocean Service, Rockville, Md.). This technical government publication discusses trends in sea level from the mid-1800s to the present based on the tide gauge records.

Shore Protection Manual, 4th ed., by the U.S. Army Corps of Engineers, 1984, 3 volumes. This is the "bible" of shoreline engineering. It outlines the various types of engineering structures, including their destructive effects. Available from the Government Printing Office, Washington, DC 20402; stock no. 008-002-00218-9.

"A 'Thumbnail Method' for Beach Communities: Estimation of Long-Term Beach Replenishment Requirements," by Orrin H. Pilkey, 1988, *Shore and Beach* 56(3):23–31. This is a short paper with tables and illustrations demonstrating that current methods of estimating long-term volume requirements for replenished beaches are inadequate. The long-term volume required can be estimated by assuming that the initial restoration volume must be replaced at prescribed intervals that depend on the geographic location. Applicable to any replenishment situation.

Chapter 7. River Flooding in Coastal Alaska

"Alaska Floods and Droughts," by R. D. Lamke, 1991, pp. 171–80 in *National Water Summary, 1988–89, Hydrologic Events and Floods and Droughts,* ed. R. W. Paulson, E. B. Chase, R. S. Roberts, and D. W. Moody, USGS Water Supply Paper 2375. Describes weather patterns responsible for floods in Alaska. Other papers in the same volume are very informative about hydrological processes and paleohydrological methods.

Alaskan Communities Flood Hazard Data, by the U.S. Army Corps of Engineers, Alaska District, 1993. This extensive inventory (335 pages plus a 7-page bibliography) of Alaska communities is a quick reference for flood hazard evaluation. Community citations include location, borough, population, number of houses, community services, maps available at the Alaska District office, and flood data. The latter includes National Flood Insurance Program information (participating or not), flood insurance studies, causes of flooding, last flood event, worst flood event, property in floodplain, and other information. Unfortunately, not all data categories are complete for each community. Available through the U.S. Army Corps of Engineers and local libraries.

Flood Characteristics of Alaskan Streams, by R. D. Lamke, 1978, USGS Water Resources Investigation 78-129, 61 pp.

Glacier Dammed Lakes and Outburst Floods in Alaska, by Austin Post and Lawrence Mayo, 1971, USGS Hydrologic Investigations Atlas HA-455. Plots the locations of 750 lakes and outlet drainages from numerous lakes. A short paper accompanying the map is packed with tidbits and useful information. Aimed at informed readers, but not overly technical.

Chapter 8. Human-Induced Hazards and Health Risks

Oil Spills, Human-Induced Hazards

Alaska Fish and Game magazine, special oil spill issue, July–August 1989. One of numerous journalistic accounts. The entire range of effects is treated in a mostly objective manner. Discussion is mostly speculative in view of the early date of the issue (just 2–3 months after the spill).

Black Tides: The Alaska Oil Spill, by Brian O'Donoghue, 1989 (Alaska Natural History Association, Anchorage). Journalistic day book with glossy photographs. Limited scientific discussion of effects; much anecdotal commentary.

"Coastal Oil Spills: Myth and Reality," by Ed Owen, 1992, *Shore and Beach* 60:2–6. Oil company consultant attempts to debunk persistent public myths about the *Exxon Valdez* spill. While mostly factual, the bias is evident.

Degrees of Disaster. Prince William Sound: How Nature Reels and Rebounds, by Jeff Wheelwright, 1994 (Simon and Schuster, New York). Controversial discussion of the science conducted after the *Exxon Valdez* spill. Argues that natural forces were more effective in cleaning the environment than human action. Very readable, written for general public; covers oceanography, health effects to cleanup workers, and subsistence consumers. Reportedly banned in Cordova.

Exxon Valdez *Oil Spill Information Packet,* State of Alaska, Office of the Governor, 1989. Provides a chronology and statistical tabulation of the oil spill's effects on fisheries and biota. Definitely polemical in places, defending the state's role. The appendix is especially useful in quantifying impacts on fisheries.

Five Years Later: 1994 Status Report on the Exxon Valdez *Oil Spill,* by *Exxon Valdez* Trustee Council, 1994; free on request from the Oil Spill Public Information Center, inside Alaska: (800) 478-7745; outside: (800) 283-7745. Succinct scientific analysis of effects on biota of the 1989 spill, including numerical estimates. Lacks references; rather long on promised goals and results of studies. Glossy.

Marine Mining on the Outer Continental Shelf, by Michael Cruikshank, 1993, Minerals Management Service Report 87-0035 (U.S. Department of the Interior, Washington,

D.C.). Revised from 1987 edition. Although oriented toward deeper water, discussion of methods is appropriate to cases in Alaska.

"Movement of Oil Spilled from the T/V *Exxon Valdez*," by J. A. Galt and D. L. Payton, 1990, pp. 4–17 in *Sea Otter Symposium, Biological Report* 90(12) (U.S. Fish and Wildlife Service, Washington, D.C.). Good scientific discussion of the physical oceanography and progress of the oil spill across southern Alaska.

Nome Harbor, Alaska: Dredged Material Disposal Environmental Assessment, by the U.S. Army Corps of Engineers, Alaska District, 1990 (U.S. Army Corps of Engineers, Anchorage).

Oil in Arctic Waters: The Untold Story of Offshore Drilling in Alaska, by Greenpeace, 1993. A polemical but well-documented discussion of issues and costs associated with oil development. General discussion of ice, biota, and effects of oil on both.

Recreational Users Guide to the Shoreline Impact Maps from the Exxon Valdez *Oil Spill*, by the Alaska Department of Environmental Conservation, *Exxon Valdez* Oil Spill Response Center, 1991. This 41-page atlas includes color photos of the types of beach pollution and shows differences between high-energy and low-energy beach settings in terms of oil persistence. Maps of Prince William Sound shorelines indicate locations of oiled sediment and oil type.

Chapter 9. Arctic Alaska

General

Beaufort Sea Coastal Erosion, Shoreline Evolution and Sediment Flux, by Erk Reimnitz, S. M. Graves, and Peter W. Barnes, USGS Map I-1182-G. Large-scale map of shoreline changes (erosion, accretion) over 30-year period based on analyses of aerial photos. Illustrated; very good. The somewhat technical but readable and informative 20-page text describes variables influencing erosion on the Arctic coast.

"Coastal Erosion Rates along the Chukchi Sea Coast near Barrow, Alaska," by John Harper, 1978, *Arctic* 31:428–33.

"Coastal Geomorphology of Arctic Alaska," by Peter W. Barnes, Stuart Rawlinson, and Erk Reimnitz, 1988, pp. 3–30 in *Arctic Coastal Processes and Slope Protection Design* (American Society of Civil Engineers, New York). Gives brief summary of geomorphic processes operating in Arctic Alaska and discusses various coastal features (spits, bluffs, etc.) and effects of ice.

The Environment and Resources of the Southeastern Chukchi Sea, by M. J. Hameedi and A. S. Naidu, 1988, Outer Continental Shelf Study MMS 87-0113 (NOAA, U.S. Department of Commerce, Washington, D.C.). Geographic limits: Kotzebue Sound; focuses on ecology of fish, sea birds, sea mammals; some discussion of sea ice, physical oceanography, marine geology, and tides.

"Storm Surge Effects on the Beaufort Sea Coast," by Erk Reimnitz and D. K. Maurer, 1979, *Arctic* 32(4):329–44.

Barrow

Bluff and Shoreline Protection Study for Barrow, Alaska, by Tekmarine, Inc., 1987. Adequate treatment of erosion rate and storm frequency; variable calculations of erosion rates. Emphasizes structural solutions to erosion problems. Available from North Slope Borough planning department.

"Bluff Erosion at Barrow and Wainwright, Arctic Alaska," by Jesse Walker, 1991, *Zeitschrift für Geomorphologie* 81:53–61. Describes the effects of the September 1986 storms, which produced considerable erosion; includes meteorological data and sea ice extent.

Mitigation Alternatives for Coastal Erosion at Wainwright and Barrow, Alaska: Active and Passive Considerations, 1989. Prepared by BTS/LCMF, Ltd., 723 West 6th Avenue, Anchorage. Brief discussion of erosion rates; concentrates on engineering solutions.

North Slope Borough Coastal Management Program, by Maynard and Partch, Woodward-Clyde Consultants, 1984, 2 volumes. Detailed discussion of oceanic hazards, including sea ice, wave erosion, and ice override. Focus is mostly on Point Hope to Point Lay; Barrow is not explicitly included. Effects of oil spills and other contaminants well covered. Much information on biologic communities, especially in lagoons.

Kivalina

City of Kivalina Relocation Study, by DOWL Engineers, Anchorage, 1994. Terse study of community survey conducted for the city. Little attention to geomorphology; maps of alternative sites plotted; basic cost estimates are probably too low. Available on request from DOWL Engineers, 4040 B Street, Anchorage, AK 99503-5999.

Shishmaref

"Geologic and Anthropological Aspects of Relocating Shishmaref," by Owen K. Mason, 1996, pp.1–35, in *Surficial Map of Part of the Shishmaref A-3, B-3, and Teller D-3 Quads,* Public Data File Report 96-7 (Department of Geological and Geophysical Surveys, State of Alaska). Available from DGGS office, Fairbanks.

Shishmaref Erosion Control Engineering Studies, by Peratrovich and Nottingham, Inc., for Alaska Department of Transportation and Public Facilities, 1982. Consideration of storm climates in southern Chukchi Sea and various mitigation alternatives, with limited discussion of consequences for the community. Read cautiously and consult chapter 6.

Shishmaref Erosion Protection Alternatives Feasibility and Cost Study, by DOWL Engineers, 1975. One of the first planning efforts; discusses relocation costs. While clearly biased toward engineering solutions, the report does warn against using unanchored sandbags and discusses problems with seawalls on sandy barriers.

Chapter 10. Southwest Alaska: Bering Sea Coast

Southwest Alaska

The Eastern Bering Sea Shelf: Oceanography and Resources, by D. W. Hood and J. A. Calder, 1982, 2 volumes (NOAA, U.S. Department of Commerce, Washington, D.C.). The first volume is most informative for hazards-related topics: includes meteorology, tides, ice dynamics, distribution of fast ice, and interactions between oil and the ocean. Technical in places. Much of the second volume covers ecological concerns, including marine mammals, and fish. Available from Government Printing Office, Washington, DC 20402.

The Yukon Delta: A Synthesis of Information, by L. K. Thorsteinson, P. R. Becker, and D. A. Hale, 1989, Outer Continental Shelf Study MMS 89-0081 (NOAA, U.S. Department of Commerce, Washington, D.C.). Covers meteorology, hydrology, geology, storm surges, biology. Limited treatment of storm surge. Readable.

Bethel

City of Bethel, Coastal Management Program Background Report, by City of Bethel Coastal Management Working Group, 1982. Discussion of most geophysical hazards is brief; erosion and flooding are reasonably well detailed, but information is not up-to-date. Available from the Coastal Zone Commission in Juneau.

Nome

City of Nome, Coastal Management Program Background Report, prepared by Environmental Services, Ltd., 1981. Comprehensive; contains numerous maps for flood and surge limits; covers geology, hazards, and biological and cultural resources (including land status, employment). Available from the Coastal Zone Commission in Juneau.

Ice Study for the Port of Nome, Alaska, by the Iowa Institute of Hydraulic Research, 1982 (University of Iowa, Iowa City). Technical; describes results of mathematical modeling and laboratory studies. Good background on sea ice in Norton Sound. Available at Rasmuson Library, University of Alaska, Fairbanks.

"Nome, City of the Golden Beaches," ed. Terrence Cole, 1984, *Alaska Geographic* 11(1). Good history of Nome, including its first 40 years. A section at the end covers storms in some detail. Well illustrated; well-documented discussion of failed attempt to retreat from the coast in late 1940s. Highly recommended.

Nome Harbor, Alaska: Dredged Material Disposal, by the U.S. Army Corps of Engineers, Alaska District, 1990. Technical report of environmental assessment. Analyzes program of dredging and also environmental consequences of offshore mining.

Nome Storm Damage, Emergency Repair, Nome-Council Road M.P. 14-33, State of Alaska Department of Transportation and Public Facilities, Northern Region, Fairbanks, 1993. Fully documented with memoranda, letters, engineering plans, and storm data. The reader can follow the bureaucratic process in one project.

Port Feasibility Study, Nome, Alaska, by Tetra Tech, Inc., Anchorage, 1980. Technical report filled with good data on dredging history, storm and ice effects, maps; includes discussion of ice override and such diverse topics as herring fishery, flooding, seismicity. Available at Rasmuson Library, University of Alaska, Fairbanks.

Naknek, Bristol Bay Borough

Bristol Bay Borough Coastal Management Program Description, prepared by Kramer, China and Mayo, Inc., 1983, 115 pp. plus 21 maps, 17 tables. Fairly detailed discussion and maps of coastal zone and its hazards. Additional information on demography, economics, and biota of borough is extensive. Available from Coastal Zone Management Office, Juneau.

Bristol Bay Borough Coastal Management Program Resource Inventory, volume 1, by Greg Peters, Tim Hostetler, and Patricia Roullier, 1984 (Northern Resource Management, Anchorage). Basically an atlas of small-scale (1:1,000,000) maps packed with data on resources, hazards, and socioeconomic information. Text is good; includes process and historical data. Available at Rasmuson Library, University of Alaska, Fairbanks.

Chapter 11. The Gulf of Alaska and South-Central Coast

South-Central Alaska

The Alaska Earthquake Series: Effects on Communities, USGS Professional Papers 542-A–F, 1966–68. Reports effects on Anchorage, Whittier, Valdez, Homer, Seward, Kodiak, and other communities (Chenega, Cordova) of the 1964 earthquake—its effects on hydrology, submarine landslides, shoreline changes, and subsidence.

"Geomorphology of a Tectonically Active, Glaciated Coast, South-Central Alaska," by Larry Ward, Thomas Moslow, and Kenneth Finkelstein, 1987, pp. 2–32 in *Glaciated Coasts,* ed. D. M. Fitzgerald and P. S. Rosen (Academic Press, New York). Somewhat technical; discusses coastal evolution in terms of size of materials on beach and supply of material to beach. Focus is on the largely uninhabited outer Kenai Peninsula coast, but applicable to much of the Gulf of Alaska coast, including southeast Alaska.

"Modern Clastic Depositional Systems of South-Central Alaska," by Miles O. Hayes and Jacqueline Michel, 1989, pp. 1–42 in *Glacial Geology and Geomorphology of North America,* volume 1, American Geophysical Union Field Trips (American Geophysical Union, Washington, D.C.). Very useful and well illustrated; shows beach profiles, sediment types, etc.; somewhat technical. Detailed discussion of climatic and oceanographic influences on shorelines of Cook Inlet, Copper River area, and Yakutat. Available from American Geophysical Union, 2000 Florida Avenue NW, Washington, DC 20009.

Anchorage

Anchorage Coastal Resources Atlas, by Municipality of Anchorage, 1980–81, 2 volumes. Includes all kinds of surficial hazards information with detailed maps. A must before purchasing Anchorage property. Available at public libraries and the city planning office.

Anchorage Snow Avalanche Zoning Analysis, prepared for the Municipality of Anchorage by Arthur Mears, Inc., Gunnison, Colo., 1982. This report, based on data from 2 years, provides an overview of the avalanche problem for the northern and southern hill slopes of the Anchorage metropolitan area. The study's principal contribution was a full set of large-scale maps; these are available only at the municipality office, not in the report. Available from the Municipality of Anchorage.

Earthquake Alaska: Are We Prepared? Proceedings of Conference on the Status of Knowledge and Preparedness for Earthquake Hazards in Alaska, November 19–20, 1992, ed. Rodney Cumbellick, Roger Head, and Randall Updike, 1994, USGS Open File Report 94-218, 192 pp. This collection of transcripts provides clear and concise discussions of seismic and tsunami hazards, with special focus on Anchorage. Distributed free by the USGS; also available from Alaska Department of Geological and Geophysical Surveys.

Engineering-Geologic Map of Southwest Anchorage, Alaska, by R. G. Updike and C. A. Ulery, 1986, Alaska Division of Geological and Geophysical Surveys Professional Report 89. This map shows areas around Anchorage where bluff slumping is a strong possibility.

A Field Guide to the Geologic Hazards of Anchorage and Turnagain Arm, Alaska, by J. M. Brown, K. J. Crossen, and J. Holzman, 1987, Alaska Geological Survey Field Guide (Alaska Geological Society, Anchorage). A critical publication for anyone, scientist or layperson, concerned with the hazards of the Anchorage area. No prospective home purchaser, teacher, rock hound, or city planning official should be without this report. Available from the Alaska Geological Society, P.O. Box 101288, Anchorage, AK 99510.

Geotechnical Hazards Assessment Study, prepared for the Municipality of Anchorage by E. C. Winterhalder, T. Williams, and J. M. England, Harding Lawson Assoc., Anchorage, 1979. Several critical plates. General review of hazards, including tsunami, wind, icing, groundwater, drainage, etc. Annotated bibliography and glossary of terms. Minimal coverage of earthquake hazards; specific discussion refers to plates that may not be included in every copy.

A Guide to the Geology of Anchorage: A Commentary on the Geotechnical and Historical Aspects of Selected Localities in the City, by R. G. Updike, C. A. Ulery, and J. L. Weir, 1986, Public Data File 86-49 (Department of Geological and Geophysical Surveys, Fairbanks). Originally prepared as an informal guide for the American Association of State Geologists in 1983, this is a road guide with some basic information. The DGGS guide includes several articles on earthquake-related slope failure and construction. Contact Department of Geological and Geophysical Surveys, 794 University Avenue, Fairbanks, AK 99709.

Slope-Stability Map of Anchorage and Vicinity, Alaska, by E. Pobrovolny and H. R. Schmoll, 1974, USGS Miscellaneous Investigations, Map I-787-E. A useful map for homeowners or potential homeowners seeking a low-risk site.

Cordova

Evaluation of Recent Channel Changes on the Scott River Near Cordova, Alaska, by Dave Blanchet, 1983 (U.S. Forest Service, Chugach National Forest, Anchorage). Discusses hydrological changes and outburst floods on glaciers upriver from Eyak Lake. Presents observations, data, and suggestions on hazard mitigation; nontechnical.

Eyak Lake AMSA, prepared by Professional Fishery Consultants, Cordova, 1985. Much of this report is concerned with fish and bird habitat, but there are good discussions of soils, hazards, and lake levels. While centered on Eyak Lake, the report has some applicability to Cordova. Available from the Coastal Zone Management Commission in Juneau.

Homer

Engineering Analysis of Beach Erosion at Homer Spit, Alaska, by Orson P. Smith, Jane M. Smith, Mary A. Cialone, Joan Pope, and Todd Walton, 1985, CERC 85-13 (Coastal Engineering Research Center, Vicksburg, Miss.). A technical summary of geological evolution of the spit, the history of shoreline stabilization measures, and possible future "solutions" to the erosion problem.

Kenai Peninsula

Geology and Geologic Hazards of the Eastern Coast of the Kenai Peninsula from Kenai to English Bay, Alaska, by J. R. Riehle, R. D. Reger, and C. L. Carver, 1977 (Alaska Department of Natural Resources, Division of Geological and Geophysical Surveys). This report is an excellent, though out-of-date, summary of the potential hazards facing people who live on the western side of the Kenai Peninsula. Hazards discussed include earthquakes, tsunamis, volcanism, coastal slumping, flooding, and coastal erosion. This may be the best single technical document available for Kenai citizens concerned with hazards.

Ground Conditions and Surficial Geology of the Kenai-Kasilof Area, Kenai Peninsula, by Thor N. V. Karlstrom, 1958, USGS Miscellaneous Geologic Investigations Map I-269.

The map shows a fairly straightforward division of sediment types and discusses land-use problems. Drainage is a principal part of the classification. Eroding bluffs are well marked. A short geological history is included.

Guide to the Geology of the Kenai Peninsula, Alaska, ed. Alexander Sisson, 1985, Alaska Geological Society Field Guide. Summarizes the geology, oceanography, and archaeology of Kachemak Bay, Turnagain Arm, and the peninsula. Includes a road log of important geologic features from Anchorage to Homer via the Seward and Sterling highways. Technical in places; covers hard-rock as well as surficial geology. Available from the Alaska Geological Society, P.O. Box 101288, Anchorage, AK 99510.

Kodiak

Geologic Effects of the March 1964 Earthquake and Associated Seismic Sea Waves on Kodiak and Nearby Islands, by George Plafker and Reuben Kachadoorian, 1966, USGS Professional Paper 543-D. Detailed description of effects of the quake; well illustrated, good maps.

Kadyak: A Background for Living, prepared by E. H. Buck et al., 1975 (Arctic Environmental Information and Data Center, Anchorage), 326 pp. An excellent, well-illustrated example of a regional synthesis of environment and resources. Summarizes the geographic, geologic, climatic, and oceanographic setting; the region's plant and animal life; human history; navigation, resource utilization, and environmental quality. Chapter 3, "Natural Disasters," reviews earthquakes, tsunamis, volcanoes, landslides, avalanches, floods, storms, and their relation to land-use planning. Recommended reading for every citizen of the region. The Kodiak City Public Library has multiple copies to encourage readership.

Kodiak Island Borough Coastal Management Program Concept Approved Draft and related resource classification maps, by the Kodiak Island Borough Community Development Department, 1983 (reprinted 1988). The report is part of the borough's coastal management program (see previous reference) and includes information on public safety, a resources analysis summary, policies and their implementation, and identifies areas meriting special attention. Property owners and developers should examine the resource classification maps: "Development with Restrictions, A" shows areas without coastal management restrictions; "Development with Restrictions, B" shows areas of high wave energy, erosion, and shoreline and property damage. Other maps include landownership and conservation. These maps are updated versions of the earlier (1981) color resource maps.

Kodiak Island Borough Coastal Management Program Progress Report, reprinted by the Kodiak Island Borough Community Development Department, 1983. Discusses the range of geologic hazards plus archaeological sites and resources. Notes the need for site-specific evaluation of earthquakes, mass wasting, and the coastal constraints of tsunamis, coastal erosion, and flooding (storm surge and riverine); recommends mitigation through site engineering and construction where possible. Discussion of volcanic ashfall hazard is limited. The report and related resource maps are available in the Community Development Department, Kodiak Island Borough, and the Coastal Zone Management Commission, Juneau.

Pillar Mountain Landslide, Kodiak, Alaska, by Reuben Kachadoorian and Willard Slater, 1978, USGS Open File Report 78-217. Analyzes in detail this major event near town.

Seward

Flood of October 1986 at Seward, Alaska, by Stanley K. Jones and Chester Zenone, 1988, USGS Open File Report 88-4278, 43 pp., 2 maps. Prepared in conjunction with Alaska Department of Transportation.

Flood Plain Information, Resurrection River and Salmon Creek, Seward Alaska, by the U.S. Army Corps of Engineers, Alaska District, 1975.

Seward Concept Plan: A Study and Discussion of the Goals and Opportunities and Problems Facing the City of Seward, with Emphasis on Land Use, Economics and Port Activity, by Kramer, Chin and Mayo, 1975.

Water Resources Data of the Seward Area, Alaska, by L. L. Dearborn, G. S. Anderson, and Chester Zenone, 1979, USGS Resources Investigations 79-11. Includes discussion of flooding history and hazards at Seward.

Valdez

Geologic Studies of Critical Areas: Valdez, Alaska, ed. R. A. Combellick and R. G. Updike, 1987, Public Data File 87-29, Division of Geological and Geophysical Surveys (Alaska Department of Natural Resources), 75 pp.

The Great Alaska Earthquake of 1964, by Frank Norton and J. Eugene Haas, 1970 (National Academy of Sciences, Washington, D.C.), pp. 357–99.

Old Town Hazards Assessment, Valdez, Alaska, by DOWL Engineers, Anchorage, 1983, 49 pp. including maps. Available at library and city office in Valdez.

Valdez Coastal Management Program, 1986, Community Development Department, City of Valdez, 239 pp. Detailed maps, some discussion of hazards. Very good for identifying slope hazards. Highly recommended. Every hazard is well mapped.

Valdez Duck Flats Area Meriting Special Attention Plan, 1992, Concept Approved Draft, City of Valdez, Coastal Management Program, Jon Isaacs and Associates, 108 pp.

"Valdez Seismicity Risk Study," appendix E in *West Mineral Creek Site Investigation,* by DOWL Engineers, Anchorage, 1982. Available at library and city office in Valdez. Details location and effects of the 11 major fault systems that can affect the area. Six major quakes, most before 1925, are described.

Yakutat

Preliminary Report on the Reconnaissance Engineering Geology of the Yakutat Area, Alaska, with Emphasis on Evaluation of Earthquake and Other Geologic Hazards, by Lynn A. Yehle, 1975, USGS Open File Report 75-529. Part of the USGS response to the 1964 quake. Surficial maps of hazards in the Yakutat area. Somewhat dated, but very readable and still informative.

Chapter 12. Southeast Alaska

"Geologic Hazards in Southeastern Alaska: An Overview," by R. A. Cumbellick and W. E. Long, 1983, Report of Investigations 83-17, Division of Geological and Geophysical Surveys (Alaska Department of Natural Resources), 17 pp. This short report describes major hazards in succinct terms. Major focus is on slope problems.

The Nature of Southeast Alaska: A Guide to Plants, Animals and Habitats, by Rita M. O'Clair, Robert H. Armstrong, and Richard Carstensen, 1992 (Alaska Northwest

Books, Anchorage). Covers everything from glacial deposits and history to bears, fish, and flowers. Well illustrated and written for the general public. Ecological orientation is clear (e.g., discussion of deer and old-growth forests). Not much on hazards to humans.

Regional and General Factors Bearing on Evaluation of Earthquake and Other Geologic Hazards to Coastal Communities in Southeastern Alaska, by R. W. Lemke and L. A. Yehle, 1972, USGS Open File Report. Introduction to USGS program of surficial mapping and hazards analysis in response to the 1964 earthquake.

Haines

City of Haines Alaska District Coastal Management Program, 1992, Concept Approved Draft prepared by City of Haines (P.O. Box 1049, Haines, AK 99827). This report contains information on environmental, geological, and sociological aspects of Haines but is rather general on geophysical hazards. Refers to a more detailed city flood and hazards map than that included in the report. Distributed by the Juneau office of the Coastal Zone Management Commission.

Reconnaissance Engineering Geology of the Haines Area, Alaska, with Emphasis on Evaluation of Earthquake and Other Geologic Hazards, by Richard Lemke and Lynn A. Yehle, 1972, USGS Open File Report. General background on southeast Alaska, using the limited database available in 1972. The discussion of underlying sediments is very useful for planning and building. Best to consult more recent seismic literature on earthquakes and tsunamis.

Juneau

Geophysical Hazards Investigation for the City and Borough of Juneau, Summary Report, by Daniel, Mann, Johnson, and Mendenhall, Inc., 1972. A very important summary document discussing earthquake, mass wasting, foundation, and snow avalanche hazards. Relative risks are rated on a series of excellent and quite detailed maps. The technical supplement (a separate document) provides the technical basis for the conclusions and recommendations. Available in the City and Borough of Juneau Community Development Office.

Juneau Area Mass Wasting and Snow Avalanche Hazard Analysis, by Doug Fesler and Jill Fredston, 1992. Report showing lot-by-lot avalanche and debris flow hazard in Juneau. A must for every banker and property owner. Available in the Juneau Public Library and the City and Borough of Juneau Community Development Office.

Surficial Geologic Map of the Juneau Urban Area and Vicinity, Alaska, by R. D. Miller, USGS Map I-885; 1 sheet, scale 1:48,000. Detailed map of glacial deposits (especially in Mendenhall Valley), debris slides, and former marine deposits now uplifted; also human-emplaced deposits. Radiocarbon ages are plotted on map. A considerable amount of fieldwork is represented in this one sheet. Very informative for planning and educational purposes.

Surficial Geology of the Juneau Urban Area and Vicinity, Alaska, with Emphasis on Earthquakes and Other Geologic Hazards, by R. D. Miller, 1972, USGS Open File Report. Summarizes the geologic setting and geophysical hazards of Juneau. An excellent but dated summary meant for the general public.

The Tsunami of the Alaska Earthquake 1964—Engineering Evaluation, by U.S. Army Corps of Engineers, 1968. Coastal Engineering Research Center, Technical Memoir 25, 444 pp.

Ketchikan

Ketchikan, Alaska, District Coastal Management Program, Public Hearing Draft October 18, 1991, by Susan Dickinson et al., Ketchikan Gateway Borough. A broad-brush effort; includes small-scale maps of diverse environmental data. Most valuable for policy statements. An updated final version is due soon. Available from the Coastal Zone Management Office in Juneau.

Reconnaissance Engineering Geology of the Ketchikan Area, Alaska, with Emphasis on Evaluation of Earthquake and Other Geologic Hazards, by Richard W. Lemke, 1975, USGS Open File Report 75-250. Thorough but somewhat dated summary of principal geophysical hazards in Ketchikan area. Much material is applicable to many southeast Alaska communities. Soil, seismic, flood, and slope hazards are emphasized; little attention to wind or avalanches.

Metlakatla

Reconnaissance Engineering Geology of the Metlakatla Area, Annette Island, Alaska, with Emphasis on Evaluation of Earthquakes and Other Geologic Hazards, by Lynn A. Yehle, 1977, USGS Open File Report 77-272. Summarizes local geology and geologic structure and potential for earthquake, storm waves, landslides, and stream floods.

Petersburg

Reconnaissance Geology of the Petersburg Area, Southeastern Alaska, with Emphasis on Geologic Hazards, by Lynn A. Yehle, 1978, USGS Open File Report 78-675.

Sitka

Reconnaissance Engineering Geology of Sitka and Vicinity, Alaska, with Emphasis on Evaluation of Earthquakes and Other Geologic Hazards, by L. A. Yehle, 1974, USGS Open File Report 74-53.

"Sitka District Coastal Management Program," 1989, City and Borough of Sitka Planning Department. A brief 4-page discussion of all hazards, not well mapped.

Skagway

Reconnaissance Engineering Geology of the Skagway Area, Alaska, with Emphasis on Evaluation of Earthquakes and Other Geologic Hazards, by Lynn A. Yehle and Richard W. Lemke, 1972, USGS Open File Report. Very detailed analysis of effects of earthquakes on varying types of sediment on the Skagway delta, but only cursory treatment of the major hazard—flooding. Discussion of tsunamis is also good. Despite its age, still a "must read" for residents.

Skagway Coastal Management Program, prepared by the City of Skagway, 1991. Brief discussion of hazards; the ongoing flood hazard is somewhat better described than the catastrophic submarine sliding hazards. The Skagway River area is extensively analyzed in a special section that includes hydraulic and engineering data. Maps are very good.

Wrangell

Reconnaissance Engineering Geology of the Wrangell Area, Alaska, with Emphasis on Evaluation of Earthquake and Other Geologic Hazards, by R. W. Lemke, 1974, USGS Open File Report 74-1062, 103 pp. One of a series of reports on coastal communities published in the decade following the 1964 Good Friday earthquake. Suggests that insufficient geological and seismological data prevented confident evaluation of earthquake probability and likely effects. Concerns include earthquake-induced subsidence and failure of fill, delta deposits, shore deposits, and muskeg due to seismic ground shaking, compaction, liquefaction, and slides and slumps. Evaluates other phenomena such as tsunamis, seiches, and flooding. Available from the USGS in Denver.

Wrangell Coastal Management Program, by Environmental Services, Ltd., for the City of Wrangell, 1981. Generally similar in its plan and implementation outline to statewide coastal zone management. Comments on hazards and is generally permissive with respect to land use. The plan exerts some restrictive control over the floodplain environment. The report is available through the planning office and the Wrangell Public Library.

Wrangell Comprehensive Development Plan, by Alaska State Housing Authority, 1968. In the 1960s, Wrangell experienced rapid economic growth that led citizens to recognize the need for planning. This plan provides a good summary of Wrangell's physical setting, history, and problems, as well as brief mentions of hazards. In particular, the upper town plat ignores natural topography, artificial fill along the shoreline could be problematic in the future, and muskeg areas require special engineering for building foundations. The report is in the Wrangell Public Library.

Chapter 13. Mitigating Wind, Snow Loading, and Permafrost Hazard Impacts through Construction

The five FEMA publications listed below are excellent sources of information on wind-related construction. The latest ones reflect the new building suggestions and techniques formulated after the recent destructive hurricanes. These publications are free and available from FEMA, P.O. Box 70274, Washington, DC 20024.

Against the Wind: Protecting Your Home from Hurricane Wind Damage, 1993, FEMA-247.

Building Performance: Hurricane Andrew in Florida, 1992, FIA-22.

Building Performance: Hurricane Iniki In Hawaii, 1993, FIA-23.

Coastal Construction Manual, 1986, FEMA-55.

Manufactured Home Installation in Flood Hazard Areas, 1985, FEMA-85.

Building Practices for Disaster Mitigation, ed. R. Wright, S. Kramer, and C. Culver, 1973, NBS Building Series 46 (U.S. Department of Commerce, National Bureau of Standards, Washington, D.C.).

Coastal Design: A Guide for Builders, Planners and Homeowners, by O. H. Pilkey Sr., O. H. Pilkey Jr., and W. J. Neal, 1983 (Van Nostrand Reinhold, New York).

Connectors for High Wind–Resistant Structures: Retrofit & New Construction, by Simpson Strong-Tie Company, Inc., 1992.

An Engineering Analysis: Mobile Homes in Windstorms, by W. Pennington and J. R. McDonald, 1978 (Texas Tech University Press, Lubbock).

How to Build Storm-Resistant Structures, by Southern Forest Products Association, P.O. Box 52468, New Orleans, LA 70152.

Hurricane-Resistant Construction for Homes, by T. L. Walton Jr. and M. R. Barnett, 1991 (Florida Sea Grant Publication, University of Florida, Gainesville).
Hurricane Resistant Construction Manual, 1986, Southern Building Code Congress International, Inc., Birmingham, Ala.
"Minimum Design Loads for Buildings and Other Structures," 1990, in ASCE 7-88 (American Society of Civil Engineers, New York). (Formerly ANSIA58.1-1982.)
Mitigating Damages in Hawaii's Hurricanes: A Perspective on Retrofit Options, by George F. Wallace, 1994, Insurance Institute for Property Loss Reduction Occasional Paper Series.
Protecting Mobile Homes from High Winds, 1974, Civil Defense Preparedness Agency Report TR-75 (U.S. Department of Defense, Washington, D.C.).
The South Florida Building Code, 1988 ed., Board of County Commissioners, Metropolitan Dade County, Miami, Fla.
Standard Building Code, 1988 ed., Southern Building Code Congress International, Inc., Birmingham, Ala.
"A Study of the Effectiveness of Building Legislation in Improving the Wind Resistance of Residential Structures," by S. M. Rogers Jr., P. R. Sparks, and K. M. Sparks, 1985, in *Proceedings of the Fifth U.S. National Conference on Wind Engineering* (Texas Tech University Press, Lubbock).
Suggested Technical Requirements for Mobile-Home Tiedown Ordinances, 1974, Civil Defense Preparedness Agency Report TM-73-1 (U.S. Department of Defense, Washington, D.C.).
Surviving the Storm: Building Codes, Compliance and the Mitigation of Hurricane Damage, 1989, All-Industry Research Advisory Council, Oak Brook, Ill.
"Wind Forces on Structures," 1961, ASCE *Transactions,* paper o. 3269 (American Society of Civil Engineers, New York).
"The Winds of Change?" by Paul Tarricone, 1994, *Civil Engineering* (January).

Permafrost

"Building on Permafrost: The Rules Are Different," by C. W. Lorell, 1989, *ASTM Standardization News* 17:24–26.
Frozen Ground Engineering, by A. Phukan, 1985 (Prentice-Hall, Englewood Cliffs, N.J.).
Permafrost, by L. L. Ray, 1993, USGS, 93-0640-P.
Permafrost and Related Engineering Problems in Alaska, by Oscar J. Ferrians Jr., Reuben Kachadoorian, and Gordon W. Greene, 1969 (Washington, D.C.: Government Printing Office).
Permafrost or Permanently Frozen Ground and Related Engineering Problems, by S. W. Muller, 1947, Edwards, Mich.

Chapter 14. Earthquake-Resistant Design and Construction

Construction Responses

APA *Residential Construction Guide,* by APA, P.O. Box 11700, Tacoma, WA 98411-0700.
Bolt-It-Down, a Homeowner's Guide to Earthquake Protection, International Conference of Building Officials, 5360 Workman Mill Road, Whittier, CA 90601-2298.
Bracing of Wood Burning Stoves and Propane Tanks, by NETAC, Task Order 2, California Governor's Office of Emergency Services, Sacramento, CA.

Buildings at Risk: Seismic Design Basics for Practicing Architects, by AIA/ACSA Council on Architectural Research, 1735 New York Avenue, NW, Washington, DC 20006.
Checklist of Nonstructural Earthquake Hazards, by BAREPP, 101 8th Street, Suite 152, Oakland, CA 94607.
Diaphragms, APA Publication 1350, APA, P.O. Box 11700, Tacoma, WA 98411-0700.
Earthquake Safeguards, by APA, P.O. Box 11700, Tacoma, WA 98411-0700.
A Guide to Repairing and Strengthening Your Home Before the Next Earthquake, by BAREPP, 101 8th Street Suite 152, Oakland, CA 94607.
Handbook to the Uniform Building Code, by Vincent R. Bush, International Conference of Building Officials, 5360 Workman Mill Road, Whittier, CA 90601-2298.
The Home Builder's Guide for Earthquake Design, 1980 ed., by Applied Technology Council, 3 Twin Dolphin Drive, Redwood City, CA 94065.
The Home Builder's Guide for Earthquake Design, by FEMA, 1992, FEMA-232.
Home Buyer's Guide to Earthquake Hazards, by BAREPP, 1989, 101 8th Street, Suite 152, Oakland, CA 94607.
The Homeowner's Guide to Earthquake Safety, by California Seismic Safety Commission, 1992, 1900 K Street, Suite 100, Sacramento, CA 95814-4186.
Reducing the Risks of Nonstructural Earthquake Damage: A Practical Guide, by FEMA, FEMA-74.
Residential Steel Framing Manual, by American Iron and Steel Institute, 1101 17th Street, NW, Washington, DC 20036-4700.
Strengthening Woodframe Houses for Earthquake Safety (ABAG P90004BAR), by BAREPP, 1990, 101 8th Street, Suite 152, Oakland, CA 94607.
1991 Uniform Building Code with 1993 Supplement, by International Conference of Building Officials, 5360 Workman Mill Road, Whittier, CA 90601-2298.
Wood Frame Construction Manual for One and Two Family Dwellings, by American Forest and Paper Association.
Wood Frame House Construction, rev. ed., by L. O. Anderson (Craftsman Book Company, Carlsbad, Calif.).

As part of the National Earthquake Hazards Reduction program (NEHRP), FEMA has developed more than 100 publications in its Earthquake Hazards Reduction series. A list of these publications, which are available free of charge, can be requested by writing to FEMA, P.O. Box 70274, Washington, DC 20024.

Preparedness in Schools and Offices

Earthquakes: A Teacher's Package for K–6 Grades, by the National Science Teachers Association, 1988, 280 pp. Available from NSTA Publications, 1742 Connecticut Avenue, NW, Washington, DC 20009, (202) 328-5800. Schools may obtain a free single copy from FEMA, Earthquake Programs, 500 C Street, SW, Washington, DC 20472.
Employee Earthquake Preparedness for the Workplace and Home, by American Red Cross, 1988, 12 pp. Send $1.00 to Red Cross Disaster Services, 1550 Sutter Street, San Francisco, CA 94109.
Living Safely in Your School Building, 1986, 9 pp., Lawrence Hall of Science, University of California, Berkeley, CA 94720, (415) 642-8718.

Making Buildings Safer

Earthquake Hazards and Wood Frame Houses: What You Should Know and Can Do, by M. Comerio and H. Levin, 1982, 46 pp., Center for Environmental Design Research, 390 Wurster Hall, University of California, Berkeley, CA 94720, (415) 642-2896.

Peace of Mind in Earthquake Country, by Peter Yanev, 1990 (Chronicle Books, San Francisco), 304 pp.

Rapid Visual Screening of Buildings for Potential Seismic Hazards: A Handbook, FEMA, 1988, 185 pp., FEMA-154.

Reducing the Risks of Non-structural Earthquake Damage: A Practical Guide, 1988, 86 pp. Order by mail from ABAG, P.O. Box 2050, Oakland, CA 94064-2050, (415) 464-7900.

Strengthening Wood Frame Houses for Earthquake Safety, 1990, 36 pp. Order by mail from ABAG, P.O. Box 2050, Oakland, CA 94064-2050, (415) 464-7900.

Chapter 15. Natural Hazards and Coastal Zone Management in Alaska

Alaska Coastal Management Program: Annual Report, Fiscal Year 1990, by the Division of Governmental Coordination, Office of the Governor, State of Alaska, 1991.

Alaska Coastal Management Program: Handbook, compiled by the State of Alaska. Includes all regulations, state and federal, pertaining to the coastal zone. Only for those with a strong stomach, will, and need to peruse the legal framework. Available from Office of the Governor, Division of Governmental Coordination, P.O. Box 110030, Juneau, AK 99811, (907) 465-3562.

Nome Storm Damage, Emergency Repair, Nome-Council Road M.P. 14-33, Alaska Department of Transportation and Public Facilities, Northern Region, Fairbanks, 1993. Fully documented with memoranda, letters, engineering plans, and storm data. Follow the bureaucratic process in one project.

Water Resources Development in Alaska 1991, by U.S. Army Corps of Engineers, Alaska District, 1991, 56 pp. Summary of the Corps' civil works projects in Alaska, including description and cost of each project. Available from U.S. Army Engineer District, Alaska, P.O. Box 898, Anchorage, AK 99506-0898.

Index

Library of Congress Cataloging-in-Publication Data
Mason, Owen K.
Living with the coast of Alaska / Owen Mason, William J. Neal,
and Orrin H. Pilkey ; with chapters by Jane Bullock . . . [et al.].
p. cm. Includes index.
Living with the shore.
ISBN 0-8223-2009-6 (cloth : alk. paper).—ISBN 0-8223-2019-3
(paper : alk. paper) 1. Natural disasters—Alaska—Pacific Coast.
2. Natural disaster hazards assessment—Alaska—Pacific Coast.
3. Coastal zone management—Alaska—Pacific Coast. 4. Human
ecology—Alaska—Pacific Coast. 5. Natural disasters—Alaska.
6. Natural disaster hazards assessment—Alaska. 7. Coastal
zone management—Alaska. 8. Human ecology—Alaska.
I. Neal, William J. II. Pilkey, Orrin H., 1934–. III. Title.
GB5010.M37 1998
363.34'09798—dc21 97–15156 CIP